T0187692

Aperiodic Structures in Condensed Matter
Fundamentals and Applications

Series in Condensed Matter Physics

Series Editor:
D R Vij
Department of Physics, Kurukshetra University, India

Series in Condensed Matter Physics

Aperiodic Structures in Condensed Matter

Fundamentals and Applications

Enrique Maciá Barber
Universidad Complutense de Madrid
Spain

CRC Press
Taylor & Francis Group
Boca Raton London New York

CRC Press is an imprint of the
Taylor & Francis Group, an **informa** business
A TAYLOR & FRANCIS BOOK

CRC Press
Taylor & Francis Group
6000 Broken Sound Parkway NW, Suite 300
Boca Raton, FL 33487-2742

First issued in paperback 2019

© 2009 by Taylor & Francis Group, LLC
CRC Press is an imprint of Taylor & Francis Group, an Informa business

No claim to original U.S. Government works

ISBN-13: 978-1-4200-6827-6 (hbk)
ISBN-13: 978-0-367-38633-7 (pbk)

Library of Congress Cataloging-in-Publication Data

Barber, Enrique Macia.
   Aperiodic structures in condensed matter : fundamentals and applications / Enrique Macia Barber.
      p. cm. -- (Series in condensed matter physics)
   Includes bibliographical references and index.
   ISBN 978-1-4200-6827-6 (hardcover : alk. paper)
      1. Condensed matter--Structure. I. Title. II. Series.

QC173.458.S75B37 2009
530.4'11--dc22   2008040940

Visit the Taylor & Francis Web site at
http://www.taylorandfrancis.com

and the CRC Press Web site at
http://www.crcpress.com

Contents

Preface

The universe is nicely ordered. There is order in the sequence of events determining the pace of evolution and the rhythms of life alike. Order can be found in all the structures unfolding around us at different scales. There is order in the arrangements of matter, in energy flow patterns, in every work of Nature interweaving space and time. This book is devoted to the study of a special kind of order referred to as aperiodic order.

Etymologically aperiodic order means order without periodicity. Accordingly, aperiodic order has nothing to do with disorder in any of its possible multiple forms. Aperiodic ordered matter exhibits long-range order in space, just as periodic orderings do. This property clearly distinguishes aperiodic structures from amorphous matter, the latter being able to display short-range correlations only. Aperiodic systems can be classified according to different criteria. For instance, certain aperiodic arrays of atoms are able to give rise to high quality x-ray or electron diffraction patterns composed of a collection of discrete Bragg spots, as periodic arrays of atoms also do. This phase of matter is referred to as quasiperiodic crystals (quasicrystals, for short) and they represent a natural extension of the periodic crystal notion. The diffraction patterns of quasicrystals are quite bizarre, unveiling the existence of unexpected symmetries which endow them with an impressive esthetical appeal. They also exhibit unusual physical properties, closely related to the fractal nature of their energy spectra. Physically this feature means that some specific fragments of the spectra appear once and again at different scales. Accordingly, we do not have periodicity but scalability. Indeed, fractal structures, characterized by their invariance under inflation/deflation operation symmetries, provide another representative example of aperiodically ordered systems.

But, in my opinion, the most important feature of aperiodic systems is their ability to encode relevant information in a way periodic order is completely unable to do. It suffices to compare a periodically arranged string of letters, namely *abcabcabcabc...*, with the preceding paragraph to immediately grasp the main point: in a periodic arrangement the information stored is limited to the basic period defining its structure (the unit cell in the case of a periodic crystal, for instance), whereas the amount of information stored in an aperiodic structure progressively increases as the system size is increased. The stacking of Watson–Crick complementary bases determining the genetic code in DNA is perhaps the most paramount example one can find in Nature. In fact, in DNA two kinds of order coexist in the same sample at just the same space scale. On the one hand, one has the aperiodic stacking of bases determining its biological information. On the other hand, one has the periodic arrangement of sugar-phosphate groups conforming the double-helix structure which preserves the physical integrity of the macromolecule at physiological conditions.

Such a blending of ordering principles can provide an inspiring guide for technological applications. For instance, one can grow layered structures consisting of a large number of films aperiodically stacked. The simplest example of such nanostructured materials is a two-component aperiodic heterostructure, where layers of two different materials (metallic, semiconductor, superconductor, dielectric, ferroelectric, ceramics) are arranged according to certain aperiodic sequence. In this way, two kinds of order are introduced in the same sample at different length scales. At the atomic level we have the usual crystalline order determined by the periodic arrangement of atoms in each layer, whereas at longer scales we have the aperiodic order determined by the sequential deposition of the different layers. This long-range aperiodic order is artificially imposed during the growth process and can be precisely controlled. Since different physical phenomena have their own relevant physical scales, by properly matching the characteristic length scales we can efficiently exploit the aperiodic order we have introduced in the system, hence opening new avenues for technological innovation. Recent works in optoelectronics and signal communication have fruitfully considered aperiodic designs in order to obtain improved devices and the very possibility of intentionally combining periodic and aperiodic materials in hybrid order composed structures has been recently explored in some detail.

Several topics on the role of aperiodic order in different domains of physical sciences and technology will be covered in this book. The first chapters address some basic notions and present the most characteristic features of different kinds of aperiodic systems in a descriptive way. In Chapter 1 we introduce different orderings of matter and describe the progressive transition from periodic to aperiodic thinking in physical sciences. In Chapter 2 the very notion of aperiodic crystal is introduced, fully describing its historical roots as well as the paramount discovery of quasicrystalline alloys and their beautiful forbidden symmetries. The study of the unusual physical properties of quasicrystalline alloys is then presented in more detail in Chapter 3, paying special attention to their intriguing electronic structure and the possible nature of chemical bonding in hierarchically arranged cluster-based solids. In Chapter 4 we introduce the basic structural properties of man-made materials consisting of aperiodic sequences of layers such as Fibonacci semiconductor-based superlattices or Cantor-like dielectric multilayers. The main mathematical features of the substitution sequences defining their growth rule are also reviewed along with the possible signatures of quasiperiodicity in their physical properties.

The two following chapters focus on some theoretical aspects and useful mathematical approaches introduced to properly study the physical systems introduced in previous chapters. Accordingly, Chapter 5 is devoted to introducing some simple models describing the fundamental physics of several aperiodic systems in one dimension. Remarkable properties of their energy and frequency spectra, such as a highly fragmented, self-similar arrangement of progressively narrower bands or the critical nature of the eigenstates, are discussed in detail by considering suitable models for different systems of interest. The impact of the peculiar energy spectra on their related transport properties is also addressed. In Chapter 6 we turn our attention to the aperiodic crystal of life, by considering some basic features of DNA molecules from the perspective of condensed matter physics. Some fundamentals on the diffraction theory by helices are first introduced. Then we discuss the electronic structure of nucleic acids and summarize what experiments say about the possible charge transfer processes in DNA. Different effective Hamiltonians aimed at describing the basic physics of these processes are subsequently introduced. On the basis of these results, the role of long-range correlations is critically analyzed from the biophysicist viewpoint.

Afterwards we shift towards more applied issues. Chapter 7 discusses how to exploit aperiodic order in different technological devices based on multilayered optical systems, photonic and phononic quasicrystals, complex metallic alloys or DNA-based nanocells. The appealing possibility of introducing novel designs based on the aperiodic order notion to achieve some specific applications is further discussed in Chapter 8 by considering not only one-dimensional systems, but also arrangements of matter in two and three dimensions. Finally, in Chapter 9 we present some useful mathematical tools which are of common use in the study of aperiodic systems.

The book is specially intended for both condensed matter physicists and materials science researchers coming into the field of aperiodic systems from other areas of research. It can also serve as a useful text for graduate students.

I am gratefully indebted to Esther Belin-Ferré, Jean-Marie Dubois, Uichiro Mizutani, Patricia A. Thiel, and An Pang Tsai for their continued support and interest in my research activities during the last decade, as well as to Victor R. Velasco, who kindly agreed to review several chapters of this book. Their illuminating advice has significantly contributed to guide my scientific research in the aperiodic order realm. It is a pleasure to thank Janez Dolinšek, Francisco Domínguez-Adame, Sergey V. Gaponenko, Carlos V. Landauro, Stephan Roche, Rogelio Rodríguez-Oliveros, and Tsunehiro Takeuchi for sharing with me their time and efforts in joint research works. I also express my thanks to Eudenilson L. Albuquerque, José Luis Aragón, Claire Berger, Arunava Chakrabarti, Gianaurelio Cuniberti, Luis Elcoro, J. César Flores, Federico García-Moliner, Didier Mayou, Gerardo G. Naumis, Juan M. Pérez-Mato, Rudolf A. Römer, Manuel Torres, Chi-Tin Shih, Jewgeni B. Starikov, Alexander Voityuk, and Chumin Wang for inspiring conversations, and to Emilio Artacho, Michael Baake, Javier García-Barriocanal, Roberto Escudero, José Reyes-Gasga, Ai-Min Guo, Roland Ketzmerick, Kazumoto Iguchi, and Ruwen Peng for sharing with me very useful materials.

The author is grateful to Taylor & Francis, and to John Navas in particular, for giving me the opportunity to prepare this book. Last, but not least, I warmly thank M. Victoria Hernández for her invaluable support, unfailing encouragement, and her continued care to the detail.

This work has been supported by the Universidad Complutense de Madrid through projects PR27/05-14014-BSCH and PR34/07-15824-BSCH.

Enrique Maciá Barber   Madrid, June 2008

1

Orderings of matter

1.1 Periodic thinking in physical sciences

The notion of periodicity allows one to easily grasp the basic order underlying certain patterns and rhythms in Nature. The essence of periodicity relies on a basic motif which is indefinitely repeated, along with a set of basic rules prescribing the way such a repetition process takes place. Periodicity can occur in time, space, or simultaneously in both of them. Periodicity in time guarantees that what is known to occur now will also occur later, and can be asserted to have already occurred before, provided that a certain relationship between those different instants is fulfilled. Let t be a real number measuring the passage of time. Then, a function satisfying the condition $f(t \pm T) = f(t)$ is periodic in time with a period T, since its value is preserved (i.e., it is invariant) under transformations describing the set of translations generated back and forth by the arrow of time by the real number T.

The existence of cyclic processes in Nature accurately obeying such a periodicity condition is the basis for the possible adoption of physical clocks (characterized by their T value). In fact, from the galactic scale down to atomic and subatomic scales, the natural world has plenty of physical systems exhibiting nearly exact periodicity in time. Most of these systems can be described, at least as a first approximation, in terms of dynamic equations of the form

$$\frac{d^2 f}{dt^2} + \omega^2 f = 0, \qquad (1.1)$$

which is usually referred to as the harmonic oscillator equation, where f is some physical magnitude (e.g., a position coordinate, the intensity of an electric or magnetic field, or the chemical concentration of a substance) and ω, the so-called natural frequency, is a quantity which depends on characteristic physical parameters of the system. For instance, in the case of a (low ampli tude) swinging pendulum we have $\omega^2 = g/l$, where l is the pendulum's length, and g measures the local intensity of the Earth's gravitational field.

Eq.(1.1) is a second order differential equation. To solve it one must find a mathematical function $f(t)$ whose second derivative coincides with (minus) itself, once properly scaled by a factor ω^2. The theory of differential equations tells us that these requirements are met by solutions of the form

$$f(t) = a\cos\omega t + b\sin\omega t, \tag{1.2}$$

where the value of the constants a and b is determined from the knowledge of a suitable set of initial conditions, and $\omega \equiv 2\pi/T$. In the particular case $a = b \equiv R$ the function given by Eq.(1.2) simply describes a uniform circular motion of radius R with angular frequency ω. Since trigonometric functions satisfy (by definition) the relations $\sin[\omega(t \pm T)] = \sin\omega t$, and $\cos[\omega(t \pm T)] = \cos\omega t$, we see that the periodicity condition $f(t \pm T) = f(t)$ is properly satisfied by Eq.(1.2). Therefore, the periodicity in time exhibited by the solutions of Eq.(1.1) naturally emerges from its basic mathematical structure. Quite remarkably, harmonic equation describes a broad collection of cyclic motions in nature, ranging from atomic vibrations in solids to population dynamics in ecosystems. The profuse appearance of this basic equation in the study of such diverse dynamical systems certainly accounts for the important role played by periodic thinking in theoretical physics, probably starting with the pioneering quest for the isochronous pendulum by Galileo Galilei (1564-1642) and Cristiaan Huygens (1629-1695) in the 17th century.[1]

Periodicity in space guarantees that what is located here must also occur over there, provided that certain geometrical relationships between "here" and "there" are fulfilled. Thus, a vector function satisfying the condition $\mathbf{f}(\mathbf{r} + \mathbf{R}_0) = \mathbf{f}(\mathbf{r})$ is periodic in space, since it is invariant under transformations describing the set of space translations generated by the vector $\mathbf{R}_0$ in the vectorial space to which the variable $\mathbf{r}$ also belongs. Periodicity in Euclidean space can involve rotations as well as translations, and can be expressed in the general form $\mathbf{Mr} + \mathbf{R}_0 = \mathbf{r}$, where

$$\mathbf{M} = \begin{pmatrix} \cos\varphi & -\sin\varphi & 0 \\ \sin\varphi & \cos\varphi & 0 \\ 0 & 0 & 1 \end{pmatrix} \tag{1.3}$$

is an orthogonal matrix describing rotations by an angle φ. Let us consider the vectors describing a lattice of points, which have the general form $\mathbf{r} = n_1\mathbf{e}_1 + n_2\mathbf{e}_2 + n_3\mathbf{e}_3$, where $\{\mathbf{e}_i\}$ is a suitable vector basis and $n_i \in \mathbb{Z}$. The periodicity condition then implies that the trace of matrix $\mathbf{M}$ must take on integer values.[2] This leads to the so-called crystallographical restriction

$$1 + 2\cos\varphi = n \in \mathbb{Z}, \tag{1.4}$$

which has played a significant role in the development of classical crystallography. The main consequence of the relationship given by Eq.(1.4) is that only a few number of rotations are compatible with the periodicity condition. Thus, only two-fold, three-fold, four-fold, and six-fold symmetry axes are allowed in periodic lattices, as it can be straightforwardly deduced from Eq.(1.4). To this end, we express the crystallographic restriction in the form

$\cos \varphi = (n-1)/2$. The condition $|\cos \varphi| \leq 1$ implies $n = \{-1, 0, 1, 2, 3\}$. By plugging these values into the former expression we obtain the solutions listed in Table 1.1.

TABLE 1.1
Allowed symmetry
axes in periodic
crystals.

n	φ	AXIS
-1	π	2-fold
0	$2\pi/3$	3-fold
1	$\pi/2$	4-fold
2	$\pi/6$	6-fold
3	0	identity

The simplest illustration of processes which are simultaneously periodic in space and time can be found in wave phenomena. For instance, sinusoidal waves of the form $\Psi(\mathbf{r}, t) = \Psi_0 \sin(\mathbf{k}.\mathbf{r} - \omega t)$, where $k = 2\pi/\lambda$ is the wave number and λ measures the wavelength, often occur in waves propagating in gases, liquids or solids as well as in electromagnetic waves propagating in vacuum. Their characteristic wave function describes a periodic pattern in space if we fix the time variable (i.e., $t \equiv t_0$). Alternatively, if we fix the space variable (i.e., $\mathbf{r} \equiv \mathbf{r}_0$), it describes a harmonic motion in time at every point of space, where the quantity $\mathbf{k}.\mathbf{r}_0$ measures the relative dephasing between the oscillations of two points separated by a distance $\mathbf{r}_0$. The double periodicity (in space and time) of wave motion can be traced back to the very structure of the corresponding wave equation, which reads

$$\nabla^2 \Psi + \frac{1}{c^2} \frac{\partial^2 \Psi}{\partial t^2} = 0, \qquad (1.5)$$

where $c = \omega/k$ is the phase velocity of the wave. The first (second) term in Eq.(1.5) describes the periodicity in space (time) of the propagating wave, while its phase velocity couples its spatial pattern to its propagation rhythm.

A key feature of sinusoidal waves, significantly contributing to pervade periodic thinking in scientific thought, is that any non-sinusoidal, periodic wave can be represented as a collection of sinusoidal ones (with different frequencies) blended together in a weighted sum of the form [cf. Eq.(1.2)]

$$f(t) = \frac{a_0}{2} + \sum_{m=1}^{\infty} [a_m \cos(\omega_m t) + b_m \sin(\omega_m t)], \qquad (1.6)$$

where $\omega_m = 2\pi m/T$, and

$$a_m = \frac{2}{T} \int f(t) \cos(\omega_m t) dt, \quad b_m = \frac{2}{T} \int f(t) \sin(\omega_m t) dt, \qquad (1.7)$$

are the so-called Fourier coefficients, after the French mathematician Joseph Fourier (1788-1830) who introduced this procedure in 1822. Closely related to this series expansion, one can consider the so-called Fourier transform, which decomposes a function into a continuous spectrum of its frequency components according to the expression

$$F(\omega) = \int\limits_{-\infty}^{+\infty} f(t)e^{-i\omega t}dt. \qquad (1.8)$$

Note that completely analogous expressions hold for periodic functions in space by simply replacing the corresponding variable in Eqs.(1.6)-(1.8). In this way, a Fourier transform can be envisioned as a linear transformation relating two different mathematical domains: that corresponding to usual time or space variables (which come closer to our everyday experience), and that corresponding to the related frequency or reciprocal space spectrum, which encloses a more abstract view of the underlying order in the considered phenomenon. Remarkably enough, there exist processes in Nature able to Fourier-transform material structures in a natural way, namely, diffraction of electromagnetic (x-ray) or matter quantum waves (electrons, neutrons) by atomic scatters in condensed phases. The resulting diffraction spectra exhibit regular arrangements of bright spots (the so-called Bragg peaks) disclosing the abstract information encoded within Fourier space to our eyes. In this way, the workings of Nature translate wave motion into geometrical patterns engraved in reciprocal space through the orchestrated interaction of matter and energy in condensed matter.

Diffraction spectra contain a lot of information about structural details which must be carefully analyzed, generally requiring a formidable task in the case of relatively complex structures. But a key, basic feature follows from the very mathematical definition of the Fourier transform: close spots in diffraction patterns correspond to scattering centers which are far apart in physical space. Accordingly, Fourier space description of crystal structures takes place in the so-called reciprocal space. In Fig.1.1 a celebrated example of diffraction pattern, ultimately leading to the elucidation of the double-helix structure of DNA, is shown for the sake of illustration. The cross-shaped arrangement of Bragg spots in reciprocal space is a characteristic telltale of the helicoidal distribution of sugar-phosphate groups in physical space. The two broad dark features located up and down the image correspond to the stacked nucleotides along the helix axis. We will study the physical implications of this impressive picture in more detail in Chapter 6.

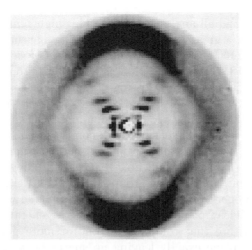

FIGURE 1.1
The so-called Photo 51 of DNA was obtained by Rosalind Franklin (1920-1958) in 1952. This image shows the diffraction spectra resulting from x-ray passing through highly hydrated fibers of calf thymus DNA samples. (Reprinted by permission from Macmillan Publishers Ltd.: *Nature* **171** 740 ©1953 [3].)

1.2   The realm of periodic crystals

Geometrical patterns appear once and again in many natural objects, ranging in size from atomic scale to the size of the galaxies, and provide us with a series of basic geometrical motives to think about. It is not surprising then that geometrical concepts play a central role in most efforts we make in order to understand and describe processes and structures in Nature, as it was pointed out by Galileo's famous metaphor of the "Book of Nature."

> "Philosophy is written in this grand book, the universe, which stands continually open to our gaze. But the book cannot be understand unless one first learns to comprehend the language and read the characters in which it is written. It is written in the language of mathematics, and its characters are triangles, circles, and other geometric figures (...)" [4]

Thus, the basic characters of this Book belong to the broad class of polygons and, more generally, polyhedra as well, which were analyzed in detail by mathematicians and natural philosophers alike from ancient times. A polygon (Greek for many angles) is a portion of a plane bounded by straight line segments, which define a closed shape. Regular polygons are those in which all sides are equal and all angles are equal. This high symmetry endows regular polygons with a great aesthetic appeal. The ancient Greeks succeeded in constructing regular polygons with large number of sides, included in the series 3,4,5,6,8,10,12,15,16 and 20. However, all efforts to construct a regular polygon of seven sides (the first missing one in the list above) failed. Proof was much later given (Carl Friedrich Gauss (1777-1855)) in 1801 and Pierre Wantzel (1814-1848) in 1837) that, if the number of sides p is a prime number, a regular polygon can be constructed with compass and straight edge if and only if $p = 2^{2^n} + 1$, $n = 0, 1, 2, ...$(the so-called Fermat primes after Pierre Fermat (1601-1665)). This result indicates that not all polygons can be treated on equal footing, but they can be grouped in different classes depending on their symmetry properties. This point can be illustrated with another example: tilings of Euclidean planes. The word tiling is generally used to describe a pattern or structure that comprises of one or more polygonal shapes of 'tiles' that pave a plane exactly leaving no spaces between them. Squares, equilateral triangles, and hexagons are particularly easy to tile with in order to achieve a periodic pattern that repeats itself at regular intervals. The obtained patterns are endowed with the characteristic symmetries – threefold, fourfold, sixfold – of the tiles they are respectively made up. In the case of a regular pentagon tile, however, no matter how hard you try, they cannot be used to fill the entire plane and form a periodic tiling pattern: unfilled gaps will always remain. In this way, we realize that (i) only a few set of regular polygons are capable of tiling the plane (namely, triangles, squares, and hexagons), and (ii) that the resulting tilings always exhibit periodic patterns.

Certainly, one may relax the original tiling rules by allowing *different* types of polygons to be simultaneously used to construct the tile. In that case one obtains the so-called Archimedean tiling patterns shown in Fig.1.2. These tiles were originally depicted by Johanes Kepler (1571-1630) in *Harmonices Mundi II* (1619), where the main requirement was that only one type of vertices is permitted at every joining point. A set of integers $(n_1.n_2.n_3.\cdots)$ denotes the vertex type in the way that n_1-gon, n_2-gon, n_3-gon,$\cdots$ different polygons consecutively meet on each vertex. For instance, the symbol $(3^2.4.3.4)$ represents a tiling in which two equilateral triangles, a square, an equilateral triangle, and a square successively gather edge-to-edge around any given vertex. In this way, uniform tilings can be listed by their vertex configuration. We note that the so-called regular or Platonic tilings are a subset of Archimedean ones corresponding to the notation (3^6), (4^4), and (6^3).

Polyhedra are solid geometric figures having polygons as their faces. The angle formed by the faces at a vertex is called polyhedral angle. If all faces are congruent and all polyhedral angles are equal, the resulting body is referred to as a regular polyhedron. Whereas there exist an infinite number of regular polygons, only five types of regular polyhedra (identical faces) are possible in three dimensions (the so-called Platonic solids), and they are shown in Fig.1.3.

The very small number of regular polyhedra could be associated with the idea that perfect harmony is scarce in real world. Quite interestingly, how-

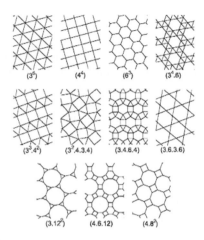

FIGURE 1.2
Archimedean tilings. When one imposes all the vertices to be of the same type only eleven types of tiling of the plane by regular polygons can be found. The three first tilings shown at the top file are composed of just one type of polygon tile and are sometimes referred to as Platonic tiles.

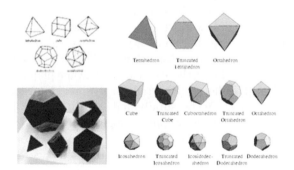

FIGURE 1.3
The five Platonic solids (left hand) and seven of the 13 Archimedean solids which can be obtained by truncation of a Platonic solid (right hand). Two additional solids can be obtained by expansion of a Platonic solid, and two further solids can be obtained by expansion of one of the previous nine Archimedean solids.

ever, Platonic shapes profusely appear in the atomic ensembles disclosed by chemists and crystallographers as the basic building blocks of matter. For instance, in the diamond crystal each atom has four neighbors, which are arranged about it at the corners of a regular tetrahedron. In the tetragonal crystal KPdCl four chlorine atoms about each Pd or Pt atom lie at the corners of a square. Octahedral complexes in cubic crystals KPtCl have their centers at the corners and the centers of a cubic unit structure. Therefore, although only a few different types of regular polyhedra exist due to the severe mathematical constraints imposed by their high symmetry degree, these basic motives are profusely adopted by atomic arrangements in Nature. This remarkable property is ultimately related to the geometrical properties of hybrid orbitals formed by atoms in solids, thereby highlighting the significant role of quantum laws in the arrangements of matter.

We are all familiar with crystalline minerals, solids bounded by naturally formed geometric faces. In fact, its regular geometrical form was a demonstration that geometry is not just an abstraction stemming from the imagination of mathematicians, for it appears spontaneously in rocks. In this way, minerals suggested a subtle link between geometry and the inner structure of solid matter. Thus, the spontaneous appearance of symmetrical shape upon cooling a saturated solution was accordingly interpreted as an indication of a regular and precise pattern of internal structure.

The earliest recorded speculation of this sort seems to have been that published by Robert Hook (1635-1703) in a book entitled *Micrographia*. Hook envisioned a crystal as a stack of spherical particles closely packed together in a regular way. Not long afterwards (1678), Cristiaan Huygens carried the idea still further. He was aware that some crystals, like calcite, mica, or Iceland spar, break most readily along certain planes, giving pieces with precisely flat surfaces, and he suggested that these so-called cleavage planes are natural lines of division between flat sheets of particles. By breaking along these planes a crystal of Iceland spar can be subdivided into little rhombohedra. Quite remarkably, the subdivision can apparently go on indefinitely, giving smaller and smaller rhombohedra, to the limits of microscopic vision. Subsequently, around 1800, René-Just Haüy (1743-1822) connected this fact with the idea that all the various natural faces of different minerals could be accounted for as various simple ways of finishing off a stack of tiny rhombohedral pieces, like those obtained from the Iceland spar crystal. In this way, faces manifest at a macroscopic scale an underlying ordered arrangement. Accordingly, the structure of minerals can be understood as regular arrangements of very small particles (say atoms) closely packed in ranks upon ranks, and the various macroscopic facets show up different ways of finishing off the stack. On the other hand, substances like glass, which do not form crystals, can be envisioned as lacking this internal regularity, so that the distinction between crystalline and amorphous substances was thought to be a fundamental one: that separating the realms of order from disorder in solid matter.

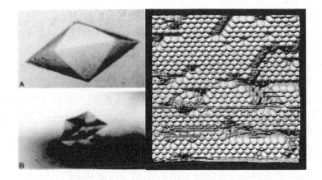

FIGURE 1.4

At a macroscopic scale (on the left) the apperance of the hexaedral bypiramid crystal of satellite tobacco mosaic virus looks like a common mineral substance, like calcite or quartz, which are constituted by inorganic molecules.[5] Upon a closer inspection (on the right), as that provided by scanning atomic microscopy, one realizes that the basic building blocks of this organic crystal are virus capsides, complex molecular structures exhibiting icosahedral symmetry.[6]

Modern condensed matter theory relies on the discrete nature of the ultimate constituents of solid bodies. This point is illustrated in Fig.1.4 where one realizes that the flat crystal surfaces one sees at a large enough scale ultimately resolve in a coarsened landscape formed by successive stacks of virus particles. The same basic principle essentially applies to any piece of condensed matter: it is composed of a huge number (of the order of magnitude of the Avogadro's number) of building blocks at the atomic/molecular scale. The way these building blocks are arranged through the space constitutes one of the most fundamental notions in solid state physics and it allows for the introduction of the useful notion of a crystal lattice: a mathematical set of material points whose positions remain fixed through the space. In addition to its undeniable mathematical convenience for a rigorous description of crystals symmetry within the framework of group theory concepts, it turns out that atoms usually behave as point-like ideal particles when they interact among themselves or with propagating energy fields.

In fact, the regular structure of crystals at the atomic level can be properly characterized by means of x-ray diffraction experiments in which the appearance of sharp spots indicates the existence of long-range translational order between different atomic distributions. The idea that x-rays might be electromagnetic waves with wavelengths of the order of magnitude of the distance between layers of atoms in crystals was originally put forward by Max von Laue (1879-1960) at the beginning of the XX century. In a series of experiments a beam of x-rays passed through a crystal of copper sulfate, and a number of spots (corresponding to diffraction peaks maxima) were observed on the photographic plates set up around the crystal. Laue then developed a set of equations, relating the direction in which diffraction maxima occur to the structure of the crystal. A few months later William Lawrence Bragg (1890-1971) developed a simpler equation to describe the diffraction maxima, given by

$$2d \sin \theta = \lambda, \tag{1.9}$$

where d is the distance that separates successive pairs of atomic plains, θ is the angle between the directions of the incident and scattered beams, and λ is the x-ray wavelength. In 1912 William Lawrence Bragg and his father William Henry Bragg (1862-1942) published several papers reporting not only the wavelengths of the x-rays but also the structures of a number of crystals: diamond, sodium chloride, fluorite. This was the start of x-ray diffraction and the determination of the structure of crystals.

It can be said that the science of crystallography first began when it was realized that in spite of existing thousands of different crystalline solids, most differing in the number and arrangement of faces, all of them can be classified in but a few basic classes. The fundamental property allowing for a simple and unified description is symmetry. Three main elements of symmetry are commonly used in crystallography, namely, center of symmetry (when a crystal shows repetition with respect to a point), axis of symmetry (repetition respect to a line), and plane of symmetry. The German scientist Johan F. C. Hessel (1796-1872) showed in 1830 that with these elements of symmetry there could be a total of 32 different crystal groups, each one characterized by a specific combination of symmetry elements, which are referred to as the crystal classes. The crystal class can be grouped, in turn, into seven crystal systems, each including a number of classes with certain symmetry elements in common. The advent of x-ray diffraction analysis revealed that atomic arrangements inside solids can introduce additional elements of symmetry in such a way that the original 32 crystal classes must be expanded into 230 space groups.

1.3   Superimposed periods

There exist materials where two or more periodicities occur. For instance, one may think of a ferromagnetic material, where one periodicity characterizes the arrangement of atoms (e.g., measured by x-ray diffraction), and other periodicity characterizes the distribution of spins (e.g., measured by neutron diffraction). Since the coupling between the magnetism and the crystal structure is generally weak, both periods are likely to be incommensurate to each other, which results in a modulated aperiodic crystal. The modulation can also appear in the density function describing the distribution of atoms themselves (Fig.1.5).

Calaverite ($AuTe_2$) and $\gamma-Na_2CO_3$ were the first examples of incommensurate modulated structures. Nowadays hundreds of compounds with a modulated structure are known in different types of materials including alloys, ionic compounds, molecular crystals, high-temperature superconductors, or ferroelectrics. For a detailed historical account the reader is referred to Refs.[2] and [7]. In incommensurate crystals the diffraction patterns are generally characterized by linear combinations of two sets of reciprocal lattice vectors, say $\mathbf{q}$ and $\mathbf{g}$, where $\mathbf{q}$ represents a basic lattice of reference and $\mathbf{g}$ represents a modulation vector. It is then clear that these incommensurate structures lack three-dimensional periodicity, so that they cannot be characterized by one of the 230 space groups. This difficulty prompted P. M. de Wolff to introduce the notion of *superspace* to account for the satellite reflections observed in the $\gamma-Na_2CO_3$ diffractograms. The basic idea is that the diffraction pattern corresponding to an incommensurate structure can be seen as the projection of a periodic lattice in four dimensions on the three-dimensional physical space. Note that this four dimensional space is an abstract, crystallographic space, and not the usual four-dimensional space-time.

Another class of materials exhibiting incommensurate periods are intergrowth compounds. For instance, the compound $Hg_{3-\delta}AsF_6$ contains a host

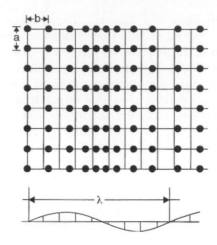

FIGURE 1.5

In an incommensurate displacively modulated crystal the atoms deviate in a periodic fashion from the positions of a conventional periodic crystal.(From ref.[2]. By permission of Oxford University Press.)

system, consisting of AsF_6 octahedra, which leave a series of void channels. Along these channels there are linear chains of Hg atoms separated by a distance which is incommensurate to the lattice parameter of the host lattice. Both the AsF_6 subsystem and the mercury subsystem show up in the diffraction spectra, and can be labelled in terms of four reciprocal lattice vectors.[2]

1.4   Order at different scales

The characteristic order exhibited in both periodic and incommensurate crystals usually manifests at the atomic scale. There exist, however, other materials where the basic building blocks define larger scale arrangements of matter. For the sake of illustration we shall briefly review some basic features of three remarkable representatives: polymers, cluster based solids, and superlattices. In the case of polymers, for instance, the basic structural motif is a molecule (or a series of a few different molecules in the case of copolymers) linked to other molecules by means of covalent bonds. The basic molecular block can be very simple, as occurs in the polyacetylene $(CH)_n$ molecule (Fig.1.6) or it may include several chemical moieties, like the different nucleotides composing DNA, which are composed of a nitrogen base (either a purine or a pyrimidine), a sugar hexose, and a phosphate group (Fig.1.7).

FIGURE 1.6

The polyacetylene molecule is composed of CH units linked together by means of unsaturated, hybrid-sp^2 covalent bonds leading to a planar geometry. Carbon (hydrogen) atoms are represented by large (small) balls.

In contrast to metals, polymers are typically insulators. However, in recent years new classes of polymers have been synthesized that are capable of carrying unusually high currents. For instance, a doped form of the polyacetylene molecule (as a consequence of an acidic treatment, for instance) was reported to have an electrical conductivity comparable to that of certain semiconducting materials ($\sim 10^3$ $\Omega^{-1}cm^{-1}$).[8] Subsequently, polymers such as polyaniline, polypyrrole, and polythiophene were found to have high electrical conductivities when chemically doped in a proper way.[9, 10, 11]

Materials assembled out of stable atomic clusters have given a new dimension to Materials Science. Thus, researchers can now use atomic clusters to form novel solids which could have dramatically different properties than ordinary materials composed of atoms. A celebrated instance of this type of solids is fullerite, the third form of stable carbon (along with diamond and graphite), which is composed of fullerene C_{60} molecules. These soccer football-shaped molecules are composed of sixty carbon atoms covalently linked together and exhibit local icosahedral symmetry. Nonetheless, when fullerene molecules crystallize they adopt a face-centered cubic arrangement, hence exhibiting conventional periodic order at a large enough scale.[12] The main difficulty in forming a cluster-based solid is related to the very structure stability of the cluster itself. Thus, the successful formation of most molecular crystals through van der Waals interaction can be attributed to the fact that intra-cluster interactions are much stronger than the intercluster ones, so that clusters keep their individuality intact.[13] One can try to form *cluster solids*

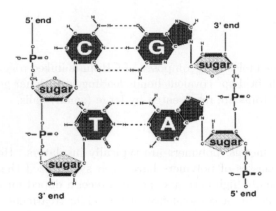

FIGURE 1.7

Double-stranded DNA molecule is composed of nucleotide units. Each nucleotide consists of a nitrogen base (**A**denine, **T**hymine, **C**ytosine, or **G**uanine), linked to a deoxiribose sugar molecule, attached to a phosphate group. The backbone is determined by the phosphorylated sugars' helical chain. Complementary bases are connected by hydrogen bonds. (Courtesy of Rudolf A. Römer.)

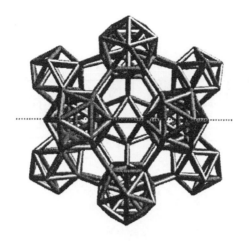

FIGURE 1.8

Structural model of a B_{156} supericosahedron cluster.(Reprinted figure with permission from Perkins C L, Trenary M, and Tanaka T 1998 *Phys. Rev. B* **58** 9980 © 1998 by the American Physical Society.)

through ionic interactions between clusters or between clusters and atoms. In the latter case, the atomic clusters serve as "superatoms" and should have their electron affinities close to atoms.[14]

An interesting example of solid structure involving successive spatial scales in its structure is provided by the β-boron phase, whose structure can be conveniently described in terms of a B_{84} unit that consists of a central B_{12} icosahedron with each vertex linked by a single boron atom to the pentagonal faces of an outer B_{60} shell having exactly the same buckyball structure as C_{60}. In this material we find an incipient hierarchical scheme which is further exploited in the case of the YB_{66} structure, described in terms of the so-called supericosahedra consisting of 13 B_{12} icosahedra. The resulting $B_{156} = B_{12}(B_{12})_{12}$ units are composed of a central B_{12} icosahedron with twelve B_{12} icosahedra bonded to each of the twelve vertices of the central subunit cluster (Fig.1.8).[15, 16] As we can see from the examples just presented, at least two different physical scales must be considered to properly understand the physical properties of these materials: at atomic level we have molecular structures from which a bulk solid progressively emerges at large enough scales. To determine the threshold scale at which a transition from typically molecular to essentially bulk solid characteristics takes place is not a simple question, and it remains as an appealing open question in most systems considered to date.

In 1970 Leo Esaki and Raphael Tsu proposed the fabrication of an artificial periodic structure consisting of alternate layers of two different semiconduct-

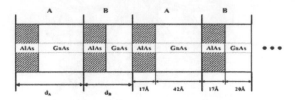

FIGURE 1.9
Schematic view of a two-component semiconductor based superlattice. Each
building block is composed of AlAs/GaAs bilayers of different widths on the
nanometer scale. The unit cell of the superstructure is given by the repetition
of *AB* couples.

ing materials, with layer thickness of the order of nanometers (Fig.1.9).[17]
They called this synthetic structure a *superlattice* because periodic order appears at two different scale lengths: At the atomic level we have the usual
crystalline order determined by the periodic arrangement of atoms in each
layer (lattice structure), whereas at longer scales we have the periodic order
determined by the sequential deposition of the different layers (superlattice
structure). The artificial periodicity length can thus be made less than the
electron free path and the de Broglie wavelength. Thus, the one-dimensional
ordering introduced along the growth direction in the manufacturing process
gives rise to novel properties by its own, such as the formation of electronic
energy bands with significantly small bandwidths, of the order of a hundred
meV, which are impossible to obtain in naturally occurring materials. The
rapid progress achieved with crystal growth technologies, like molecular beam
epitaxy or chemical vapor deposition, has made it possible to grow artificial
structures using different materials (metallic, semiconductor, superconductor, dielectric, ferroelectric, ceramics) and chemical composition modulations
along the growth direction. Accordingly, the physical properties of superlattices are determined by the chemical nature of the constituent bulk materials
as well as the layer thicknesses, so that these structures can be currently
grown to tailor a number of physical properties as required, as we will discuss
in Chapters 4, 7, and 8.

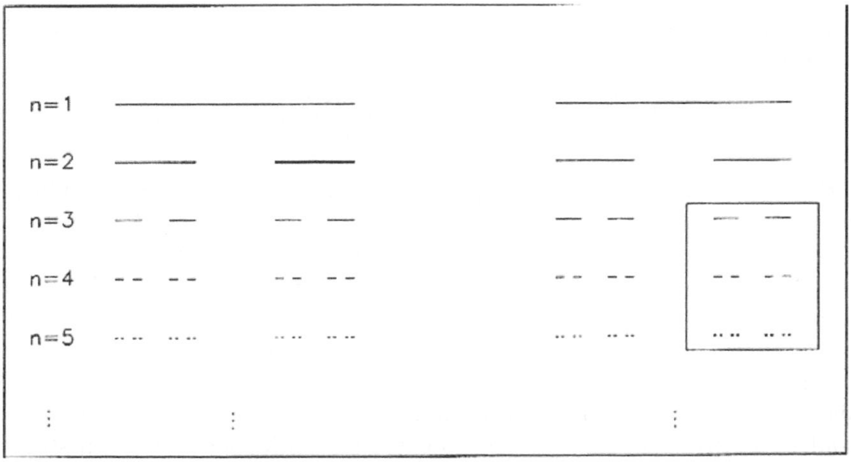

FIGURE 1.10
Successive stages defining the generation of a triadic Cantor set. A characteristic property of these sets is its self-similarity, which is illustrated in the box.

1.5 Scale invariance and fractal geometry

During the last twenty years it has been realized that a large amount of objects and processes in Nature can be characterized by a class of geometric entities called fractals, which were originally introduced by Benoit Mandelbrot.[18] The concept of fractal relies on the notion of dimensionality. In particular, a fundamental property of fractal structures is that they assume fractional dimension values, which distinguish them from homogeneous euclidean objects. This basic property stems from their peculiar scaling behavior referred to as self-similarity. By this we mean that the global structure of the system as observed on one length scale is repeated on successively smaller (or larger) scales. This property is illustrated in Figs.1.10 and 1.11

A Cantor set (after Georg Cantor 1845-1918) is defined as what is left from a finite segment after removing parts of it according to some iterative procedure. For the sake of illustration, let us consider the so-called triadic Cantor set, a geometrical structure which can be defined by means of the iterative process depicted in Fig.1.10 Let us start with a closed segment of length $L = 1$ and divide it in three equal intervals taking away the middle one. After repeating the process n times we get 2^n identical intervals of length $L/3^n$. If we repeat the process an infinite number of times we are left with an infinite and nonconnected set of isolated points having a null measure. This characteristic property of Cantor-like sets can be easily seen by evaluating the total length

of its complement, that is, the part of the original interval taken away. This
length is

$$\frac{L}{3} + \frac{2L}{3^2} + ... + \frac{2^n L}{3^{n+1}} + ... = \frac{L}{3} \sum_{n=0}^{\infty} \left(\frac{2}{3}\right)^n = L, \tag{1.10}$$

so that the spatial extent of the part removed equals that corresponding to
the original segment! Nevertheless, the points belonging to the Cantor set are
spread over the space, separated by empty voids. Accordingly, our intuition
prevents us to simply accept that they can be regarded as a unique point
measure. The solution of this seeming paradox requires the extension of the
very notion of spatial dimension in order to include dimensions which are
intermediate between that of a segment (unity) and that of point (zero). To
this end, the dimensionality of a Cantor set is defined as the number $N(\lambda)$ of
smallest intervals of size λ needed to cover the entire set at a given stage of
the iterative process, through the relationship

$$D = - \lim_{\lambda \to 0} \frac{\ln N(\lambda)}{\ln \lambda}. \tag{1.11}$$

By inspecting Fig.1.10 we obtain $D = \ln 2 / \ln 3 = 0.6309...$ for a Cantor set,
indicating that its coarse-grained structure is compatible with some degree of
spatial extension. This value is usually referred to as the Hausdorff dimension
of the set (after Felix Hausdorff 1868-1942). We observe that D no longer
corresponds to an integer number as it is usual in dealing with Euclidean geometrical figures. This non-integer dimensionality is a typical feature of fractal
structures, geometrical entities intermediate between single points ($D = 0$),
line segments ($D = 1$), planes ($D = 2$), and regular solids ($D = 3$).

The notion of fractal dimension can be extended to higher dimension structures. An illustrative example is provided by the Koch star, which is obtained
following the iterative process sketched in Fig.1.11. One starts with an equilateral triangle of side L. Successive equilateral triangles of sides $L/3^n$ are
added at the middle of the sides, growing a figure which resembles a snow
flake appearance. At a given iteration stage the perimeter of the Koch star is
obtained by multiplying the total number of sides by its effective length,

$$P_n = (3 \times 4^n)\frac{L}{3^n} = P_0 \left(\frac{4}{3}\right)^n, \tag{1.12}$$

where $P_0 = 3L$ is the perimeter of the original triangle. Since the ratio of
the series given by Eq.(1.12) is larger than unity it grows indefinitely as n
increases. As a result, the perimeter of the Koch curve tends to infinity. On
the other hand, the area of the Koch star is given by

$$A_n = A_0 \left[1 + \frac{3}{9} \sum_{k=1}^{n} \left(\frac{4}{9}\right)^{k-1}\right], \tag{1.13}$$

FIGURE 1.11
Five order Koch curve illustrating the way an ever increasing perimeter polygon can be contained in a finite portion of the plane.

where $A_0 = \sqrt{3}L^2/4$ is the area of the original equilateral triangle. The sum inside the parenthesis is the partial sum of a geometric series with ratio $4/9 < 1$, and the sum converges as n goes to infinity, so that one obtains $A_* = \lim_{n\to\infty} A_n = 8A_0/5$. In this way, we realize that the Koch star is a polygon with an infinite perimeter which encloses a finite area. Finally, making use of Eq.(1.11) we get $D = \ln 4/\ln 3 = 1.2619...$which is greater than the dimension of a line but less than that of a plane. The notion of Cantor set can also be extended to higher dimensions, and the corresponding Hausdorff dimension is given by $D_N = N \times \ln 2/\ln 3$.

As we can see, the recourse to the self-similarity allows to evaluate D in a straightforward manner. Unfortunately, fractals found in nature, such as colloidal aggregates, lightning strikes, dendritic particles, spatial distributions of cracks in solids, or filamentary arrangements of galaxy clusters, differ from ideal, mathematical fractals, because they exhibit only statistical self-similarity in a limited range of spatial scaling lengths. In fact, most naturally occurring fractal structures can be regarded as an ensemble of diverse fractals of different dimensions characterized by different weights. Such objects are called multifractals and more elaborated computation approaches are required in order to obtain their fractal dimensions. Making use of these techniques the fractal dimension of Britain ($D = 1.24$) and Norway ($D = 1.52$) coastlines have been measured.[19] The notion of fractal dimension also applies to diverse biological representatives like broccoli ($D = 2.66$), surface of human brain ($D = 2.79$)[20], or alveoli of a lung ($D = 2.97$).[19] The question about why Nature gives rise to fractal structure so profusely still remains. This will require the formulation of suitable models of fractal growth based on physical phenomena and subsequent understanding of their mathematical structure.

1.6 Hierarchical architecture of biomolecules

Biological macromolecules are a prerequisite for all forms of life. Attending to their structural order we can group them into two broad sets. On the one hand, we have macromolecules consisting of a linear periodic arrangement of monomers. On the other hand, we have macromolecules composed of several kinds of monomers aperiodically ordered instead. Most of structural biomolecules, like amylose, cellulose, or collagen belong to the first class, whereas nucleic acids and most proteins belong to the second one. Nucleic acids can be further separated into biological and synthetic. In general, synthetic nucleic acids are oligonucleotides where a few (5-60) base pairs are periodically arranged. These structures are quite different from the biological ones where thousands to millions of base pairs, including four different nucleotides, are aperiodically distributed. In fact, it is precisely on account of their aperiodic order that nucleic acids and proteins can carry genetic information or perform specific catalytic functions, respectively. The aperiodic order present in the structure of proteins can be understood in terms of a *hierarchical scheme,* as it is illustrated in Fig.1.12. In the first level we find periodically ordered structures, like $\alpha-$helix or $\beta-$sheets, interspersed with random sections, where the protein chain exhibits enhanced flexibility. As a result, the macromolecule adopts its biologically functional ternary structure. In some cases, several ternary units associate to build up a quaternary, supramolecular structure. Accordingly, different kinds of order predominate at different scales in proteins.[21]

An even richer hierarchy of ordering schemes can be found in nucleic acids. In the short range, a sequence periodicity of 3 base pairs indicates the presence of protein coding sequences, a feature which can be used to distinguish coding and noncoding DNA regions.[23] Initial analyses from human genome show that protein coding regions constitute less than 3% of the total genome. In contrast, about 50% of the human genome consists of repetitive sequences.[24, 25] These repeats are approximate copies of patterns of nucleotides of various lengths interspersed throughout the genome. Thus, sequence periodicities of about 10 base pairs reflect DNA bendability, much as the secondary structure of proteins.[26] On the next length scale, correlations in the order of 100 base pairs can explain the nucleosomal structure in eukaryotes.[27] Finally, compositional heterogeneities in the range $10^2 - 10^6$ base pairs are related to the presence of wide specific domains characterized by long-range correlations obeying power laws.[28, 29]

FIGURE 1.12

Crystal structure of human hemoglobin at 1.73 Å resolution.[22]

1.7 The role of correlated disorder

Up to now we have mainly considered perfectly ordered systems, which can only describe a highly idealized situation. In actual samples different types of structural defects usually appear ranging in size from the atomic dimension scale to bulk fractures in the macroscopic domain. Among the different sources of disorder in real crystals we found isotopic substitution (when a given atom is replaced by some isotope, hence locally changing some physical properties, but leaving the structure chemically invariant), substitutional impurities (when an atom of a given valence state replaces other atom of identical valence in the original lattice, for instance $B^{+3} \rightarrow Al^{+3}$), missing atoms at one or more lattice sites (vacancies), atoms which force their way into a hole between lattice sites (interstitial), or extra half-planes of atoms that go part way through the lattice (dislocations). In addition to these structural defects any crystalline structure at finite temperature is subjected to thermal fluctuations which separate their constituent atoms from their ideal equilibrium positions, introducing dynamical disorder effects.

Disorder manifests itself in terms of several destructive effects in many instances. Thus, as a consequence of the breaking of the translational symmetry, Paul W. Anderson showed that, in tight-binding models with independent random interactions, the spatial distribution of electronic states exhibits a finite localization length, outside which the probability of finding the electron is negligible.[30] Correspondingly, the charge is mainly localized around

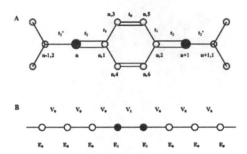

FIGURE 1.13

A typical quinoid defect in emeraldine polymer (A) can be exactly mapped into a dimer-like defect in a linear chain (B) by renormalizing the carbon atoms comprised between the two nitrogen atoms (black circles) at the end positions n and $n + 1$.

atomic positions and the electron diffusion coefficient vanishes in the considered system, which becomes an insulator. Localization by disorder was also found in vibrations of glasslike materials around the same time.[31] These results led to the view that disorder induces the localization of all eigenstates in one-dimensional chains, even for an infinitesimal amount of disorder.[32] However, it was subsequently realized that disorder effects can also be of creative nature, playing a significant role in the emergence of complex phenomena, provided that some of the following ingredients (or simultaneously both) are also present in the considered system: correlations and non-linearity.

Disorder correlations imply that the neighbor random parameters are not independent within a certain length – the so-called correlation length. This property naturally introduces a short-range order scale in the system, so that the competition between the long-range disorder and this short-range correlation can eventually suppress the localization effect, allowing for materials with better than expected transport properties.

Inspired by this physical scenario, D. Dunlap, along with H. L. Wu and P. Phillips, proposed a model aimed at studying the role of disorder in one dimensional systems where the existence of short-range correlations was explicitly considered.[33] This model, known in the literature as random dimer model, is characterized by the presence of randomly distributed impurities along a chain, so that they can only appear as dimers. In this way, the correlation length coincides with the lattice spacing. They showed that for a certain energy the reflection coefficient of a single dimer vanished, and that this resonance was preserved when a finite concentration of dimers were randomly placed in the chain. This gives rise to a set of delocalized states whose number increases as the square root of the number of sites, hence affecting the transport properties of the system. In a series of subsequent works, this model was successfully applied to describe the possible mechanism of electronic transport in diverse conducting polymers, such as polyaniline and polyparaphenylene.[34] Polyaniline refers to a general class of aromatic rings of benzenoid and quinoid character connected by nitrogen atoms. In Fig.1.13 a typical quinoid building block of polyaniline molecule is shown. Such a unit is composed of an unsaturated ring of six carbon atoms flanked by two nitrogen atoms. As it is illustrated in this figure, this basic structure can be exactly mapped into a simpler dimer-like effective molecule. The resulting system (referred to as renormalized chain) can then be properly described in terms of the random dimer model. In this way the interest of this model was extended from the fundamental research field to play an active role in the domain of experimentally testable models.

The suitability of the random dimer model to describe other systems of physical interest has been demonstrated in the case of superlattices as well. To this end, three GaAs–AlGaAs based superlattices were grown by molecular beam epitaxy and their relative thicknesses tailored to obtain a periodic, random, or random dimer arrangement of layers, respectively. The electronic properties of the different kinds of superlattices were studied by means of photoluminiscence and vertical resistance measurements. The obtained results clearly demonstrated that the intentional introduction of short-range correlations in a disordered semiconductor superlattice inhibits localization and gives rise to extended states, as theoretically predicted.[35]

1.8 Long-range correlations: The DNA case

From a fundamental viewpoint the study of the random dimer model properly illustrates that, as soon as short-range dimer correlations among basic structural units are introduced in an otherwise random linear structure, a significant number of extended states appear, efficiently contributing to the electrical transport.[36] It is then reasonable to expect that self-similar systems, exhibiting richer correlation patterns among their basic building blocks (as those we will introduce in Chapters 2, 3, and 5), are able to support extended electronic states as well.[37, 38, 39, 40, 41] Generally speaking, the presence of correlations (whose main effect is to reduce the degree of disorder) will favor the existence of resonant extended states in the energy spectrum, hence contributing to enhance the transport properties. In fact, it was theoretically shown that chain models with certain long-range correlations among the energy of atomic sites (ε_n) exhibit a phase of extended electronic states.[42] The considered disorder distribution obeys a power-law spectral density of the form $S(k) = 1/k^\alpha$, with $\alpha > 2$, where $S(k)$ is the Fourier transform of the two-point correlation function $\langle \varepsilon_i \varepsilon_j \rangle$, and k is related to the wavelength λ of the undulations in the random parameter landscape by $k = 1/\lambda$. At variance with the results obtained for random n-mer models (where delocalization is observed only at particular resonance energies) this model exhibits a true Anderson transition, with mobility edges separating localized and extended states. The existence of mobility edges in one-dimensional potentials in the presence of specific long-range correlations was further analyzed in terms of the relation between the correlation function and the localization length.[43] In particular an algorithm to numerically construct potentials with mobility edges was provided and such a mobility edge was indeed observed in a waveguide with intentionally introduced correlated disorder.[44]

During the past few years, the nature of long range correlations in DNA sequences has been the subject of intense debate.[45] Scale invariant properties in complex genomic sequences with thousands of nucleotides have been investigated in particular with wavelet analysis, and have been argued to play a crucial role in gene regulation and cell division. Indeed, a precise understanding of DNA-mediated charge migration should have a strong impact on the description of damage recognition processes and protein binding, or in engineering biological processes.[46, 47] Biological and artificial DNA molecules significantly differ in size, chemical complexity, and the kind of structural order. Consequently, one can hardly expect that results obtained from the study of the oversimplified synthetic molecular systems may be directly extrapolated to understand the physical properties of complex DNA molecules of biological interest. In fact, both the sugar-phosphate backbone and the nucleotide bases sequence are periodically ordered in, say, polyG-polyC chains, whereas in biological DNA the nucleotide bases are aperiodically ordered instead. From general principles one expects the aperiodic nature of the nucleotide sequence distribution would favor localization of charge carriers in biological nucleic acids, reducing charge transfer rate due to backscattering effects. Nevertheless, this scenario must be refined in order to take into account correlation effects among nucleotides reported in biological DNA samples, since these correlations can enhance charge transport via resonant effects.[48, 49, 50] Besides its fundamental importance for the progress of biological condensed-matter theory, several properties of biological interest may be directly related to the presence of sequence correlations in genomic DNA, including gene regulation, cell division, or damage recognition processes due to DNA-mediated charge migration.[46, 47] In this sense, certain aperiodic systems may be regarded as useful model prototypes, able to mimic relevant features related to long-range correlation effects in natural DNA samples, as we will discuss in Chapters 5 and 6.

References

[1] Matthews M R 2005 *The Pendulum: Scientific, Historical, Philosophical and Educational Perespectives* (Springer-Verlag, Berlin)

[2] Janssen T, Chapuis G, and de Boissieu M 2007 *Aperiodic Crystals: From Modulated Phases to Quasicrystals* (Oxford University Press, Oxford)

[3] Franklin R E and Gosling R G 1953 *Nature* **171** 740. A complete explanation of most details enclosed in this diffraction picture can be learnt at http://www.pbs.org/wgbh/nova/photo51/

[4] Galileo G 1623 *The Assayer*

[5] Koszelak S, Day J, Leja C, Cudney R and McPherson A 1995 *Biophys. J* **69** 13

[6] Goddard T http://www.rbvi.ucsf.edu/Research/afm/stmv/afmaverage.html

[7] Chapuis G 2003 *Crystal Engineering* **6** 187

[8] Skotheim T A 1986 *Handbook of Conducting Polymers* (Dekker, New York)

[9] Chen J, Heeger A J, and Wudl F 1986 *Solid State Commun.* **58** 251

[10] Bredas J L and Street G B 1985 *Acc. Chem. Res.* **18** 309

[11] Ofer D, Crooks R M, and Wrighton M S 1990 *J. Am. Chem. Soc.* **112** 7869

[12] Kroto H 1997 *Rev. Mod. Phys.* **69** 703

[13] Khanna S N and Jena P 1992 *Phys. Rev. Lett.* **69** 1664; ibid. 1995 *Phys. Rev. B* **51** 13705

[14] Ashman C, Khanna S N, Liu F, Jena P, Kaplan T, and Mostoller M 1997 *Phys. Rev. B* **55** 15868

[15] Perkins C L, Trenary M, and Tanaka T 1998 *Phys. Rev. B* **58** 9980

[16] Mackay A L 1998 *Nature (London)* **391** 334; 1962 *Acta Crystallogr.* **15** 916

[17] Esaki L and Tsu R 1970 *IBM J. Res. Develop.* **14** 61

[18] Mandelbrot B B 1982 *The Fractal Geometry of Nature* (Freeman, San Francisco)

[19] http://en.wikipedia.org/wiki/List_of_fractals_by_Hausdorff_dimension

[20] http://www.heise.de/tr/artikel/54311/2/0

[21] Volkenshtein M V 1985 *General Biophysics* (Academic Press, New York)

[22] Dasgubta J, Sen U, Choudhury D, Chakrabarti A, Chabrabarty S B, Chakrabarty A, Dattagupta J K 2004 *Biochem. Biophys. Res. Comm.* **303** 619. Picture downloaded from the *Image Library of Biological Macromolecules* database (http://www.imb-jena.de/IMAGE.html)

[23] Staden R and McLachlan A D 1982 *Nucleic Acids Res.* **10** 141

[24] Lander E et al. 2001 *Nature (London)* **409** 860

[25] Venter J C et al. 2001 *Science* **291** 1904

[26] Herzel H, Weiss O, and Trifonov E N 1999 *Bioinformatics* **15** 187

[27] Audit B, Thermes C, Vaillant C, d'Aubenton-Carafa Y, Muzy J F, and Arneodo A 2001 *Phys. Rev. Lett.* **86** 2471

[28] Peng C K et al. 1992 *Nature (London)* **356** 168

[29] Voss R F 1992 *Phys. Rev. Lett.* **68** 3805

[30] Anderson P W 1958 *Phys. Rev.* **109** 1492

[31] Dean P 1964 *Proc. Phys. Soc.* **84** 727

[32] Ziman J M 1979 *Models of Disorder* (Cambridge University Press, London)

[33] Dunlap D H, Wu H L, and Phillips P W 1990 *Phys. Rev. Lett.* **65** 88

[34] Phillips P W and Wu H L 1991 *Science* **252** 1805; Wu H L and Phillips P W 1991 *Phys. Rev. Lett.* **66** 1366

[35] Bellani V, Diez E, Hey R, Toni L, Tarricone L, Parravicini G B, Domínguez-Adame F, and Gómez-Alcalá R 1999 *Phys. Rev. Lett.* **82** 2159; Parisini A, Tarricone L, Bellani V, Parravicini G B, Diez E, Domínguez-Adame F, and Hey R 2001 *Phys. Rev. B* **63** 165321

[36] Sánchez A, Maciá E, and Domínguez-Adame F 1994 *Phys. Rev. B* **49** 147

[37] Maciá E 1998 *Phys. Rev. B* **57** 7661

[38] Maciá E and Domínguez-Adame F 1996 *Phys. Rev. Lett.* **76** 2957

[39] Kumar V 1990 *J. Phys. Condens. Matt.* **2** 1349

[40] Chakrabarti A, Karmakar S N, and Moitra R K 1992 *Phys. Lett. A* **168** 301; Oh G Y, Ryu C S, and Lee M H 1992 *J. Condens. Matter* **4** 8187; Chakrabarti A, Karmakar S N, and Moitra R K 1994 *Phys. Rev. B* **50** 13276

[41] Oviedo-Roa R, Pérez L A, and Wang C M 2000 *Phys. Rev. B* **62** 13805; Sánchez V and Wang C M 2004 *Phys. Rev. B* **70** 144207

[42] de Moura F A B F and Lyra M L 1998 *Phys. Rev. Lett.* **81** 3735

[43] Izrailev F M and Krokhin A A 1999 *Phys. Rev. Lett.* **82** 4062

[44] Kuhl U, Izrailev F M, Krokhin A A, and Stöckmann H J 2000 *Appl. Phys. Lett.* **77** 633

[45] Carpena P, Galván P B, Ivanov P Ch, and Stanley H E 2002 *Nature (London)* **418** 955; 2003 *Nature (London)* **421** 764

[46] Braun E, Eichen Y, Sivan U, and Ben-Yoseph G 1998 *Nature (London)* **391** 775

[47] Treadway C, Hill M G, Barton J K 2002 *Chem. Phys.* **281** 409

[48] Roche S, Bicout D, Maciá E, and Kats E 2003 *Phys. Rev. Lett.* **91**, 228101; Roche S, Bicout D, Maciá E, and Kats E 2004 *Phys. Rev. Lett.* **92**, 109901(E).

[49] Roche S and Maciá E 2004 *Modern Phys. Lett. B* **18** 847

[50] Bagci V M K and Krokhin A A 2007 *Chaos, Solitons and Fractals* **34** 104

2

The notion of aperiodic crystal

2.1   Order without periodicity

Our current understanding of basic properties of most physical, chemical, and biological systems relies on the very notion of order. In most cases, however, we usually restrict ourselves to mainly consider *periodic order*, as we have seen in Chapter 1. Such is the powerful attraction the notion of a regular pattern exerts on our minds. Notwithstanding this, advances in several research fields have led to a progressive change of paradigm during the last decades, as the concept of *order without periodicity* has emerged to properly describe an increasing number of complex systems. In this way, the study of phenomena exhibiting aperiodic arrangements of matter in space and/or time has been spurred, and the basic knowledge gained from these studies has naturally extended to applied research domains. Thus, new designs based on the aperiodic ordering of different building blocks are paving the way for promising technological applications as well.

From a historical perspective, the mathematics of *aperiodic order* has been an active branch for a long time,[1, 2] where it was well known that some functions had the property of being *"almost periodic"* (see Section 9.1). The mathematical bases for the study of aperiodic functions were laid in 1933 by Harald Bohr (1887-1951). In particular, quasiperiodic functions, a subset of the broader family of almost periodic functions, will play a major role in the contents described in this book. Conversely, such a notion was scarcely considered in other scientific domains. In fact, as we have reviewed in Chapter 1, the study of the structure of matter has been traditionally based upon the notion of a regular arrangement of atoms in space, which can be generated by periodic translations of a basic motif or building block. Thus, the recourse to the translational and rotational symmetries, stemming from the periodic order present in the underlying substrate, has allowed for a proper understanding of most relevant properties of crystals. The notion of a periodic arrangement of atoms in space also introduces a quite natural classification scheme of matter into two broad categories: crystalline matter and amorphous matter. The dichotomy implied by this classification scheme led to the progressive consideration of crystalline matter as a paradigm of *order* in solid state physics.

This view naturally pervaded life sciences as well. During the 1930s, DNA was considered to be merely a tetranucleotide of the form, namely, GACT, composed of one unit each of deoxy-adenylic, -guanilyc, -thymidylic, and cytidilic acids (the particular order of appearance of each of these bases being considered as irrelevant). Even when it was subsequently realized, in the early 1940s, that the molecular weight of DNA is actually much higher, it was still widely believed that the tetranucleotide unit was the basic repeating building block of the large DNA polymer, in which the four different kinds of nucleotides recur in periodic sequence, i.e., $(GACT)_n$.[3] Thus, DNA was originally viewed as a trivially periodic macromolecule, unable to store the amount of information required for the governance of cell function.

The conceptual difficulty of assigning a genetic role to a periodically arranged DNA did not escape the first researchers in the field, who stated that "nucleic acids must be regarded as possessing biological specificity, the chemical basis of which is as yet undetermined."[4] The mystery of the nature of the genetic material attracted some physicists to genetics. In 1935 Max Delbrück (1906-1981) published a speculative paper in which he pointed out that one of the most striking aspects of the gene is its long-term stability.[5] To account for that stability, he proposed that the gene is a molecule whose constituent atoms are fixed in their mean positions and its electronic properties are determined by the electronic states according to the basic concepts put forward by the (then) recently introduced quantum mechanics. According to such a theory, whenever the DNA molecule is excited by acquiring an energy amount greater than the activation energy required to change its particular state, a discontinuous change would occur, corresponding to a mutation.

Ten years later, Erwin Schrödinger (1887-1961) extended this seminal idea in order to account for the notion of the gene as an information carrier. To this end, he suggested that its information is stable because the chromosome in which it is embedded consists of a long sequence of a few repeating elements exhibiting a well defined order *without* the recourse of periodic repetition, and illustrated the vast combinatorial possibilities of such a crystal. In this way, the notion of a one-dimensional *aperiodic crystal* was introduced,[6] and we can consider Schrödinger was the first person to put forward the notion of a linear genetic code.[3] Subsequent, more refined, chemical analysis of DNA samples from diverse biological sources by Erwin Chargaff (1905-2002) showed that the tetranucleotide hypothesis was indeed flawed, since the relative contents of the four nucleotide bases adenine (A), guanine (G), cytosine (C), and thymine (T) vary over a wide range.[7] Hence, Schrödinger's proposal of considering DNA as a one-dimensional aperiodic *chain*, with four different nucleotides arranged in a way able to store the required genetic information, was progressively incorporated into dominant biological thinking. Nevertheless, the very notion of a full-fledged three-dimensional aperiodic *crystal* remained dormant for almost four decades until the discovery of a new type of solids, which defy standard crystallographic classification, awoke the condensed matter community.

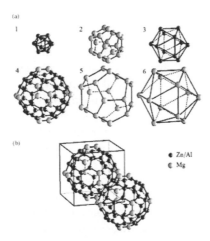

FIGURE 2.1
(a) Successive atomic shells of the six-cell Bergman cluster. (b) Body-centered packing of the Bergman clusters, sharing a hexagonal face of the fourth shell produce the structure of the $(Al,Zn)_{49}Mn_{32}$ Bergman phase. ([8] Courtesy of Janez Dolinšek.)

2.2 Bringing in icosahedral thought

2.2.1 A hierarchy of platonic shells

The very possibility of explaining orderings in nature by means of a hierarchical geometrical construction based on the systematic use of Platonic solids can be probably traced back to Johannes Kepler (1571-1630) who tried to explain the relative size of orbits of the six planets known at the time by successively inscribing an octahedron, icosahedron, dodecahedron, tetrahedron, and hexahedron of proper sizes in a series of concentric spheres. This cosmological model was published in his book *Mysterium Cosmographicum* (1597). Since then, the basic idea – introducing geometrical constraints in the orderings of matter – has reappeared in several scientific domains. In particular, the possible existence of materials containing units of icosahedral symmetry has received some attention in a variety of different contexts.

Chemists became interested in icosahedra because several clusters of atoms have been found to be closely related to the icosahedron point group symmetry, the celebrated case of fullerene being a paramount example of this. Another instance is provided by icosahedral arrangements of 13 atoms, which are common in gas phase metal clusters or the structures of many boron-rich solids containing icosahedral arrangements of boron atoms,[9, 10] sometimes hierarchically structured in terms of an icosahedron of icosahedra building blocks (see Fig.1.8 in Section 1.2).

On the other hand, in solid state physics, the interest on the possible role of icosahedral structural units in some condensed phases can be traced back at least to the 1950s when the icosahedral thought came up in several contexts. There are many known structures of intermetallic compounds that involve icosahedral coordination. These structures are usually complex, with 20, 52, 58, 162, 184, or even more atoms in the unit cell. Relatively simple cubic crystals based on icosahedral building blocks are provided by $MoAl_{12}$, WAl_{12}, or $MnAl_{12}$ alloys. In their structures, at each lattice point of a body-centered cubic lattice, there is a regular icosahedron of twelve aluminum atoms about the smaller transition metal atom which occupies the central position.[11] Linus Pauling (1901-1994) and co-workers successfully described a complex phase corresponding to $(Al,Zn)_{49}Mn_{32}$ alloy containing 162 atoms in its unit cell (Fig.2.1).[12] The structure is based on a body-centered lattice. At each lattice point there is a small atom (Zn, Al) which is surrounded by an icosahedron of twelve atoms. This group is then surrounded by 20 atoms, at the corners of a pentagonal dodecahedron, each atom lying directly out from the center of one of the centers of the pentagonal faces of the icosahedron. Twelve more atoms lie out from the centers of the pentagonal faces of the dodecahedron. At this stage, the resulting cluster is composed of 45 atoms, the outer 32 of which lie at the corners of a rhombic triacontahedron, a polyhedron with 30 rhombic faces which can be obtained as the union of an icosahedron and a dodecahedron. The next shell consists of 60 atoms, each directly above the center of a triangle that forms one-half of each of the 30 rhombic faces of the underlying triacontahedron (Fig.2.2). These 60 atoms then lie at the corners of a truncated icosahedron, which has 20 hexagonal faces and 12 pentagonal faces. Finally, twelve additional atoms are located out from the center of 12 of the 20 hexagonal faces completing the impressive series of closed shell atomic clusters based on icosahedral symmetry.

Following a different line of thought John D. Bernal (1901-1971) considered the icosahedral coordination as a key to understand the structure of liquid water, precisely on the basis that objects of icosahedral symmetry cannot fill three dimensional space, which naturally prevents periodic crystallization. In fact, J. D. Bernal was extremely keen on hierarchy as a principle of building things and generalizing crystallography,[14] and this guiding principle inspired Alan L. Mackay to construct an atomic cluster based on an arrangement of three successive icosahedral symmetry shells (Fig.2.3): an inner icosahedron, a double-sized icosahedron, and an icosidodecahedron. The atomic cluster introduced by Mackay has played a relevant role in the field of quasicrystal research. In fact, the first thermodynamically stable quasicrystal found exhibits a triacontahedral growth habit (see Fig.2.10 in Section 2.3.4) and a slightly modified version of the Mackay icosahedron plays a fundamental role in some structural models proposed for icosahedral quasicrystals (see Section 3.2.6).

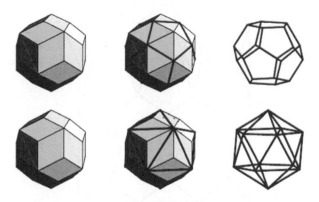

FIGURE 2.2

The short diagonals of the faces of the rhombic triacontahedron give the edges of a dodecahedron (top row), while the long diagonals give the edges of an icosahedron.

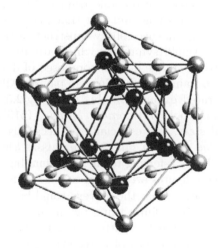

FIGURE 2.3

A Mackay cluster contains 54 atoms arranged within three successive shells: 12 atoms at the vertices of an inner icosahedron (in black), 12 atoms at the vertices of a larger icosahedron (in dark gray), and 30 atoms on the two-fold axes of an outer icosidodecahedron (in gray). The diameter of this atomic cluster is about 0.96 nm. (From ref.[13]. By permission of Oxford University Press.)

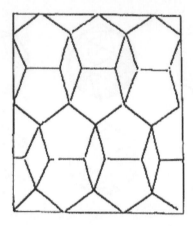

FIGURE 2.4
Finite tiling containing pentagons and rhombi.

2.2.2   Fivefold tilings in plane and space

In Section 1.2 we considered the possible tilings of the Euclidean plane by different sorts of regular polygons and saw that pentagons are not allowed. Notwithstanding this, one can successfully construct certain particular, *finite* tiles exhibiting global fivefold symmetry which are able to cover the plane without gaps. A relatively simple design, composed of double pentagons and rhombi (Fig.2.4), was obtained by the German painter Albrecht Dürer (1471-1528). A more elaborated example of this sort (Fig.2.5) was published by Johannes Kepler in his book *Harmonices Mundi* (1519), and it is formed by pentagons, fivefold stars, and especially by decagons and double tiles of combined decagons (*duo decagoni inter se commissi*) which were termed by Kepler *monstrum*, probably referring to their unusual big size. As we will see in Section 2.4.2, all these figures are related to the Golden mean.

Although Kepler was aware that the pattern could be continued, the extension of Kepler's patch to an infinite tiling has no unique solution.[16] Thus, the key to obtain fivefold symmetric patterns was to relax the periodicity condition in order to obtain collections of polygons capable of covering a plane with neither gaps nor overlaps in such a way that the resulting overall pattern lacks any translational symmetry, that is, to construct two-dimensional aperiodic tilings.

The existence of aperiodic sets of shapes capable of tiling an *infinite* plane only in an aperiodic fashion was proved in the 1960s, involving the presence of a huge number of different geometrical tiles.[17, 18] These early results spurred an intense research activity among mathematicians which ultimately led to the discovery of the *simpler aperiodic set* by Roger Penrose in 1974[19],

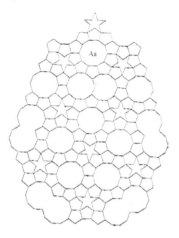

FIGURE 2.5

Kepler's Aa tiling: a finite patch containing different decagonal and pentagonal tiles. (Adapted from ref.[15, 16].)

which consists of just two different tiles: a flat and a thick rhombus, as it is illustrated in Fig.2.6.

In 1991 Sergei E. Burkov realized that a quasiperiodic tiling of the plane can be obtained by using just a single decagonal tile if the tiles are allowed to *overlap*, hence relaxing a basic property of previously considered tilings.[20] Five years later, Petra Gummelt proved rigorously that the Penrose tiling can be obtained by using a single "decorated" decagon combined with a specific overlapping rule.[21] Although the recourse to overlapping may seem somewhat tricky from a purist mathematical viewpoint, it turns out that it makes physical sense if one considers the presence of atomic clusters sharing a number of common atoms in actual quasicrystals models (see Section 3.2).

Instead of using plane tiles to cover a plane one may think of using three-dimensional bricks to fill up the space, hence extending Penrose's approach to higher dimensions. This enterprise was undertaken by the mathematician Robert Ammann (1946-1994), who in 1976 discovered a pair of squashed and stretched blocks, referred to as Ammann rhombohedra (Fig.2.7), that can fill up the space with no gaps.[22] The pattern that emerges when properly ensembling these building blocks is nonperiodic in nature and exhibits the operation symmetries of the icosahedral group, a forbidden symmetry in classical crystallography. In this way, we are led to another instance of a geometrical construction which allows for the emergence of a forbidden order by simply abandoning the periodicity paradigm.

At the early 1980s, Dov Levine was starting his Ph.D. under the supervision of Paul Joseph Steinhardt with a project concerning whether metallic

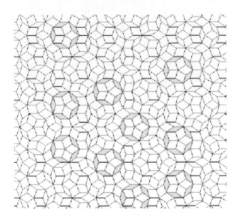

FIGURE 2.6
Penrose tiling illustrating the local isomorphism characteristic of Conway's theorem. The entire tile is composed of two basic tiles.

glasses have any inherent symmetry. He was (no wonder) fascinated by Penrose tilings and their possible generalization to higher dimensions. Following the Ammann steps they realized that the most relevant feature of Penrose construction was not the fact it is aperiodic, but that it possessed a kind of translation order, and it is capable of producing sharp diffraction peaks.[14] To understand this key property the intuitive concepts of order and symmetry need to be suitably extended beyond some of the traditional views related to the concept of periodicity. Thus, the notion of repetitiveness, typical of periodic arrangements, should be replaced by that of local isomorphism, which expresses the occurrence of any bounded region of the whole tiling infinitely often across the tiling, irrespective of its size. In the particular case of Penrose tilings, John Horton Conway obtained a theorem which states that given any local pattern having a certain diameter, an identical pattern can be found within a distance of two diameters.[23] This interesting result is illustrated in Fig.2.6. In this way, a purely mathematical search looking for higher dimensional analogous to the Penrose lattice paved the way to the very notion of quasiperiodic crystals.

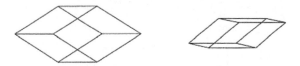

FIGURE 2.7
The Ammann rhombohedra.

2.3 Beautiful forbidden symmetries

"as I was studying rapidly solidified aluminum alloy which contained 25% manganese by transmission electron microscopy, something very strange and unexpected happened. There were 10 bright spots in the selected area diffraction pattern, equally spaced from the center and from one another. I counted them and repeat the count in the other direction and said myself: 'There is no such animal.' In Hebrew: 'Ein Chaya Kazo.' I then walked out to the corridor to share it with somebody, but there was nobody there..." (Dan Shechtman commenting on his April 8, 1982 finding [14])

2.3.1 Ein chaya kazo

As it was mentioned in Section 1.2, modern crystallography started in 1912 with the introduction of x-ray crystallography. Since then, all the crystals studied were periodically ordered, and thus a paradigm evolved that all crystals *must* be periodic. Consequently, crystallography textbooks properly stated that the allowed rotational symmetries are twofold, threefold, fourfold, and sixfold. The 14 Bravais lattices, along with the 230 space groups, provided the basic tool for crystal classification and the *International Tables of Crystallography* was the ultimate classification catalog for crystals. Accordingly, for many decades, it had been believed that five-fold rotational symmetry could not exist in stable condensed phases.

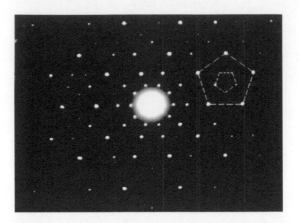

FIGURE 2.8

Electron diffraction pattern corresponding to an AlCuFe icosahedral quasicrystal. A 5/10 fold symmetry axis around the origin can be clearly appreciated. A Pythagorean pentagram is shown on the upper right corner.
(Courtesy of José Reyes-Gasga).

Mineralogists have discovered materials containing icosahedral units, either
isolated or linked, but these materials exhibited global symmetries proper of
periodic order. Thus, the pentagon symmetry, which is widely found in the
world of the living, was excluded from the mineral kingdom, until the existence of long-range quinary symmetry was first reported by Dan Shechtman,
Ilan Blech, Denis Gratias, and John W. Cahn in a paper entitled "Metallic
phase with long-range orientational order and no translational symmetry,"
published in *Physical Review Letters* on 12 November 1984.[24] In this paper
the existence of a metallic solid which diffracts electrons like a single crystal
does but has icosahedral point group symmetry, which is inconsistent with
periodic lattice translations, was first reported. This unexpected symmetry is
illustrated in the electron diffraction pattern shown in Fig.2.8, where a conspicuous 5/10-fold symmetry axis is clearly visible. When the specimen is
tilted and viewed from other directions, patterns of two and threefold symmetry can be observed. The complete account of symmetry rotation axes yields
15 twofold axes, 10 threefold axes, and 6 fivefold axes, amounting to the 120
symmetry elements characteristic of the icosahedral group.

The remarkable sharpness of the diffraction spots indicates a high coherency
in the spatial interference, comparable to the one usually encountered in ordinary crystals. Thus, these alloys exhibit well-defined Bragg peaks, characteristic of long-range order in their diffraction patterns, but one which is
explicitly incompatible with periodic translational one, since the crystallo-

graphic restriction theorem (see Section 1.1) indicates that fivefold axes are inconsistent with translational order.

2.3.2 The twinning affair

"I have found it hard to believe that any single crystal with 5-fold axes could give reasonably sharp diffraction patterns, resembling those given by crystals" [25] Linus Pauling (Nobel Prize in Chemistry 1954, Nobel Peace Prize 1963)

Faced with an apparent paradox, the very nature of these materials was early disputed by prominent scientists. A representative example of the conservative position is clearly illustrated by the above quotation. This attitude, somewhat surprising coming from one of the scientists who did more to bring in the icosahedron symmetry into crystallography, can be understood on the basis that the only existing data were in reciprocal space, which lacked a simple real-space structural interpretation, and led him to interpret the icosahedral symmetry exhibited by the diffraction patterns as due to directed multiple twinning of ordinary cubic crystals with large edges (within the range 23-28 Å). The resulting structures were very complex, each basic structural unit containing about one thousand atoms and requiring these units to be precisely arranged to produce an aggregate with approximate icosahedral symmetry.[25]

Certainly, Shechtman was completely aware of such a possibility and he took care of disregarding it from the very beginning.[14, 24] To this end, he generated a series of dark-field images. In these experiments one takes an image from a diffracted point such that all the information that passes through the plate is contained in this beam only. The obtained images were almost identical, which means that the same part of the crystal produced *all* the diffracted beams. A complementary experiment in the electron microscope is the convergent-beam experiment, in which a diffraction pattern can be taken by focusing the beam onto a very fine spot. In this case one gets again a ten-fold diffraction pattern, and this means that if they were twins, they had to be smaller than the diameter of the convergent beam spots (about 40 nm). These observations strongly suggested that the sample was not twinned at all. In fact, as more data on better quasicrystalline specimens were progressively accumulated the reliability of the twinning hypothesis was fading down in favor of the advocates of a new paradigm based on a substantial revision of classical crystallography, a revision which appeared necessary in order to incorporate the very possibility of *ordered aperiodic* structures in condensed matter physics.

By 1982 Alan L. Mackay had published a simulated diffraction electron pattern which turned out to exhibit a certain resemblance to Shechtman's experimental observation. He had also obtained an optical Fourier transform of a Penrose tiling keeping an even closer similarity with Shechtman's diffraction.[26] However, at that time he was mainly interested in the study of hierarchic patterns, rather than in aperiodicity as such, and the interesting possibility of extended structures exhibiting fivefold symmetry in condensed matter remained mainly unnoticed by the physicist community.[14] Fortunately, Dov Levine and Paul Joseph Steinhardt were actively working on 3D generalizations of the Penrose tiling at the time and, spurred by the Shechtman's report, the notion of a new class of ordered solids was promptly introduced by them in a paper entitled "Quasicrystals: a new class of ordered structures."[27] Published in *Physical Review Letters* on 24 December 1984, the notion of a quasiperiodic crystal was introduced as the natural extension of the concept of crystal to structures where the periodic spatial order is replaced by a kind of translational order described in terms of *quasiperiodic functions* instead (see Section 9.1). For this reason these alloys were referred to as *quasicrystals* (QCs), a shorthand for quasiperiodic crystal. In retrospective, this semantics may appear somewhat misleading since the prefix "quasi" suggests an intermediate position between crystalline and amorphous matter rather than a well defined kind of ordered spatial arrangement. By the light of subsequent developments in crystallography a better term would probably be *hypercrystals* [28, 29].

2.3.3   The sequel

Shortly after the publication of the two seminal papers by D. Shechtman and co-workers, and by D. Levine and P. J. Steinhardt, a number of works reporting the existence of new QCs characterized by the presence of different types of non-crystallographic symmetry axes appeared in rapid succession. Thus, a new ordered state of matter exhibiting a twelvefold symmetry axis (dodecagonal) was found in small particles of a Ni-Cr alloy.[30] Quasicrystals with an eightfold rotational axis were reported in rapidly solidified (V,Cr)-Ni-Si alloys, and were referred to as representatives of the octagonal phase.[31] A phase in rapidly solidified Al-Mn alloys with higher manganese content (18-22 at.%) than those originally studied by Shechtman (10-20 at.%), which had been initially labeled as the T-phase, was subsequently realized to be another example of quasicrystal: the so-called decagonal phase.[32, 33] In this regard, it is interesting to note that, in 1978, it was reported the observation of a diffraction pattern with a pseudo-pentagonal symmetry in an investigation by transmission electron microscopy (the same technique used by Shechtman) of rapidly solidified AlPd alloys. Unfortunately, not enough attention was paid to this system, which was later recognized to belong to the decagonal phase. QCs belonging to this phase are characterized by the presence of a tenfold rotation axis along a decaprism axis (Fig.2.9). Cross sections of the decaprism show ten faceted planes containing aperiodically arranged clusters of atoms. These quasiperiodic planes, in turn, are periodically stacked along the prism axis, so that both kinds of order, periodic and quasiperiodic, coexist in the same sample. This interesting finding suggested that other kinds of mixed order

FIGURE 2.9
Scanning electron images showing the growth morphology of two representative quasicrystalline alloys. (right) A regular dodecahedral habit corresponding to the AlCuFe icosahedral phase. (left) A decagonal prism habit corresponding to the AlNiCo decagonal phase. (Courtesy of An Pang Tsai).

structures may exist. In fact, a one-dimensional QC derived from decagonal phase representatives was found in rapidly solidified AlNiSi, AlCuMn, and AlCuCo alloys. In addition to the periodic translation along the tenfold axis, the translation along one of the twofold axes normal to the tenfold one also becomes periodic in these materials.[34]

As it can be seen, all the different classes of QCs described in this account were obtained by means of far from equilibrium techniques and exhibited small sizes. In fact, they were not stable and returned to a usual crystalline state after heating at moderate temperatures. Also, the diffraction lines were rather broad, resembling those observed with highly faulted crystals. Thus, during the early years of the field, there were some speculations that these new solids, exhibiting symmetries forbidden for ordinary crystals, were located at a somewhat intermediate position between crystals and amorphous materials so that they might be inherently unstable. Those expectations proved, however, to be wrong.

2.3.4 A new phase of matter

Using a Debye-Sherrer method of power diffraction, Hardy and Silcock identified in 1955 the intermetallic compounds in equilibrium with the aluminum solid solution in the AlLiCu system. At $500°$ C they found several compounds corresponding to hexagonal and cubic structures, in addition to a

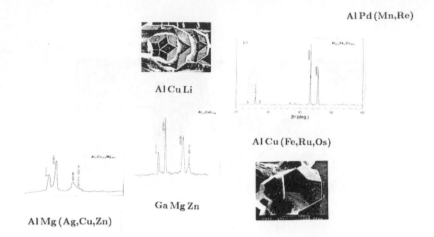

FIGURE 2.10

Comparison between the x-ray diffraction patterns corresponding to different families of thermodynamically stable quasicrystals. The scanning electron micrographs corresponding to the triacontahedral (dodecahedral) growth morphologies of AlCuLi (AlCuRu), respectively, are shown for the sake of illustration.

phase called T2, with composition close to Al_6CuLi_3, which exhibited Bragg peaks escaping any crystalline indexation.[35, 36] Notwithstanding this, the oddity remained dormant for more than three decades, until B. Dubost and co-workers proved that this phase actually belongs to the icosahedral family. In fact, their report announced the discovery of first thermodynamically stable quasicrystals, so that large grains could be grown to millimeter size by conventional solidification techniques in close to equilibrium conditions.[37] In this way, large quasicrystalline grains, exhibiting the growth morphology of a triacontahedron (Fig.2.10), were demonstrated in $Al_{56}Cu_{11}Li_{33}$ alloys by several groups. Shortly after, a second stable quasiperiodic crystal exhibiting a dodecahedral solidification morphology was reported in the $Ga_{20}Mg_{37}Zn_{43}$ system.[38]

The structural quality of a sample can be estimated from the so-called correlation length. This magnitude can be obtained from diffraction spectra through the relation $2\pi/FWHM$, where FWHM is the full width at half maximum of the diffraction peak. By inspecting Fig.2.10 we see that $Ga_{20}Mg_{37}Zn_{43}$ icosahedral phases do not show a very good structural quality, showing correlation lengths smaller than 50 nm. Following the identification of these phases, however, a series of thermodynamically stable quasicrystalline alloys of high structural quality in the icosahedral AlCu(Fe,Ru,Os), AlPd(Mn,Re) and Cd(Yb,Ca) systems,[39, 40, 41] and decagonal AlCo(Cu,Ni) system[42, 43] were discovered by A. P. Tsai and co-workers. Titanium based icosahedral quasicrystals in the TiZrNi alloy were also reported by K. F. Kelton and co-workers.[44] Currently, more than a dozen different quasicrystalline compounds have been reported to be thermodynamically stable up to their respective melting points and to exhibit Bragg peaks of extraordinary quality, comparable to those observed for the best monocrystalline samples ever grown. For the sake of illustration more representative compounds belonging to the icosahedral phase are listed in Table 2.1.

TABLE 2.1
Main representatives of high structural quality,
thermodynamically stable QCs.

Al - based	Zn - based	Ti - based	Cd - based
$Al_{70}Pd_{20}Mn_{10}$	$Zn_{80}Sc_{15}Mg_5$	$Ti_{45}Zr_{38}Ni_{17}$	$Cd_{85}Yb_{15}$
$Al_{63}Cu_{25}Fe_{12}$	$Zn_{60}Mg_{30}RE_{10}$	$Ti_{40}Hf_{40}Ni_{20}$	$(Ag,In)_{85}Ca_{15}$
$Al_{56}Li_{33}Cu_{11}$	$Zn_{43}Mg_{37}Ga_{20}$		

The discovery of this set of large, high quality samples has opened promising avenues in the study of physical properties of quasicrystals, allowing for detailed experimental studies of their related transport properties, which we will discuss in great detail in Section 3.1.2.

2.4 Unveiling Pythagorean dreams

"He himself could hear the harmony of the Universe, and understood the music of the spheres, and the stars which move in concert with them, and which we cannot hear because of the limitations of our weak nature" (Porphyry ca. 232-304 A. D. in his *Life of Pythagoras* ca. 570- B. C.)

2.4.1 The Pythagorean pentagram

If you connect all the vertices of a regular pentagon by diagonals you obtain the so-called Pythagorean pentagram (Fig.2.11), named after the symbolic use the members of this school gave to this figure. At their intersecting points the diagonals form a smaller pentagon at the center, and the diagonals of this pentagon form a new pentagram enclosing a yet smaller pentagon. This progression can be continued ad infinitum, creating smaller and smaller

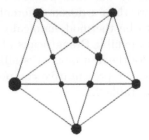

FIGURE 2.11

Sketch showing the first step in the construction of a nested series of Pythagorean pentagrams. The different sizes of the circles at the vertices and crossing points reflect the different intensities of the corresponding Bragg peaks in actual diffraction spectra (see Fig.2.8).

pentagons and pentagrams in an endless succession exhibiting a self-similar nesting characteristic of the fractal structures discussed in Section 1.4.

Surprisingly enough this geometrical construction can be easily recognized in the electron diffraction pattern shown in Fig.2.8. In fact, the diffraction pattern of icosahedral QCs is characterized by a series of pentagonal arrangements of Bragg peaks. In the ideal case these peaks will densely fill reciprocal space, though most of them are extremely weak in real samples, making it possible to distinguish individual spots. In turn, the self-similar property leads to the appearance of a hierarchy of intensities extending several orders in the diffraction pattern, which is illustrated by the nested pentagons highlighted on the upper right corner of Fig.2.8. In this way, we find a nice example of the Pythagorean pentagram engraved in the inner structure of a new phase of condensed matter. The contemplation of this diffraction pattern then provides an impressive example of novel geometrical patterns in reciprocal space, ultimately emerging from interference processes interweaving matter and waves at the atomic scale.

Certainly, the presence of a potentially infinite series of nested sets of Bragg spots endows QCs diffraction patterns with a peculiar beauty, unveiling the aesthetic fingerprint of the characteristic quasiperiodic order underlying their atomic arrangement. But it also precludes direct application of some basic tools of classical crystallography. In fact, the diffraction pattern of an ordinary crystal essentially reduces to one of the Bravais lattices so that in order to index such patterns one simply must obtain a set of three primitive basis vectors that generate the lattice. In that case, a physical length scale is easily deduced from the minimum separation of Bragg peaks in the three directions. Conversely, since the pattern of peaks is in principle dense for QCs, it is not even clear what sets the physical scale in this case, and such a procedure no longer applies to QCs.

2.4.2 The Golden mean

The diffraction pattern shown in Fig.2.8 illustrates several main features related to the characteristic inflation symmetry and self-similarity properties of quasicrystals. Let us start by measuring the ratios between successive diffraction spots along the radial directions measured from the centre. The obtained values are listed in the first column of Table 2.2 and they clearly follow a non-periodic series. Nevertheless, underlying these apparently unconnected number series we can appreciate a clear correspondence with successive powers of a certain number: the so-called Golden mean τ (from the Greek $\tau o\mu\eta$, which means "section"), as it is seen in the second and third columns.

TABLE 2.2
The main spots in the diffraction pattern of icosahedral QCs can be arranged according to a power series related to the golden mean.

	d_n/d_1	τ^m	m
d_1	1.00	1.00	0
d_2	1.42	1.62	1
d_3	2.63	2.62	2
d_4	4.25	4.24	3
d_5	6.87	6.85	4

The Golden mean has been largely known from the ancient times, since it is of frequent occurrence in the various pentagonal polygons and polyhedra and many geometrical figures can be constructed making explicit use of the Golden mean. Thus, for instance, a dodecahedron with an edge length of one has a total surface area of $15\tau/\sqrt{3-\tau}$, and a volume of $5\tau^3/(6-2\tau)$. Similarly, an icosahedron with a unit length edge has a volume of $5\tau^5/6$. In plane geometry the radius of the circle that circumscribes a decagon with a side length of a unit is equal to τ. In a similar way a triangle with a ratio of side to base of τ is known as a Golden triangle, whereas those later triangles with a ratio of side to base of τ^{-1} are sometimes called Golden gnomons. A unique property of both types of triangles is that they can be indefinitely dissected into smaller triangles that are also Golden triangles and Golden gnomons. Such a property underlies the construction of the Penrose tiling shown in Fig.2.6. The so-called Golden rectangle is a rectangle in which the ratio of length to width is given by τ. When you snip off squares from a Golden rectangle a series of nested Golden rectangles is obtained. If one connects the successive points where these squares divide the sides in Golden ratios, you obtain a logarithmic spiral that coils inward. Due to these plethora of interesting properties the Golden mean has been profusely implemented in arts such as painting, sculpture, and architecture.[45]

This number can be defined in several ways. For example, a segment is said to be divided in the Golden mean if the ratio of the whole segment to the larger part is equal to the ratio of the larger to the smaller part. If we take the smaller segment as unit and label the larger part as the unknown x, this geometrical definition can be expressed as

$$\frac{x}{1} = \frac{x+1}{x}, \tag{2.1}$$

which leads to the algebraic equation $x^2 = x + 1$, whose positive solution is given by the irrational number $x_+ = \tau = (\sqrt{5}+1)/2 = 1.6180339887....$ The other solution of the quadratic algebraic equation is negative, and can be expressed as $x_- = -\tau^{-1}$. Accordingly, we get the following expressions relating the Golden mean, its square, and its reciprocal

$$\tau^2 = \tau + 1, \quad \tau^{-1} = \tau - 1. \tag{2.2}$$

To our purposes, the Golden mean can be also obtained from the Pythagorean pentagram shown in Fig.2.8, where τ now appears as the ratio between the diagonal and the side of the original pentagon. One can readily check that the size of the original pentagon is related to the size of the smaller one by the scale factor τ^2. Since this process can be recursively applied to define smaller pentagons inside the previous ones, we realize that the entire diffraction patterns is hierarchically arranged according to a principle of scale invariance symmetry. In fact, the so-called inflation symmetry is a characteristic feature of the quasilattice of Bragg reflections from a quasicrystal. Once a vector basis for its indexation have been chosen, an alternative basis of diffraction vectors can be found with vectors parallel to the original ones and an irrational number as scale factor. Let us consider, for instance, the basic relation $\tau^2 = \tau + 1$. By successively multiplying it by τ we get

$$\tau^3 = \tau^2 + \tau = 2\tau + 1,$$
$$\tau^4 = 2\tau^2 + \tau = 3\tau + 2,$$
$$\tau^5 = 3\tau^2 + 2\tau = 5\tau + 3,$$
$$...$$
$$\tau^n = \tau F_n + F_{n-1}, \tag{2.3}$$

where we have introduced the so-called Fibonacci numbers, which are defined by the sequence $F_n = \{1, 1, 2, 3, 5, 8, 13, 21, ...\}$, where each number in the sequence is just the sum of the preceding two. Accordingly, any power of τ can be expressed as a linear combination of two successive Fibonacci numbers. Making use of Eq.(2.3) along with Table 2.2 one can properly order the different diffraction peaks along the radial directions of the diffraction pattern shown in Fig.2.8 in terms of appropriate linear combinations of Fibonacci numbers with the Golden mean. Thus, the translation symmetry characteristic of classical crystals is replaced by an inflation/deflation symmetry in the case of quasiperiodic crystals. Their characteristic scale factor is given by the Golden mean, which sets in this way a new physical scale in the arrangements of matter.

2.4.3 The Fibonacci world

Fibonacci numbers are ubiquitous, unexpectedly emerging in different places. In chemistry, for example, one finds Fibonacci numbers when counting the number of Kekulé structures, k, in zigzag fused linear chains of benzoic rings. Thus, we have $k = 5$ for phenanthrene, $k = 8$ for chrysene, $k = 13$ for pirene, $k = 21$ for fulminene, and so on.[46] Most of these polycyclic aromatic hydrocarbons are widespread through diverse astrophysical environments like dense molecular clouds, planetary nebulae, evolved stars, protoplanetary nebulae, meteorites, comets, or interplanetary dust particles, so that we can find the blueprints of Fibonacci numbers in the electronic structure of these compounds everywhere in the universe.[47] In biology, the widespread appearance of ordering patterns based on the Fibonacci sequence is well established in many botanical arrangements.[48] Curiously, the basic methods of study in *phyllotaxis* ("leaf arrangement" in Greek), like the representation of leaf distribution as a point-lattice on a cylinder, were introduced in 1837 by Auguste Bravais (1811-1863), one of the founders of classical crystallography.[49]

In Western civilization Fibonacci numbers were introduced, along with the modern notation for the so-called Arabic (truly Hindu) numerals, by Leonardo Pisano (ca. 1170-1240) in his celebrated book *Liber Abacci* (1201),[1] although the sequence $F_n = \{1, 1, 2, 3, 5, 8, 13, 21, ...\}$ was not referred to as the Fibonacci series up to nineteenth century (1877) by the French mathematician Edouard Lucas (1842-1891).[50] The terms in this sequence are obtained from the recursive equation $F_n = F_{n-1} + F_{n-2}$, starting with $F_1 = 1$ and $F_2 = 2$. Thus, the sequence is perfectly ordered, but the rule used to generate it has nothing to do with periodicity. On the contrary, it is closely related to certain inflation properties connecting these numbers with the Golden mean. The close relationship between Fibonacci numbers and the Golden mean is nicely illustrated by the so-called Binet formula (after the French mathematician

Jacques Phillipe Marie Binet, 1786-1856)

$$F_n = \frac{1}{\sqrt{5}} \left[\tau^n - (-1)^n \tau^{-n} \right], \qquad (2.4)$$

which allows one to find the value of a Fibonacci number if its place in the sequence is known, hence avoiding the cumbersome procedure of recursively calculating all the precedent numbers in the sequence. This is a quite remarkable formula, since it indicates that a countable infinite set of integer numbers (the Fibonacci ones) can be exactly obtained by simply adding suitable integers powers of irrational numbers (the Golden mean and its reciprocal). Another mathematical relationship between the Fibonacci series and the Golden mean is given by the asymptotic limit

$$\lim_{n \to \infty} \frac{F_n}{F_{n-1}} = \tau, \qquad (2.5)$$

which succinctly accounts for the property (discovered in 1611 by Johannes Kepler) that the ratio of two successive larger and larger Fibonacci numbers comes closer and closer to the Golden mean. However, the process of convergence is very slow, since the Golden mean is the "most irrational" among the irrational numbers. This property can be grasped by considering the continued fraction given by the expression

$$\tau = 1 + \cfrac{1}{1 + \cfrac{1}{1 + \cfrac{1}{1 + \cfrac{1}{1 + \dots}}}}, \qquad (2.6)$$

which is entirely composed of ones, so that it very slowly converges towards its limiting value. In order to determine this value we can proceed the direct way, by a series of successive approximations, interrupting the continued fraction farther and farther down. In so doing, we get the approximant series

$$1, \ \frac{2}{1}, \frac{3}{2}, \frac{5}{3}, \frac{8}{5}, \frac{13}{8}, \dots \qquad (2.7)$$

which, according to Eq.(2.5) precisely converges towards the Golden mean.

The symbolical analog of the Fibonacci sequence, constructed using two types of letters, say A and B, can be obtained from the substitution rule $A \to AB$ and $B \to A$, whose successive application generates the sequence of letters $A, AB, ABA, ABAAB, ABAABABA, \dots$ and so on. In this way, we get a perfectly ordered "word" which is not periodic at all. We note that the construction process guarantees that the total number of letters of a Fibonacci word is a Fibonacci number and the ratio of A's over B's approaches the Golden mean as the size of the Fibonacci word is progressively increased. It also guarantees that no BB pairs appear in the entire word. In this regard, it is worth noting that a Fibonacci lattice can be regarded as a particular class of random dimer alloy (see Section 1.7) where strong repulsions exist between B species, such that two B species do not appear in succession. A very remarkable property of this string of letters is that it endows self-similar properties, so that any pattern of letters we arbitrarily choose in the original sequence is found in the sequence on another scale. Another interesting property is related to the Conway's theorem, stating that given an arbitrary portion of a Fibonacci word we will find a replica of it at a distance smaller than twice is length. This property is illustrated for the short words *ABAB* and *AABABAA* in the longer Fibonacci word

$$AB\mathbf{AABAB}AAB\mathbf{AABABAABAB}AAB\mathbf{AABABAAB}AAB$$
$$AB\mathbf{AABABA}ABAABAB\mathbf{AABABA}ABAABABAABAAB.$$

As we will see in the following chapters of this book, a considerable number of works have focused on systems whose relevant physical parameters are arranged according to the Fibonacci sequence. Several reasons can be invoked to account for the interest spurred by the Fibonacci lattice (and other closely related self-similar lattices) in the condensed matter community during the last twenty years. We can highlight the following ones:

- From a conceptual point of view, the Fibonacci lattice can be considered as the one-dimensional analogue of some celebrated quasiperiodic arrangements in higher dimensions like, for example, the Penrose or Ammann tilings in two and three dimensions, respectively.

- The recourse to orderings based on the successive application of an inflation rule is particularly well suited to grow superlattices made of a sequential deposition of alternating layers of different materials.

- A number of rigorous mathematical results have been restricted to a few basic aperiodic sequences, including the Fibonacci sequence as a paradigmatic case.

2.5 Crystallography in six dimensions

2.5.1 Hyperspace

Since its very beginning in 1912, in order to determine the structure of crystals traditional crystallography relies primarily on x-ray diffraction, rather than electron diffraction, because x-ray diffraction techniques provide a good quantitative tool. Conversely, it is quite difficult to precisely measure crystalline length parameters by electron microscopy. Accordingly, the first high-resolution x-ray scattering measurements of the AlMn quasicrystalline phase discovered by Shechtman were welcomed with the greatest expectation.[51]

Those measurements showed that the obtained diffraction patterns can be indexed to a mixture of conventional face-centered cubic Al and a new phase whose reciprocal vectors can be described by a sum of the form,

$$\mathbf{q} = \sum_{i=1}^{6} n_i \mathbf{q}_i \qquad (2.8)$$

where n_i are integers and $\mathbf{q}_i$ are vectors pointing to the vertices of an icosahedron given by

$$\mathbf{q}_1 = (1, \tau, 0), \quad \mathbf{q}_2 = (1, -\tau, 0), \quad \mathbf{q}_3 = (0, 1, \tau),$$
$$\mathbf{q}_4 = (0, 1, -\tau), \quad \mathbf{q}_5 = (\tau, 0, 1), \quad \mathbf{q}_6 = (-\tau, 0, 1), \qquad (2.9)$$

where τ is the Golden mean and the components of the $\mathbf{q}_i$ vectors are referred to a suitable Cartesian reference frame. Thus, just as the Bragg lattice of ordinary crystals is generated by a basis of three vectors, the icosahedral patterns can be also generated by a suitable basis, although six in number. In this way, all the diffraction peaks could be expressed as an integer linear combination of these six vectors. For instance, the second strongest peak in the diffraction pattern reported in Ref.[51] can be indexed as (110001), which leads to $\mathbf{q} = (2 - \tau, 0, 1)$.

It was subsequently realized that it is useful to think of a six dimensional *periodic* lattice, with basis vectors $\mathbf{e}_i$, which are properly projected into the usual three-dimensional space in the form $\mathbf{e}_i = (\mathbf{e}_\parallel, \mathbf{e}_\perp)$, where $\parallel$ denotes the physical space and $\perp$ denotes the so-called perpendicular space.[52] For instance, to achieve an icosahedral QC structure a six-dimensional simple cubic lattice is rotated around the origin by the 6×6 matrix,[13]

$$\mathcal{R} = \frac{1}{\sqrt{4 + 2\tau}} \begin{pmatrix} \tau & \tau & 0 & -1 & 0 & 1 \\ 0 & 0 & 1 & \tau & 1 & \tau \\ 1 & -1 & -\tau & 0 & \tau & 0 \\ \tau & -\tau & 1 & 0 & -1 & 0 \\ -1 & -1 & 0 & -\tau & 0 & \tau \\ 0 & 0 & \tau & -1 & \tau & -1 \end{pmatrix}. \qquad (2.10)$$

The second three coordinates, $\mathbf{e}_\perp$, of the resulting 6D $\mathbf{e}_i$ vector determine whether or not a lattice point is projected. If positive, then the first three coordinates, $\mathbf{e}_\parallel$, represent the real-space position of this point. In this way, icosahedral structures can also be described by space-group operations consisting of translations combined with rotations, as in the case of periodic crystals. The translations, however, take place in a 6D space, and the point-group describes generalized six-dimensional rotations and reflections. In this regard the considered structure is quite different from the usual incommensurate structures, previously studied in Section 1.3, which involve at least two independent lengths.[53]

A general Bragg vector can be then written as

$$\mathbf{g}_{\|} = \frac{\pi}{a} \sum_{i=1}^{6} n_i \mathbf{e}_{\|}^i, \tag{2.11}$$

where a represents the "quasilattice" constant. For each Bragg vector given by Eq.(2.11), there corresponds a unique partner in orthogonal space

$$\mathbf{g}_{\perp} = \frac{\pi}{a} \sum_{i=1}^{6} n_i \mathbf{e}_{\perp}^i. \tag{2.12}$$

Accordingly, the diffractions spots can be labeled by six generalized Miller indices and the diffraction pattern is spanned by six linearly independent reciprocal-lattice vectors $\mathbf{e}_i$, according to expressions (2.11) and (2.12). The main theme is well known from algebra. When you have a great number of unknowns in a mathematical problem, you must manage to find out a suitable number of equations relating them. Otherwise, no closed solution is generally possible. Analogously, the most fruitful way to deal with a complex problem some times consists in introducing new degrees of freedom in order to get a suitable description of complex physical systems with the aid of a more elaborated mathematical framework. The systematic use of complex numbers in different domains of mechanics, acoustics, electronics, etc. provides a nice example of this approach. The generalization of Miller's indexing scheme from three to six dimensions just represents another instance of this fruitful approach in crystallography. Although the physical meaning of the vectors given by Eq.(2.12) is not immediately clear, it turns out that they play a significant role in quasicrystallography. In fact, a close inspection of the electron diffraction pattern shown in Fig.2.8 reveals another unusual feature of quasicrystals crystallography: the intensity of the diffracted beams does not decrease as a function of the centre, as it does for a periodic crystal according to the atomic scattering amplitude.

This feature can be easily understood within the superspace projection framework sketched in Fig 2.12 if one assumes that the intensities decay with the magnitude of $g\perp$. Current model structures of QC's are usually carried out in the framework of hyperspace crystallography, a mathematical recipe that treats a QC as a periodic structure embedded in a higher (than three) dimension space. Thus, for example, the icosahedral quasilattice in three dimensions can be generated from a cubic lattice defined in six dimensions.[54] Starting with the square lattice in the plane a line $g_{\|}$ with an irrational slope is drawn (which defines the so-called parallel space), surrounded by a parallel strip of finite width, W, measured in the so-called perpendicular space $g_{\perp}$. This window defines the acceptance domain, since only those square lattice points inside the strip are projected onto $g_{\|}$. If experimental resolution is represented by the pair of dashed lines shown in Fig.2.12 then only the lattice vectors between these lines give rise to observable peaks. Increasing the resolution of the spectrometer has the effect of moving the dashed lines outward

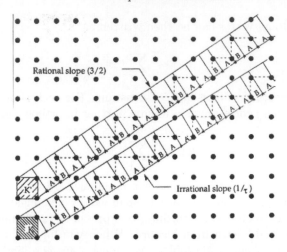

FIGURE 2.12

Construction of the Fibonacci chain and the 3/2 rational approximant through the cut and project algorithm. (Courtesy of Stephan Roche).

to a larger $|g\bot|$, thus filling in the regions between peaks on the $g_\parallel$ axis with weaker peaks. Then, if the $g_\parallel$ axis is incommensurate with the underlying lattice, the projections will be dense. We note that two projections with very nearly equal values of $g_\parallel$ can arise only at the expense of having their corresponding lattice vectors widely separated in the orthogonal $g\bot$ direction. In this way, the main qualitative features observed in the QC diffraction patterns can be nicely visualized.

The atomic structure determination of QCs is best achieved in the context of hyperspace crystallography where the periodicity is recovered in higher dimensions. For icosahedral QCs the periodic space has six dimensions and decomposes into two orthogonal 3D subspaces: the parallel (physical space) and the perpendicular (complementary) space. The 6D unit cell is decorated by 3D objects known as "occupation domains," which describe the local atomic environment in the physical space. This approach allows modelling and refinement of the structure against experimental diffraction data in a way similar to that achieved for usual 3D periodic crystals.[55] Nevertheless, the amount of observed diffraction data is rather limited in general, which precludes a detailed refinement of the chemical order in ternary QCs. From this perspective, the discovery of the first stable binary icosahedral QC in the Cd(Yb,Ca) family was a significant breakthrough, due to its simple chemical composition.

A lot of information is encoded in the mathematical recourse to higher dimension crystallography to describe ideal QC structures. A half of this information is used to describe the three dimensional appearance of QCs in the physical space through the matrix projection from hyperspace. What about the information remaining in the perpendicular space? As we will see, this extra information is related to two new concepts directly related to higher dimensional crystallography: the phasonic defects and the notion of crystalline approximants.

2.5.2   The notion of phason defects

Within the cut-and-project framework we can introduce kinds of defects which are specific of quasiperiodic systems. Let us consider the construction depicted in Fig.2.12, but now describing a distribution of atoms rather than an ensemble of Bragg spots in the reciprocal superspace. In the particular case of a 2D superspace, if the slope is τ^{-1} and the width of the acceptance domain coincides with the projection of the fundamental unit cell K, we obtain a Fibonacci lattice, characterized by the presence of long (A) and short (B) segments arranged according to the Fibonacci sequence. Analogously, the celebrated Penrose tiling can also be obtained via projection of a five-dimensional periodic lattice onto a planar surface.[13] On the other hand, if the slope is a rational approximant of the golden mean $(3/2$, for instance$)$ we obtain a periodic lattice, whose unit cell is given by the sequence $ABAAB$. As we see, a translation of the square lattice along the $E_\perp$ direction causes some original lattice points to move outside the acceptance window and some others to move into it. As a consequence, a number of local rearrangements in the original projected sequence take place. For example, some local strings of the form ABA turn to the form AAB in the Fibonacci sequence. These sort of rearrangements are called phason flips and can give rise to novel physical properties. In fact, while a distortion of the lattice along the $E_\parallel$ space corresponds to a phonon propagation, a distortion of the lattice along the $E_\perp$ space defines a kind of diffusive motion which is specific to QCs. Note that square lattice points near the edges of the acceptance window fall very easily outside it, while points close to its center are very robust against phason effects, and are responsible for the system stability.[56] In this sense, one may think of introducing phason flips by changing the geometrical properties of the acceptance window on their own. For example, by defining it in terms of a sinusoidal function instead of a straight line we can introduce a periodic modulation in the overall phason distribution.

An interesting physical consequence of the phason flip notion is that one may describe a continuous transformation of a one-dimensional lattice from quasiperiodic to periodic order by simply changing the slope of the $E_\parallel$ line. In this way, one realizes that the very notion of defect can be addressed from a completely different viewpoint. In fact, while one usually considers the presence of defects as destroying periodic order, in this case one starts with a quasiperiodically ordered structure and, by increasing the number of phasonic defects, one actually improves the periodic order in the system. Quite interestingly, the progressive transition from quasiperiodic to periodic order could be assessed, at least in principle, performing a systematic experimental study of the acoustic (or optical) response of certain aperiodic structures as a number of planar defects (mimicking phasonic defects) is progressively introduced in its way towards the periodic limit.[57]

2.5.3 Approximant phases

In every alloy system the true QC is accompanied by compositionally related classical crystals, having huge unit cell sizes, often forming micro-twinned networks with nearly aperiodic symmetries. These crystals not only have very similar compositions, but also structures closely resembling that of the true QC, from which they can nevertheless be distinguished. For these reasons such crystals are called approximants. In some systems, however, only the approximants are thermodynamically stable and the QCs need to be produced by rapid cooling. Nevertheless, the definition of such an approximant is still a useful one. Approximant phases are not to be confused with giant-unit-cell intermetallics, sometimes also termed "complex metallic alloys," exhibiting complex structures that contain some hundred up to several thousand atoms in the unit cell. Examples are the $Mg_{32}(Al,Zn)_{49}$ compound discussed in Chapter 1, with 162 atoms in the unit cell, orthorhombic ξ'-$Al_{74}Pd_{22}Mn_4$ (258 atoms in the unit cell),[58, 59] λ-Al_4Mn (586 atoms in the unit cell),[60] cubic β-Al_3Mg_2 (1168 atoms in the unit cell),[61] and the heavy-fermion compound $YbCu_{4.5}$, comprising as many as 7448 atoms in the supercell.[62] These giant unit cells contrast with elementary metals and simple intermetallics whose unit cells in general comprise from single up to a few tens atoms only. The giant unit cells with lattice parameters of several nanometers provide translational periodicity of the crystalline lattice on the scale of many interatomic distances, whereas on the atomic scale, the atoms are arranged in clusters with polytetrahedral order, where icosahedrally-coordinated environments play a prominent role. The structures of complex metallic alloys thus show duality; on the scale of several nanometers, these alloys are periodic crystals, whereas on the atomic scale, they resemble cluster aggregates. The high structural complexity of complex metallic alloys together with the two competing physical length scales—one defined by the unit-cell parameters and the other by the cluster substructure—may have a significant impact on the physical properties of these materials, such as the electronic structure and lattice dynamics. On this basis, complex metallic materials are expected to exhibit novel transport properties, like a combination of metallic electrical conductivity with low thermal conductivity, and electrical and thermal resistances tunable by varying the composition.

2.6   IUCr: A new definition of crystal

> "Then came the quasicrystals, and I realized that we didn't
> even understand what order is, from a mathematical point of view"
> (Marjorie Senechal, 1997)[14]

In the interval spanning from 1984 to 1990 an increasing number and variety
of quasicrystals were progressively found. Most of them were thermodynamically stable samples exhibiting a high degree of structural perfection, comparable to that observed in the best periodic crystals. Accordingly, the arrangement of atoms in QCs, which was originally regarded as some sort of defective
order, occupying an intermediate position between the short-range correlation typical of amorphous materials and the long-range-order characteristic of
periodic crystals, eventually was regarded as a new kind of long-range order.
Accordingly, a number of tentative definitions for the novel quasicrystalline
materials was introduced in the literature, generating some terminological
confusion. Thus, QCs were successively referred to as "a metallic phase with
icosahedral point group symmetry and no translational symmetry,"[24] "materials with point-group symmetries incompatible with crystal translational
symmetry,"[35] or "an arrangement of atoms which in a diffraction experiment produce infinitely sharp Bragg peaks in a pattern which exhibits overall
icosahedral symmetry."[63]

This unsatisfactory state of the affairs prompted the International Crystallographic Union to approve, in April 1991, the establishment of a Commission
on Aperiodic Crystals with the membership of J. M. Pérez-Mato (Chairman),
G. Chapuis, M. Farkas-Jahnke, M. L. Senechal, and W. Steurer. According
to their terms of reference:

> In the following by 'crystal' we mean any solid having an es-
> sentially discrete diffraction diagram, and by 'aperiodic crystal'
> we mean any crystal in which three-dimensional lattice periodic-
> ity can be considered to be absent.[64]

In the new definition, the essential attribute of crystallinity is transferred
from real space to reciprocal space. Consequently, within the crystalline family we can now distinguish between periodic crystals, which are periodic on the
atomic scale, and aperiodic crystals, lacking lattice periodicity in full agreement with the earlier Shrödinger's proposal (see Section 2.1). This broader
definition reflects our current understanding that microscopic periodicity is a
sufficient but not necessary condition for crystallinity. At the same time, it
suggests that the definition of QC *should not include* the requirement that
they possess an axis of symmetry that is *forbidden* in periodic crystals. Thus,
the presence of a mathematically well defined, long-range atomic order should
be properly regarded as the *generic* attribute of solid state matter rather

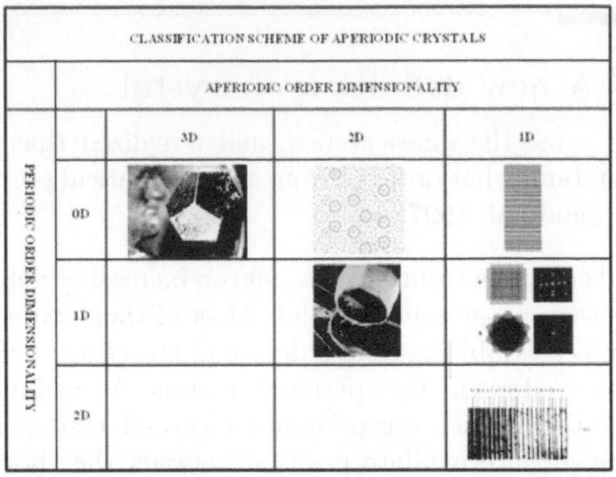

FIGURE 2.13
Classification of different types of aperiodic solids attending to the dimensionality of their aperiodic order. From left to right and from top to bottom we have representatives of an icosahedral quasicrystal, the Penrose tiling, a decagonal quasicrystal, a Fibonacci code bar,[66] an aperiodic photonic crystal,[66] and a dielectric Fibonacci multilayer (more details in the text).

than mere periodicity. In fact, in the last decades many examples of aperiodic crystals have been fully characterized by diffraction techniques, and we can currently distinguish three types of aperiodic crystals: incommensurately modulated crystals, composites, and quasicrystals.[13, 65]

The key question regarding the very nature of any possible crystal (either a periodic or aperiodic one) can then be formulated in the following terms: which are the necessary and sufficient conditions for a given arrangement of points to exhibit an essentially discrete Fourier transform?

2.7   Aperiodic crystals classification schemes

At this stage we realize that, in principle, a condensed matter phase can exhibit different kinds of order (namely, periodic or aperiodic) along different directions, so that we can properly speak of *isotropic or anisotropic aperiodic crystals*.

In order to illustrate this notion, in Fig.2.13 we present a classification of aperiodic crystals attending to their aperiodic order dimensionality, i.e., the number of spatial dimensions in which the aperiodic order occurs. For any given solid its spatial dimension, nD, is derived from the sum of its periodic dimension (labelled by the corresponding row number) plus its aperiodic dimension (labelled by the corresponding column number). Allowed entries are restricted by the condition $n \leq 3$. In addition, the coordinates of a given solid in the chart indicate the relative importance of aperiodic versus periodic order in its structure. For example, the icosahedral quasicrystal shown at the left upper corner, with coordinates (3,0), exhibits quasiperiodic order along the three dimensions of space, so that it can be regarded as a suitable example of isotropic aperiodic crystal. On the contrary, decagonal quasicrystals located at the middle of the chart, with coordinates (2,1), exhibit quasiperiodic order in the planes perpendicular to the decagonal axis, but are periodically arranged along this axis, so that they must be regarded as a suitable example of anisotropic aperiodic crystals. In this case, we have a condensed matter phase where two different kinds of order coexist, namely 1D periodic order along with 2D quasiperiodic order. Quite interestingly, the hybrid nature of the arrangement of matter in these alloys has measurable physical effects, which we will discuss in Chapter 3. At the right lower corner of Fig.2.13 we have another interesting instance of a hybrid-order system, a Fibonacci heterostructure, with coordinates (1,2). In this system a series of 2D periodic layers are stacked following a quasiperiodic 1D pattern. As we will see in Chapter 4, these kind of structures have been artificially grown and, as such, they properly illustrate the technological potential related to the very notion of aperiodic order in materials science which we will explore in more detail in Chapters 7 and 8.

Following the line of reasoning inspiring the classification scheme just introduced one may think of DNA as a sort of hybrid order system exhibiting 1D aperiodic order (as determined by the base pairs sequence) superimposed to 3D periodic order (corresponding to the sugar-phosphate backbone helix). The apparent extra-dimension in this case is related to the fact we are really dealing with two separate subsystems in the DNA helix, namely the nucleotide subsystem and the backbone system. Although both subsystems are strictly speaking 3D, in practice the planar Watson-Crick conjugate nucleobases are usually regarded as an effective one-dimensional chain as far as one is mainly concerned with the coding properties of the macromolecule. We will study the physical properties of double-stranded DNA helices in great detail in Chapter 6.

2.8 Some milestones in the aperiodic crystal route

- 1944 E. Schrödinger's proposal of aperiodic crystals to describe the nature of the gene

- 1952 W. Cochran, F. H. C. Crick, and V. Vand present the full theory of x-ray diffraction by helices

- 1953 Helical structure of DNA is proposed by J. Watson and F. H. C. Crick on the basis of x-ray diffraction studies by R. Franklin and M. H. F. Wilkins

- 1972 de Wolff introduces the higher-dimensional crystallography approach to describe the structure of incommensurate modulated phases

- 1974 R. Penrose obtains a quasiperiodic tiling of the plane containing just two different tiles

- 1982 A. L. Mackay reports on optical diffraction patterns obtained from Penrose tiling

- 1982 D. Shechtman obtains five-fold electron diffraction patterns from AlMn alloys

- 1984 D. Shechtman, I. Blech, D. Gratias, and J. W. Chan report on the finding of alloys exhibiting both long-range translational order and five-fold symmetry

- 1984 The notion of quasiperiodic crystal is introduced by D. Levine and P. J. Steinhardt

- 1985 R. Merlin and co-workers grow the first Fibonacci superlattice based on semiconductor materials

- 1985 First representative of the dodecagonal quasicrystalline phase is reported by T. Ishimasa, H. U. Nissen, and Y. Fukano

- 1985 First representative of the decagonal quasicrystalline phase is reported by L. Bendersky

- 1986 B. Dubost and co-workers announce the discovery of the first thermodynamically stable icosahedral phase in the AlCuLi system exhibiting a triacontahedral growth morphology

- 1987 First representative of the octagonal quasicrystalline phase is reported by N. Wang, H. Chen, and K. H. Kuo

- 1987 Stable GaMgZn quasicrystals with dodecahedral solidification morphology are reported by W. Ohashi and F. Spaepen

- 1987-1990 High structural quality icosahedral phases containing transition metals are reported by A. P. Tsai and co-workers in the alloy systems AlCu[Fe,Ru,Os] and AlPd[Mn,Re]

- 1991 The Commission on Aperiodic Crystals of the International Crystallographic Union broadens the definition of crystal in order to include both periodic and aperiodic crystals

- 1994 Fibonacci dielectric multilayers are first grown and the self-similarity of their optical spectral portraits is experimentally shown by W. Gellermann, M. Kohmoto, B. Sutherland, and P. C. Taylor

- 1995 High structural quality icosahedral phases containing rare-earth atoms in the alloy system ZnMg(RE) are reported by A. P. Tsai and co-workers

- 2000 The first binary icosahedral phase is found in the Cd(Y,Ca) alloy system by A. P. Tsai and co-workers

- 2005 A photonic quasicrystal with centimeter scale cells, exhibiting sizeable stop bands for microwave radiation, is constructed using stereolithography by W. Man, M. Megens, P. J. Steinhardt, and P. M. Chaikin

References

[1] Sigler L E 2002 *Fibonacci's Liber Abaci: A Translation into Modern English of Leonardo Pisano's Book of Calculation Sources and Studies in the History of Mathematics and Physical Sciences* (Springer, New York)

[2] Moody R V 1995 *The Mathematics of Long-Range Aperiodic Order,* Nato Science Series C: Mathematical and Physical Sciences vol. 489 (Kluwer, Dordrecht)

[3] Stent G S 1995 *DNA: The Double Helix Perspective and Prospective at Forty Years*, Ed. Chambers D A, Ann. N. Y. Acad. Sci. vol. 758, 25

[4] Avery O T, MacLeod C N, and McCarty M 1944 *J. Exp. Med.* **79** 137

[5] Timofeef-Ressovsky N W, Zimmer W K, and Delbrück M 1935 *Nach. Ges. Wiss. Goettingen Math-Phys.* Kl. Fachgruppe VI **13** 190

[6] Schrödinger E 1945 *What is life? The Physical Aspects of the Living Cell* (Cambridge University Press, New York)

[7] Chargaff E 1950 *Experientia* **6** 201

[8] Smontara A, Smiljanic I, Bilušić A, Jagličić Z, Klanj šek M, Roitsch S, Dolinšek J and Feuerbacher M 2007 *J. Alloys Comp.* **430** 29

[9] Boustani I, Quandt A, and Kramer P 1996 *Europhys. Lett.* **36** 583

[10] Perkins C L, Trenary M, and Tanaka K 1996 *Phys. Rev. Lett.* **77** 4772

[11] Adam J and Rich J B 1954 *Acta Cryst.* **7** 813

[12] Bergman G, Waugh J L T, and Pauling L 1952 *Nature (London)* **169** 1057; 1957 *Acta Cryst.* **10** 254.

[13] Janssen T, Chapuis G, and de Boissieu M 2007 *Aperiodic Crystals: From Modulated Phases to Quasicrystals* (Oxford University Press, Oxford)

[14] Hargittan I 1997 *The Chemical Intelligencer* October, pp.25-48

[15] Kepler J (Joannis Keppleri) 1519 *Harmonices Mundi Libri V* (Lincii Austriae, Sumptibus Godofredi Tampachhii, Franccof.)

[16] Lück R 2000 *Mater. Sci. Eng. A* **294-296** 263

[17] Wang H 1961 *Bell Syst. Tech. J.* **40** 1

[18] Berger R 1966 *Mem. Amer. Math. Soc.* **66** 1

[19] Penrose R 1974 *Bull. Inst. Math. Appl.* **10** 266

[20] Burkov S E 1991 *Phys. Rev. Lett.* **67** 614

[21] Gummelt P 1996 *Geometriae Dedicata* **62** 1

[22] Ammann R, Grünbaum B, and Shephard G C 1992 *Discrete Comput. Geom.* **8** 1

[23] Gardner M 1977 *Sci. Am.* **236** (1) 110

[24] Shechtman D, Blech I, Gratias D, and Cahn J W 1984 *Phys. Rev. Lett.* **53** 1951

[25] Pauling L 1985 *Nature (London)* **317** 512; 1987 *Phys. Rev. Lett.* **58** 365

[26] Mackay A L 1982 *Physica A* **114** 609

[27] Levine D and Steinhardt P J 1984 *Phys. Rev. Lett.* **53** 2477

[28] Maciá E 2006 *Rep. Prog. Phys.* **69** 397

[29] Lifshitz R 2003 *Foundations of Phys.* **33** 1703

[30] Ishimasa T, Nissen H U, and Fukano Y 1985 *Phys. Rev. Lett.* **55** 511

[31] Wang N, Chen H, and Kuo K H 1987 *Phys. Rev. Lett.* **59** 1010

[32] Bendersky L 1985 *Phys. Rev. Lett.* **55** 1461

[33] Fung K K, Yang C Y, Zhou Y Q, Zhao J G, Zhan W S and Shen B G 1986 *Phys. Rev. Lett.* **56** 2060

[34] He L X, Li X Z, Zhang Z, and Kuo K H 1988 *Phys. Rev. Lett.* **61** 1116

[35] Guyot P 1987 *Nature (London)* **326** 640

[36] Hardy H K and Silcock J M 1955 *J. Inst. Metall.* **24** 423

[37] Dubost B, Lang J M, Tanaka M, Sainfort P, and Audier M 1986 *Nature (London)* **324** 48

[38] Ohashi W and Spaepen F 1987 *Nature (London)* **330** 555

[39] Tsai A P, Inoue A, and Masumoto T 1988 *Jpn. J. Appl. Phys.* **27** 1587

[40] Tsai A P, Inoue A, and Masumoto T 1990 *Mater. Trans. Jpn. Inst. Metal.* **31** 98

[41] Tsai A P, Guo J Q, Abe E, Takahura H, and Sato T J 2000 *Nature* **408** 537

[42] He X L, Li X Z, Zhang Z, and Kuo K H 1988 *Phys. Rev. Lett.* **61** 1116

[43] Tsai A P, Inoue A, and Masumoto T 1989 *Mater. Trans. Jpn. Inst. Metal.* **30** 463

[44] Kelton K F, Kim Y J, and Stroud R M 1997 *Appl. Phys. Lett.* **70** 3230

[45] Livio M 2002 *The Golden Mean* (Broadway Book, New York); reviewed by Maor E 2003 *Science* **299** 1016

[46] Randil M, Morales D A, and Araujo O 1996 *J. Math. Chem.* **20** 79

[47] Bernstein M P, Sandford S A, Allamandola L J, Gillette J S, Clemett S J, and Zare R N 1999 *Science* **283** 1135

[48] Adler I, Barabé D, and Jean R V 1997 *Ann. Bot.* **80** 231

[49] Bravais L and Bravais A 1837 *Ann. Sciences Naturalles Botanique* **7** 42

[50] The Indian mathematician Acharya Hemachandra (1089-1172) reported on this series around 1150, about 50 years before Fibonacci. The interested reader is referred to the paper by Singh P 1985, *Historia Mathematica* **12** 229

[51] Bancel P A, Heiney P A, Stephens P W, Goldman A I, and Horn P M 1985 *Phys. Rev. Lett.* **54** 2422

[52] Duneau M and Katz A 1985 *Phys. Rev. Lett.* **54** 2688

[53] Bak P 1985 *Phys. Rev. B* **32** 5764

[54] Elser V 1985 *Phys. Rev. B* **32** 4892

[55] Boudard M and de Boissieu M 1999, in *Physical Properties of Quasicrystals*, Ed. Stadnik Z M, Springer series in Solid-State Sciences, vol. 126 (Springer, Berlin)

[56] Naumis G G 2005 *Phys. Rev. B* **71**, 144204

[57] Aynaou H, Velasco V R, Nougaoui A, El Boudouti E H, and Bria D 2002 *Superlatt. Microstruct.* **32** 35; Aynaou H, Velasco V R, Nougaoui A, El Boudouti E H, Djafari-Rouhani B, and Bria D 2003 *Surf. Sci.* **538** 101

[58] Boudard M, Klein H, de Boissieu M, Audier M, and Vincent H 1996 *Philos. Mag. A* **74**, 939

[59] Dolinšek J, Jeglič P, McGuiness P J, Jagličić Z, Bilušić A, Bihar Ž, Smontara A, Landauro C V, Feuerbacher M, Grushko B, and Urban K 2005 *Phys. Rev. B* **72** 064208

[60] Kreiner G and Franzen H F 1997 *J. Alloys Compd.* **261** 83

[61] Samson S 1965 *Acta Crystallogr.* **19** 401

[62] Cerny R, Francois M, Yvon K, Jaccard D, Walker E, Petricek V, Cisarova I, Nissen H-U, and Wessiken R 1996 *J. Phys.: Condens. Matter* **8** 4485

[63] Bak P and Goldman A L 1988 *Introduction to Quasicrystals* (Academic Press, New York)

[64] ICrU Report of the Executive Committee for 1991 *Acta Cryst. A* 1992 **48** 922. See also http://www.iucr.ac.uk/iucr-top/comm/capd/terms.html.

[65] Chapuis G 2003 *Crystal Engineering* **6** 187

[66] Ferralis N, Szmodis A W, and Diehl R D 2004 *Am. J. Phys.* **72** 1241

3

Quasiperiodic crystals

3.1 Quasicrystalline alloys

3.1.1 Quasicrystal paradoxes

From a macroscopic point of view crystals are characterized by regular shapes and flat surfaces. According to these phenomenological criteria thermodynamically stable QCs, exhibiting regular shapes and flat surfaces, should be regarded as common crystals by all standards.

This point is illustrated in Fig.3.1 where we compare different representatives of both periodic and aperiodic crystal families. As we can see, all samples nicely meet the phenomenological criteria for a piece of matter to be considered a "crystal." Accordingly, had quasicrystalline alloys been found to spontaneously occur in nature from ancient times, then they would most probably have been regarded as another instance of mineral crystals representatives, for they exhibit all the basic features most of them show up.

Nevertheless, quasicrystalline alloys were not discovered this way. On the contrary, they were unexpectedly found, once the theoretical framework aimed to understand the nature of crystal kingdom from a microscopic viewpoint had been completely elaborated. That theory is mathematically rigorous and very successful in accounting for the structure of every form of matter found to date. And such a rigorous and successful theory explicitly forbids the growth morphologies exhibited by quasicrystalline alloys!

Thus the discovery of quasiperiodic crystals was a terrible shock for the crystallographical, solid state, and condensed matter communities, demanding an appropriate answer for this apparent paradox. It was a hectic period of intellectual storm, as we have described in Section 2.3. The final solution fortunately arrived soon. It was as simple as every piece of wisdom usually uses to be: our theoretical understanding was entirely based on the very notion of periodic arrangements of atoms in space. But what about aperiodic ones? We had simply missed them in constructing classical crystallography. Consequently, all the conceptual troubles stemming from the discovery of this new form of matter were rooted on epistemological shortcomings rather than pointing to any sort of pathological state of matter. The message from Nature was that matter can display well-ordered patterns which are more complex than those imposed by mere periodicity. Thus, the notion of aperiodic crystal

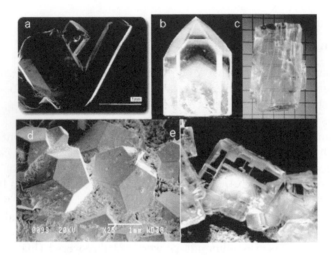

FIGURE 3.1

Periodic and aperiodic crystals gallery. (a) AlCuCo quasicrystalline alloy exhibiting a decaprismatic growth habit, (b) Quartz (SiO_2) crystal exhibiting a hexagonal prismatic habit, (c) Calcite ($CaCO_3$) crystal exhibiting a hexagonal growth morphology, (d) AlCuRu quasicrystalline alloy exhibiting a dodecahedral growth habit, and (e) Fluorite (CaF_2) crystal exhibiting a cubic growth morphology.

generalizes the one of periodic crystal and lifts the prohibitions about the rotational symmetries that can leave the material invariant. In this way, the puzzle due to the crystallographical restriction theorem was eventually fixed, forcing new progress in our understanding of the atomic arrangements in solids.[1]

Nevertheless, shortly after the discovery of thermodynamically stable quasicrystalline alloys of high structural quality, it was progressively realized that these materials occupy an odd position among the well-ordered condensed matter phases. In fact, since QCs consist of metallic elements one would expect they should behave as metals. Nonetheless, as we will describe below, it is now well established that transport properties of stable QCs are quite unusual by the standard of common metallic alloys, as most of their transport properties resemble a more semiconductor-like than metallic character. In this way, we were led to face a second quasicrystal paradox: that of a peculiar class of metallic alloys which do not behave as common metallic alloys usually do.

3.1.2 Unusual properties of quasicrystals

"If real quasicrystals exist, as suggested by Shechtman et al., they are sure to possess a wealth of remarkable new structural and electronic properties" (D. Levine and P. J. Steinhardt, 1984) [2]

"I point out that there is no reason to expect these alloys to have unusual physical properties" (L. Pauling, 1987) [3]

Once the existence of a new kind of condensed matter phase was established (Section 2.3.4), the question naturally arose regarding the possible influence of its characteristic quasiperiodic order on its physical properties. Unfortunately, the first obtained QCs were metastable, preventing a significant study of several physical properties, in particular the temperature dependence of their transport properties. Even the first thermodynamically stable QCs, obtained in the systems AlCuLi and GaMgZn, were unsuitable to this end, since they were usually contaminated with small crystalline inclusions and exhibited a relatively large number of structural imperfections (see Fig.2.10). The discovery of thermodynamically stable quasicrystalline alloys of *high structural quality* in the AlCu(Fe,Ru,Os), AlPd(Mn,Re), ZnMg(RE), and Cd(Yb,Ca) icosahedral systems, as well as the AlCo(Cu,Ni) decagonal system, allowed for detailed experimental studies of *intrinsic* transport properties of quasicrystals. In this way, unusual behaviors in the temperature dependence of electrical conductivity, Hall and Seebeck coefficients, specific heat, and thermal conductivity were progressively reported.[4, 5, 6, 7, 8, 9, 10, 11]

3.1.2.1   dc conductivity

Some characteristic electrical transport anomalies are graphically summarized in Figs.3.2 and 3.3. Fig.3.2 gives the electrical resistivities, ρ, measured at

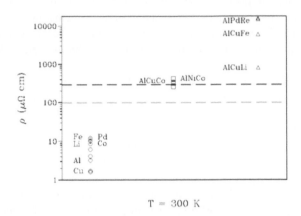

$$T = 300 \text{ K}$$

FIGURE 3.2

Room temperature resistivity of icosahedral (triangles) and decagonal (squares) QCs is compared to that of their respective constituent elements.

room temperature for different QCs and their constituent (pure) elements. It can be seen that the pure elements are good metals, possessing low electrical resistivities. On the contrary, quasicrystalline alloys composed of these elements show electrical resistivities which are higher by several orders of magnitude, and much higher than typical values for conventional metallic alloys (both crystalline and amorphous), whose representative values fall in the region between the horizontal dashed lines in Fig.3.2.

In Fig.3.3 the striking behavior of the electrical resistivity of quasicrystalline alloys is further highlighted by considering its variation with temperature. For typical metals resistivity decreases as the temperature is decreased and it can even completely vanish at low enough temperatures for those materials reaching the superconducting state. Such a behavior is illustrated for the case of aluminum (the main constituent of an important class of QCs) in Fig.3.3. Conversely, the electrical resistivity of QCs progressively increases as the temperature is decreased, suggesting the possibility of reaching a metal-insulator transition in high-quality icosahedral quasicrystals at low temperatures.[12, 13, 14, 15] On the other hand, the electrical conductivity steadily increases as the temperature increases up to the melting point, and its value very sensitively depends on minor variations of the sample stoichiometry, as it is illustrated in Fig.3.4.

Accordingly, attending to their electrical properties quasicrystalline phases are marginally metallic and should be properly located at the *border line* between metals and semiconductors.[16, 17, 18] To illustrate this point, the room temperature electrical resistivities of several classes of materials are

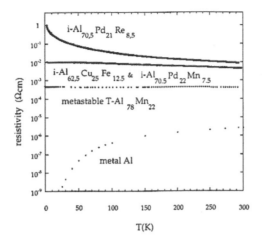

FIGURE 3.3

The general temperature dependence of electrical resistivity for icosahedral QCs based on aluminium is compared with that corresponding to aluminium, its main alloying element. (Courtesy of Claire Berger).

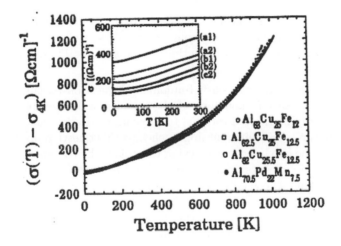

FIGURE 3.4

Temperature dependence of the electrical conductivity for four diferent quasicrystalline samples up to 1000 K. The inset illustrates the sensitivity of the residual conductivity value to minor variations in the sample composition.([7] Courtesy of Claire Berger. Reprinted figure with permission from Mayou D, Berger C, Cyrot-Lackmann F, Klein T, and Lanco P 1993 *Phys. Rev. Lett.* **70** 3915 © 1993 by the American Physical Society.)

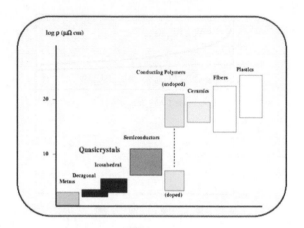

FIGURE 3.5

Room temperature electrical resistivity is compared for different materials of technological interest. Quasicrystals are located at the border line between metals and semiconductors.

compared in Fig.3.5. It can be seen that quasicrystals fill the gap between metals and semiconductors, exhibiting electrical resistivity values comparable to those reported for doped conducting polymers, metallic macrocycles, and fullerenes.

Underlying these results a basic fundamental question remains, concerning whether the purported anomalies in the quasicrystals transport properties should be mainly attributed (or not) to the characteristic *quasiperiodic order* of their structure. Two different approaches to this question are found in the literature: on the one hand, those trying to explain the transport properties of QCs in terms of concepts originally developed to describe amorphous solids; on the other hand, those advocating for specific treatments, aimed to exploit the physical implications of the quasiperiodic order notion. To the present, both approaches have obtained partial successes in describing different experimental data, thus spurring the interest for a suitable theory of quasicrystalline matter. A clear indication on the significant role played by the kind of order (i.e., periodic or quasiperiodic) in the transport properties of the underlying structure was provided by a series of studies on the transport properties of decagonal quasicrystals belonging to the AlCo(Cu,Ni) system. The atomic arrangement in these thermodynamically stable quasicrystals is periodic in the tenfold growth direction, and quasiperiodic in the plane perpendicular to it (Fig.2.9). Therefore, the study of their physical properties allows for the comparison between the transport properties in the quasiperiodic plane and those in the periodic direction in the *same sample*. Quite interestingly, a remarkable

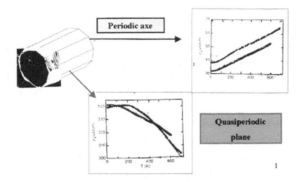

FIGURE 3.6
The electrical resistivity temperature dependence along two different directions exhibits quite differerent behaviors in the decagonal QC AlCuCo.[19] Note that, at any given temperature, the resistivity values differ by about an order of magnitude depending on the kind of order present in the underlying substrates.

anisotropy in the electrical resistivity, thermopower, and thermal conductivity was reported for high-quality, single-grained decagonal quasicrystals.[19, 20] Thus, for example, when measured along the decagonal axis (periodic order) the electrical resistivity increases with temperature, as usually occurs in metals. On the contrary, when the electrical resistivity is measured along the quasiperiodic plane it decreases with temperature (Fig.3.6). In a similar way, the electronic contribution to the thermal conductivity appears to be almost completely suppressed in the quasicrystalline plane, whereas the heat transport along the periodic direction behaves like that observed in usual periodic crystalline metals.[19] These transport anisotropy measurements provide compelling evidence on the existence of physical effects intrinsically related to the quasiperiodic order of the lattices.

Additional evidence on the role played by the novel kind of order present in these materials is that their electrical conductivity significantly *decreases* as the structural quality of the sample is improved (e.g., by annealing), in striking contrast with periodic crystals, whose electrical transport properties improve when structural imperfections are removed upon heating. Consequently, neither the notion of metal nor that of semiconductor can be applied to QCs, clearly demanding the introduction of a more adequate concept to describe them. To this end, the possible existence of general trends, allowing for a systematic classification of QCs according to their related transport coefficients, appears as a very promising starting point.

3.1.2.2   Inverse Matthiessen rule

Strong evidence on a possible qualitative *universal behavior* of the electrical conductivity of icosahedral QCs was reported from the observation that the conductivity curves of four different quasicrystalline samples are nearly parallel up to about 1000 K (see Fig.3.4), so that one can write[7]

$$\sigma(T) = \sigma(0) + \Delta\sigma(T), \tag{3.1}$$

where $\sigma(0)$ measures the sample dependent residual conductivity, and $\Delta\sigma(T)$ is proposed to be a general function. According to this expression the contribution to the sample conductivity due to different sources of scattering seems to be additive. This is just the opposite to what happens to normal metals, where the *resistivities* due to different sources of disorder are additive. This remarkable behavior, referred to as *inverse Matthiessen rule*,[7] is further illustrated for a broader collection of QC samples in Fig.3.7. It has been also observed in quasicrystalline approximants,[21] and even in amorphous phases prior to their thermally driven transition to the QC phase (see Fig.3.11).[22]

These findings indicate that the inverse Matthiessen rule may be a quite general property of structurally complex alloy phases closely related to quasicrystalline compounds. Then, the question arises concerning the possible existence of a suitable physical mechanism supporting the presumed universality of the $\Delta\sigma(T)$ function. In fact, the parallelism of the $\sigma(T)$ curves is difficult to understand in terms of a classical thermally activated mechanism, since the temperature dependence of $\sigma(T)$ does not follow an exponential law of the form $\exp(-E_g/k_BT)$, where k_B is the Boltzmann constant. The inadequacy of this fitting implies the absence of a conventional semiconducting-like gap in QCs.[25] Additional evidence comes from the fact that, for the heavily doped semiconductors, the $\sigma(T)$ curve decreases at high enough temperatures when all the impurity levels have become ionized. No evidence of such a limiting threshold has been observed in QCs.

On the other hand, signatures of electron-electron scattering, spin-orbit interaction, chemical disorder effects, and quasiperiodicity effects have been inferred from the temperature dependence of σ, although their relative role is still awaiting for a precise experimental and theoretical clarification.[26] Consequently, one would expect that different fits to the experimental data may be more or less adequate depending on the temperature ranges considered, since the relative importance of different physical mechanisms at work will depend on their own temperature scales.

3.1.2.3   Current-voltage curves

In this regard it is noteworthy to mention that characteristic current-voltage (I-V) curves of AlCuFe icosahedral quasicrystals exhibit a perfect Ohmic behavior at low temperatures ($T \simeq 4$ K) for bias voltages which vary by seven

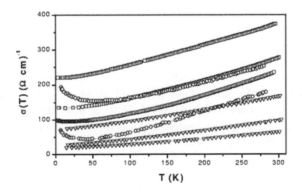

FIGURE 3.7

Diagram comparing the electrical conductivity temperature dependences for different quasicrystalline samples belonging to the AlCuFe ($\square$), AlCuRu (∇), and AlPdMn ($\bigcirc$) families. From top to bottom their chemical compositions read as follows: $Al_{63}Cu_{24.5}Fe_{12.5}$, $Al_{62.8}Cu_{24.8}Fe_{12.4}$, $Al_{70}Pd_{20}Mn_{10}$, $Al_{62.5}Cu_{25}Fe_{12.5}$, $Al_{70}Pd_{20}Mn_{10}$, $Al_{65}Cu_{21}Ru_{14}$, $Al_{65}Cu_{20}Ru_{15}$, and $Al_{65}Cu_{19}Ru_{16}$. Data for AlCuFe samples were kindly provided by Claire Berger. Data for AlCuRu and AlPdMn samples after Refs. [23] and [24], respectively.

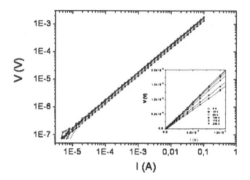

FIGURE 3.8

Double logarithmic I-V plots of an icosahedral $Al_{63}Cu_{25}Fe_{12}$ phase sample (kindly provided by Jean Marie Dubois) at $T = 9, 45, 65, 100, 175$, and 230 K. The inset shows the linear representation of the same data. (Courtesy of Javier García-Barriocanal).

orders of magnitude.[27] Such a linear behavior holds as the sample temperature is progressively increased (Fig.3.8), suggesting that a linear I-V behavior may be a common property of quasicrystals. This behavior lends support to the possible presence of relatively extended states close to the Fermi level and can be understood in the light of the electronic structure of the considered samples (see Sec.3.3.1).

In fact, it is generally accepted that the electronic structure of most icosahedral QCs is characterized by the presence of three relevant energy scales close to the Fermi level. First, the Hume-Rothery stabilization mechanism gives rise to a broad pseudogap on the energy scale of about 1 eV. Second, hybridization effects among d-states and sp-states give rise to the presence of a narrow dip of about 0.1 eV close to the Fermi level. Finally, resonant effects due to the quasiperiodic distribution of transition metal clusters along with possible chemical bonding effects are related to the existence of finer structure features in the density of states (DOS), on the scale of about 0.01 eV.[28] Accordingly, one would expect to observe some nonlinearity related to the presence of these spectral features as soon as the energy change of the charge carriers involved in the measurement process is in the range 0.01 - 1 eV. Now, the highest electric fields applied in these experiments are in the range $E = 50 - 100$ V/cm, so that we get the electron energy $\varepsilon \simeq eEL_0 \simeq 10^{-5}$ eV, where $L_0 \simeq 20$ Å is a rough estimate of the electronic mean free path in these materials.[27] Certainly, this figure is small enough to play a subsidiary role in the considered I-V measurements. In other words, stronger electric fields should be applied in order to observe the expected quasiperiodic effects in these materials.

3.1.2.4 Optical conductivity

The study of optical properties, performed over a very broad spectral range, is a powerful experimental tool for identifying the spectrum of excitations in a solid. In this way, several intrinsic parameters, such as the plasma frequency, relevant excitations due to phonons, or the strength of interband transitions, can be evaluated. To this end, one experimentally obtains the reflectivity curve as a function of the incoming electromagnetic radiation frequency, $R(\omega)$, and derives from it the optical conductivity curve $\sigma(\omega)$ by means of the so-called Kramers-Krönig transformation of the reflectivity spectrum. This transformation requires the knowledge of the optical responses at very low and very high frequencies, which are generally obtained from suitable extrapolations.

Several contributions are involved in the $\sigma(\omega)$ curve of simple metals. One is due to intraband transitions of conduction electrons and can be analyzed using the Drude model for free electrons

$$\sigma(\omega) = \frac{\sigma(0)}{1 + (\omega\tau)^2}, \tag{3.2}$$

where $\sigma(0)$ is the dc conductivity and τ is the relaxation time. This contribution becomes dominant at low frequencies and results in a characteristic Lorentzian function centered at the zero frequency, known as the Drude peak, followed by a rapid decay of the optical conductivity at large frequencies. A second contribution (located at the far-infrared region of the spectrum) is related to the presence of optical phonon modes, which are activated when the incoming radiation frequency is equal to or exceeds the necessary excitation energy. Additional contributions come from transitions involving both the valence and conduction bands (interband transitions) in the visible spectral range. Good conductors show a reflectance close to 100% at frequencies below the onset of absorption due to interband transitions and a characteristic sudden decay of $R(\omega)$ (known as the plasma edge) as the frequency increases approaching the value

$$\omega_p^2 \equiv \frac{ne^2}{m\varepsilon_0}, \tag{3.3}$$

referred to as the plasma frequency, where n is the number of electrons per unit volume and ε_0 is the vacuum dielectric constant. This frequency defines a threshold value. At low frequencies (i.e., $\omega < \omega_p$), the free electrons can couple to the oscillating electromagnetic field of incoming photons giving rise to a collective motion referred to as plasma oscillation. Accordingly, no radiation can propagate and the radiation field falls exponentially inside the solid. On the other hand, at large enough frequencies (i.e., $\omega > \omega_p$), the electromagnetic

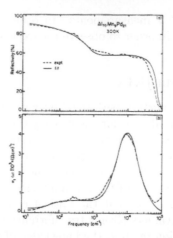

FIGURE 3.9

(a) The reflectivity spectrum at room temperature of i-AlPdMn (b) The optical conductivity obtained by Kramers-Krönig transformation of the reflectivity spectrum is compared to a phenomenological fit.(From ref.[30]. With permission from Elsevier.)

wave can propagate and the medium becomes transparent. Thus, a metal is basically transparent to light for wavelengths smaller than the plasmon cut-off $\lambda_p = c/\omega_p$, and absorbing and reflecting above.

Generally speaking, the optical conductivity of icosahedral QCs studied so far is quite different from that of either a metal or a semiconductor. Thus, reflectance of high quality icosahedral samples was found to be significantly small (about 60%) in a wide wavelength region from about 300 nm (UV region) to 15 μm (IR region), and several unusual features were observed in the optical conductivity:

1. The far infrared $\sigma(\omega)$ is very weak and no Drude peak appears (Fig.3.9), though extrapolation to the zero frequency at the low-frequency region yields conductivity values in good agreement with the measured dc conductivity.[29, 30, 31, 32, 33] Two different explanations have been proposed to account for the unusual absence of a Drude peak. In a first approach, the low $\sigma(\omega)$ was assigned to an extremely low density of states at the Fermi level due to the presence of a pseudogap in the band structure of QCs, hence leading to a substantially small value of $\sigma(0)$ in Eq.(3.2).[34] Another approach is based on the localization of charge carriers due to the quasiperiodicity of the structure, which leads to an anomalous diffusion mechanism. In that case, Drude's formula for

the optical conductivity may adopt the form

$$\sigma(\omega) = Ae^2 N(E)\Gamma(2\beta + 1)\left(\frac{\tau}{1 - i\omega\tau}\right)^{2\beta-1}, \qquad (3.4)$$

where A is a constant, Γ is the Gamma function, $N(E)$ is the density of states, and β is a diffusion exponent which depends on the energy.[35] The real part of this expression reduces to Eq.(3.2) in the case $\beta = 1$, while values as low as $\beta = 0.07$ and $\beta = 0.03$ were found from fitting analysis of the $\sigma(\omega)$ curves of AlCuFeB and AlPdMn QCs, respectively.[33]

2. All the studied QCs exhibit a typical absorption feature overlapping the low frequency tail of the far-infrared region. This relatively broad feature (which splits into two separate contributions at about 25 and 35 meV in most cases) is ascribed to phonon effects. At higher energies (~ 0.4 eV) the optical conductivity progressively rises reaching a peak at about 0.7 eV (i-ZnMgY, i-ZnMgTb), $1.2 - 1.5$ eV (i-AlCuFe, i-AlPdMn), or $2.6 - 2.9$ eV (i-AlPdRe), after which the conductivity decreases. This absorption feature is commonly ascribed to excitations across a characteristic pseudogap related to the Hume-Rothery stabilization mechanism (Section 3.3.1). Note that the location of the pseudogap absorption feature correlates with the width and depth of the related pseudogap for the different quasicrystalline families.

In summary, unlike disordered metals (where a Drude model on the strong-scattering limit is applicable) or semiconductors (with a well developed conductivity gap), the reflectivity spectra of icosahedral phases display low optical conductivity on the far-infrared energy range and a marked absorption in the visible.

These characteristic features are also observed in typical approximant phases, such as 1/1 AlMnSi, indicating that unusual optical properties are not specific to long-range quasiperiodic order. Nevertheless, for decagonal phases different behaviors of the $\sigma(\omega)$ curve can be clearly established between the quasicrystalline and the periodic directions (Fig.3.10).[36] In fact, a Drude peak is present when light is irradiated within a narrow area parallel to the periodic axis, whereas no peak is detected in a plane perpendicular to it. The analysis of the optical data shows that contrarily to the case of icosahedral QCs, there is no clear evidence for the presence of a marked pseudo-gap at the Fermi level.

3.1.2.5 Thermoelectric power

Thermoelectric power describes the electric response of a sample due to the application of an external temperature gradient through the relationship $\Delta V = S(T)\Delta T$, where $S(T)$ is the so-called Seebeck coefficient. During the last

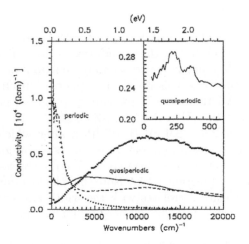

FIGURE 3.10

The optical conductivity of the decagonal AlCoCuSi quasicrystal for the periodic (short-dashed line) and the quasiperiodic (solid line) directions is compared with the conductivity of the icosahedral AlCuFe studied in Ref.[29] (solid dots). In the inset the quasiperiodic conductivity in the far-infrared part of the spectrum is shown. (Reprinted figure with permission from Basov D N, Timusk T, Barakat F, Greedan J, and Grushko B 1994 *Phys. Rev. Lett.* **72** 1937 © 1994 by the American Physical Society.)

decade the thermoelectric power of samples belonging to different icosahedral families has been measured. Reported data refer to a broad range of
stoichiometric compositions and cover different temperature ranges in the
interval from 1 K to 900 K. From the collected data the following general conclusions can be drawn for high-quality QCs containing transition
metals.[37, 38, 39, 40, 41]

- Room temperature thermoelectric power usually exhibits large values
  (50-120 μVK^{-1}) when compared to those of both crystalline and disordered metallic systems (1-10 μVK^{-1}).

- The temperature dependence of the Seebeck coefficient usually deviates
  from the linear behavior, exhibiting pronounced curvatures (either positive or negative) at temperatures above $\sim 50-100$ K. This behavior is at
  variance with that exhibited by ordinary metallic alloys where the $S(T)$
  curve is dominated by electron diffusion yielding a linear temperature
  dependence.

- Small variations in the chemical composition (of just a few atomic per-
  cent) can give rise to sign reversals in the thermopower curve.

- The $S(T)$ curves exhibit well-defined extrema in several cases. Both the
  magnitude and position of the extrema observed in the thermoelectric
  power curves are extremely sensitive to minor variations in the chemical
  stoichiometry of the sample.

On the other hand, thermopower measurements of rare-earth bearing QCs
in the system i-ZnMg(Y,Tb,Ho,Er) exhibit markedly linear temperature dependences above ~ 50 K.[42] An analogous behavior has been reported for the
thermodynamically stable CdYb QC, which also contains rare-earth atoms.[43]
Such different behaviors among the i-AlCu(Fe,Ru,Os) and i-AlPd(Mn,Re)
families (bearing transition metals) and the i-ZnMg(RE) and i-CdYb families (bearing rare-earth atoms), strongly suggest that chemical effects may be
playing a significant role.

3.1.2.6 Hall coefficient

Let us consider the following experimental situation: an external electric
field, E_x, is applied to a wire extending in the x-direction so that a current
density j_x flows. If a transverse magnetic field B_z pointing in the positive
z-direction is applied the resulting Lorentz force $\mathbf{j} \times \mathbf{B} = -j_x B_z \hat{j}$ acts to
deflect the charge carriers in the negative y-direction. This gives rise to an
electrostatic field E_y in the y-direction (the so called Hall field) that opposes
the charge motion. In the equilibrium the Hall field will balance the Lorentz
force, and current will flow only in the x-direction. The Hall coefficient is then

defined as[44]

$$R_H \equiv \frac{E_y}{j_x B_z} = -\frac{1}{nqc} .$$ (3.5)

According to Eq.(3.5) the value of the Hall coefficient measures the density of carriers and its sign indicates their nature (i.e., electrons $q = -e$, or holes $q = +e$). Experimental measurements of the Hall coefficient of samples belonging to different icosahedral families, covering a broad range of stoichiometric compositions, have been reported during the last decade. [6, 15, 25, 38, 40, 45, 46, 47, 48, 49, 50, 51, 52, 53, 54] In this way, several anomalous properties have been reported in the temperature dependence of the Hall coefficient, $R_H(T)$, within the interval 4.5 K to 300 K.

- Low temperature values suggest small carrier densities, within the range $n \simeq 10^{20} - 10^{21}$ cm^{-3}, which is an unusually low figure for alloys made of good metals.

- Positive or negative values at low temperatures have also been observed. Moreover, in some cases after increasing the temperature, a sign reversal is observed.

- In some instances, $R_H(T)$ and $S(T)$ change their signs at closely related temperatures (Table 3.1).

- The Hall coefficient is strongly sensitive to minor variations in the sample stoichiometry, and annealing conditions.

- A linear correlation between the temperature dependence of the Hall coefficient and the electrical resistivity, extending from 4 K up to room temperature, was reported for i-AlCuFe samples with varying resistivities and exhibiting Hall coefficients of both signs.[50]

TABLE 3.1
Comparison between the temperature values at which the Hall and Seebeck coefficients change their sign for a given sample. The mark * indicates extrapolated values.

Sample	T_0^H (K)	T_0^S (K)	Reference
$Al_{62.5}Cu_{25}Fe_{12.5}$	392*	349	[40]
$Al_{63}Cu_{25}Fe_{12}$	389*	393*	[41]
$Al_{65}Cu_{20}Ru_{15}$	137	162	[6]
$Al_{70}Cu_{15}Ru_{15}$	370*	285	[6]

3.1.2.7 Thermal conductivity

The thermal conductivity of QCs belonging to different families has been measured, covering different temperature ranges, and the following general conclusions can be drawn from the collected data:

- Although most metallic alloys are good heat conductors, the thermal conductivity of QCs is unusually low, even lower than that observed for thermal insulators of extensive use in aeronautical industry, such as titanium carbides or nitrides, doped zirconia, or alumina. For example, in AlPdMn icosahedral phases the thermal conductivity at room temperature is comparable to that of zirconia ($1\,\mathrm{Wm^{-1}K^{-1}}$), and this value decreases to about $10^{-4}\,\mathrm{Wm^{-1}K^{-1}}$ below 0.1 K.[55]

- Assuming that QCs obey the Wiedemann-Franz law (see Section 3.1.2.8) one estimates that the contribution of electrons to the thermal transport is, at least, one order of magnitude lower than that due to phonons over a wide temperature range ($0.1\,K \leq T \leq 200\,K$).[56]

- The thermal diffusivity of these alloys is extraordinarily low, even lower than that of zirconium oxide.[57]

The low thermal conductivity of QCs can be understood in terms of two main facts. In the first place, the charge carrier concentration is low (see Section 3.3.1), so that heat must propagate by means of atomic vibrations (phonons). In the second place, in the energy window where lattice thermal transport is expected to be most efficient the frequency spectra of quasiperiodic systems is highly fragmented. As a consequence (see Section 5.5), the corresponding eigenstates become more localized and thermal transport is further reduced. Physically, this effect can be attributed to the fact that quasicrystal lattices have a fractal reciprocal space, lacking a well defined lower bond as that provided by the lattice parameter in the case of periodic crystals. Consequently, the transfer of momentum to the lattice is not bounded below, which gives rise to a significant degradation of thermal current through the sample.[58]

3.1.2.8 Wiedemann-Franz law

The Wiedemann-Franz law links the electrical conductivity, $\sigma(T)$, and the charge carriers' contribution to the thermal conductivity, $\kappa_e(T)$, of a substance by means of the relationship $\kappa_e(T)/\sigma(T) - L_0 T$, where T is the temperature and $L_0 = (k_B/e)^2\,\eta_0$ is the Lorenz number, where k_B is the Boltzmann constant, e is the electron charge, and η_0 depends on the sample's nature. Thus, for metallic systems $\eta_0 = \pi^2/3$, and we get the Sommerfeld's value $L_0 = 2.44 \times 10^{-8}$ $\mathrm{VK^{-2}}$, while for semiconductors we have $\eta_0 \simeq 2$.[60] The Wiedemann-Franz law expresses a transport symmetry arising from the fact that the motion of the carriers determines both the electrical and thermal currents at low temperatures. As the temperature of the sample is progressively increased, the validity of Wiedemann-Franz law will depend on the nature of the interaction between the charge carriers and the different scattering sources present in the solid. In general, the Wiedemann-Franz law applies as far as elastic processes dominate the transport coefficients, and usually holds for arbitrary band structures provided that the change in energy due to collisions is small compared with $k_B T$.[44] Accordingly, one expects some appreciable deviation from Wiedemann-Franz law when electron-phonon interactions, affecting in a dissimilar way to electrical and heat currents, start to play a significant role.[61] On the other hand, at high enough temperatures the heat transfer is dominated by the charge carriers again, due to umklapp phonon scattering processes, and the Wiedemann-Franz law is expected to hold as well.

Since QCs consist of metallic elements one should expect they would behave as metals do, hence reasonably obeying the Wiedemann-Franz law. This working hypothesis is routinely assumed when studying the thermal transport properties of these materials in order to estimate the phonon contribution to the thermal conductivity, $\kappa_{ph}(T)$, by subtracting to the experimental data, $\kappa_{mes}(T)$, the expected electronic contribution according to the expression $\kappa_{ph} = \kappa_{mes} - L_0 T \sigma$. Nonetheless, since most transport properties of stable QCs are quite unusual by the standard of common metallic alloys, it seems quite convenient to check up on the validity of this law for QCs, since our understanding of thermal properties in these materials should be substantially revised if it does not hold.[9] In fact, according to the phenomenological approach discussed in Section 3.4 one expects significant deviations of the ratio $\kappa_e(T)/\sigma(T)$ from its ideal behavior above ~ 50 K, due to electronic structure effects.[62]

A suitable experimental measure of the Wiedemann-Franz law validity over a given temperature range can be gained from the study of the magnitude $\kappa_{mes}(T)/\sigma(T) = L(T) + \varphi(T)$, where the so-called Lorenz function is defined by the relationship $L(T) \equiv \frac{\kappa_e(T)}{T\sigma(T)}$, and $\varphi(T)$ accounts for the phonon contribution to the heat transport. A study of the temperature variation of the κ_{mes}/σ ratio in several intermetallic compounds showed that the experimental data may be fitted by a linear temperature dependence of the form $\kappa_{mes}/\sigma = LT + B$ over the temperature range 350-800 K.[1, 63] By comparing the slopes obtained for pure aluminum and icosahedral AlCuFe samples the ratio $L_{QC}/L_{Al} \simeq 1.21$ was obtained, hence indicating an enhanced Lorenz number for QCs at high temperatures. In a similar way, room temperature $L(T)$ values larger than L_0 have been experimentally reported, ranging from $L_{300}/L_0 = 1.15$,[64] to $L_{300}/L_0 = 1.43$.[65] Therefore, the available experimental information indicates an enhancement of the Lorenz number value in the high temperature regime, in agreement with some theoretical results.[66]

3.1.2.9 Magnetic properties

The response of a material in the presence of an external magnetic field $\mathbf{H}$ is determined by the value of the magnetization vector $\mathbf{M}$ through the relationship $\mathbf{M} = \chi \mathbf{H}$, where χ measures the magnetic susceptibility of a given material. Diamagnetic materials are characterized by negative values of χ, indicating that the magnetic particles in the material act against the applied magnetic field. On the contrary, paramagnetic materials are characterized by positive values of χ. In a metal composed of non-magnetic atoms (i.e., atoms with no intrinsic magnetic moments) the magnetic susceptibility is determined by two main contributions: the Lenz response of ion core electron orbitals to the external field (Larmor's diamagnetism, χ_L) and the conduction electron's contribution χ_e, which, in turn, can be split into the effect of spin electrons aligning in a direction parallel to $\mathbf{H}$ (Pauli paramagnetism, χ_p) and the Lenz response of free electrons (Landau diamagnetism, χ_l). In the free electron approximation the Pauli's contribution is proportional to the density of states at the Fermi level, $\chi_p = \mu_B^2 N(E_F)$, where μ_B is the Bohr magneton, and the Landau's contribution amounts $\chi_l = -\chi_p/3$, so that $\chi_e = \chi_p + \chi_l = 2\chi_p/3 > 0$, and the conduction electron's contribution becomes paramagnetic. The core electron's contribution is more difficult to determine but in most metals and metallic alloys the temperature independent contribution to the magnetic susceptibility, $\chi_0 = \chi_L + \chi_e$, takes on positive values typical of a paramagnetic response. For instance, one gets $\chi_0 = +0.6 \times 10^{-6}$ emu/g for both aluminum metal and the β−AlCuFe alloy phase.

At variance with this typically metallic behavior, i-AlCuFe QCs and approximant phases are diamagnetic in a broad temperature range, with magnetic susceptibility values comprised within the interval $[-0.6, -0.4] \times 10^{-6}$ emu/g. A similar behavior has been reported for i-AlPdRe, i-GaMgZn, and i-MgZnY representatives.[67] It is interesting to note that the more diamagnetic samples are generally also the more resistive ones. The emergence of a diamagnetic behavior in these QCs is attributed to a weak Pauli term contribution due to the existence of a pseudogap close to the Fermi level and to an anomalously strong Landau term $\chi_l = -(m/m_*)^2 \chi_p/3$, resulting from a peculiar band structure, characterized by flat bands with large effective masses m_*.[68]

The temperature dependence of the magnetic susceptibility of i-AlCuFe samples can be fitted to the form $\chi(T) = \chi_0 + AT^2$, where the parabolic term is ascribed to a temperature dependent Pauli susceptibility. In fact, the Pauli paramagnetic contribution is temperature independent as long as the DOS at the Fermi level does not vary. However, the chemical potential slightly shifts with the temperature (see Eq.(5.42)) and this variation can have a significant influence in those systems exhibiting sharp features in the DOS near the Fermi level. This is precisely the case of most i-AlCuFe samples (see Section 3.3.2), so that the Pauli susceptibility can be expressed as

$$\chi_p(T) = \mu_B^2 N(E_F) \left\{ 1 + \frac{\pi^2}{6}(k_B T)^2 \left[\frac{1}{N(E)}\frac{d^2 N}{dE^2} - \left(\frac{1}{N(E)}\frac{dN}{dE}\right)^2 \right]_{E=E_F} \right\},$$
(3.6)

which properly accounts for the experimental data reported for AlCuFe.[69] Nevertheless, this expression can not account for the linear dependence $\chi(T) = \chi_0 + AT$ reported for i-AlPdRe samples, which indicates the existence of substantial differences in the electronic structure of both quasicrystalline families.

Quite remarkably a diamagnetic behavior is also observed in QCs containing magnetic atoms such as i-AlPdMn, i-AlSiMn, and ZnMg(Ho,Yb,Tb,Er). The temperature dependence of the magnetic susceptibility in these phases obeys a Curie-Weiss law

$$\chi(T) = \chi_0 + \frac{C}{T - \theta},$$
(3.7)

where χ_0 is comprised within the interval $[-0.6, -0.4] \times 10^{-6}$ emu/g, and the second term describes the contribution due to the presence of ions with incomplete orbitals giving rise to a net angular momentum J (Curie paramagnetism). The Curie constant is usually expressed as

$$C = \frac{g^2 \mu_B^2 J(J+1)N_m}{3k_B},$$
(3.8)

where g is the Landé factor ($g = 2$ for Mn atoms) and N_m measures the number of magnetic atoms. In the case of the i-AlPdMn phases the analysis of the obtained measurements indicates that only a minor fraction (i.e., 0.04% - 4%) of the Mn atoms present in the QC carry a magnetic moment.[70] This fraction increases rapidly with the Mn concentration in the alloy QC. These Mn atoms are an intrinsic feature of the QC phase, although the precise location of the magnetic Mn sites could not be identified because of both their small number and the uncertainties in the chemical decoration of the quasilattice structure. Thus the magnetic momentum formation on Mn atoms is very sensitive to environmental effects determined by atomic distances, coordination number, and the kinds of atoms around Mn ones. For instance, Mn sites with a low Al coordination experience weaker sp-d hybridization effects, which favour the appearance of a magnetic moment.[71]

Whereas the magnetic properties of QCs containing Mn atoms are mainly determined by the number of Mn atoms carrying a magnetic moment, the magnetic behavior of ZnMgRE QCs containing rare-earth atoms with particularly strong magnetic moments such as Ho,Yb,Tb, and Er is strongly influenced by the nature of the magnetic ordering of these atoms. In ferromagnetism, for instance, all the magnetic moments point in the same direction, whereas in antiferromagnetism neighboring atoms point in alternate directions. The presence of relatively large, negative values of θ (from -5 K to -26 K) in Eq.(3.7) indicates the existence of dominant antiferromagnetic exchange interactions between magnetic atoms. In fact, the direct confirmation of the presence of short-range spin antiferromagnetic correlations in i-ZnMgHo[72] spurred the interest of theoretical works predicting that long-range ferromagnetic order is possible in QCs.[73] However, previous claims of both antiferromagnetic,[74] and ferromagnetic,[75] QC phases may be probably related to the presence of secondary magnetic crystalline phases in the considered samples.[76]

Evidences of the so-called spin glass transitions have been observed in both i-AlPdMn and i-ZnMgRE QCs which take place at relatively low temperatures of a few kelvin degrees (as compared to those usually observed in conventional alloys which are one order of magnitude larger).[67] In fact, due to the icosahedral symmetry, atoms in a QC find themselves in a variety of different local environments, which means that magnetic interactions often become "frustrated." In other words, there is no possible configuration that allows magnetic moments to align in their preferred directions. A well-known example is that of antiferromagnetic spins on a triangular lattice: the three spins cannot be arranged so that all neighboring spins are antiparallel. The low value of the transition temperature is explained by the relatively small fraction of magnetic atoms present in QC phases. The very existence of such a transition indicates that quasiperiodic order is sufficient to effectively coupling the magnetic atoms (via delocalized d electrons through the so-called Ruderman-Kittel-Kasuya-Yoshida interactions) in order to induce a transition leading to the spin glass state. Magnetic properties of decagonal phases in AlCuCo and AlPdMn systems also indicate the presence of anisotropy effects. Thus the value of the local magnetic moments for the i-AlPdMn is about twice as large as those for the d-AlPdMn.[77]

3.1.2.10 Mechanical and tribological properties

Quasicrystals are noteworthy for their hardness (comparable to that of silica), low surface energy, and low friction.[78] The influence of commensurability on friction has been examined by a number of experimental and theoretical studies.[79, 80] In the ideal case, when two workpieces with incommensurate lattices are brought in contact, the minimal force required to achieve sliding (known as the static frictional force) vanishes, provided the two substrates are stiff enough.[81] Thus, it has been observed that friction becomes negligible for incommensurate surfaces sliding under conditions of elastic contact. In real situations, however, physical contact between two (uncontaminated) surfaces is generally mediated by third bodies acting like a lubricant film. In that case the sliding interface should be properly described in terms of three characteristic lengths, corresponding to the periods of both substrates and the lubricant layer.

Scanning tunneling microscopy studies have revealed the presence of structures closely related to the Golden mean in quasicrystalline surfaces. For example, on the surface of AlPdMn icosahedral quasicrystals atomic terraces are separated by steps of three different heights whose values are related, not by a simple integer, but by the irrational number $\tau \simeq 1.618$.[82, 83] This means that the spacing between similar planes of the bulk structure has not one but two dominant spacings, and this is reflected in the step heights observed on the surface. In this way, the bulk quasiperiodic order of the sample naturally emerges to its surface, hence suggesting that quasicrystal surfaces can act as templates for the growth of thin films having quasicrystalline order as well.[84, 85]

However, recent theoretical studies indicate that the best low-friction regime is achieved for incommensurabilities related to cubic irrational numbers rather than to quadratic irrationals, like the Golden mean.[86] A suitable example of cubic irrational number is provided by the so-called spiral mean, which satisfies the equation $\omega^3 - \omega - 1 = 0$. Its rational approximants are generated by the recursion relation $G_{n+1} = G_{n-1} + G_{n-2}$ with $G_{-2} = G_0 = 1$ and $G_{-1} = 0$, leading to the sequence $G_n = \{1, 0, 1, 1, 1, 2, 2, 3, 4, 5, 7, 9, 12, 16, 21, 28, ...\}$ whose terms satisfy the asymptotic limit, $\lim(G_{n+1}/G_n) = \omega \simeq 1.3247....$ According to these results, the low friction observed in QCs cannot be simply justified in terms of their characteristic Fibonacci-based surface ordering. Quite interestingly, experimental studies on friction anisotropy of a clean, decagonal AlNiCo QC, whose surface terminations exhibit periodic as well as Fibonacci-like atomic ordering along different directions, reveal a strong connection between interface atomic structure and the mechanisms by which energy is dissipated.[87] This result suggests that electronic and phononic contributions probably play a significant role in the tribological properties of QCs. Other attractive properties of QC surfaces, which are currently intensively explored, include oxidation resistance,[88, 89] low surface energy,[90] and catalytic activity.[91, 92, 93, 94]

3.1.3   On the nature of chemical bond

> "It is the metallic bonding that makes it possible. I am doubtful whether they will ever make quasicrystals out of anything other than metals" (John W. Cahn 1997) [95]

As it is well known, metallic substances exhibit a number of characteristic physical properties which are directly related to the presence of a specific kind of chemical bond among their atomic constituents: the so-called metallic bond. In fact, it is precisely the existence of the metallic bond which accounts for both the atomic structure and physical properties of metallic compounds.[96] For the sake of comparison in Table 3.2 we list a number of representative physical properties of both metals and QCs. By inspecting this Table one realizes that quasicrystalline alloys significantly depart from metallic behavior, resembling either ionic or semiconducting materials (respectively labeled I or S in Table 3.2). Thus, QCs are an intriguing example of solids made of typical metallic atoms which do not exhibit any of the physical properties usually signaling the presence of metallic bonding.

TABLE 3.2
Comparison between the physical properties of QCs and typical metallic systems.

PROPERTY	METALS	QUASICRYSTALS
MECHANICAL	ductility,malleability	brittle (I)
	Young modulus	
TRIBOLOGICAL	relatively soft	very hard (I)
		low friction coefficient
	easy corrosion	corrosion resistant
ELECTRICAL	high conductivity	low conductivity (S)
	resistivity increases with T	decreases with T (S)
	small thermopower	large thermopower (S)
MAGNETIC	paramagnetic	diamagnetic
THERMAL	high conductivity	very low conductivity (I)
	high specific heat values	low specific heat values
	high melting points	
OPTICAL	Drude peak	IR absorption (S)

Therefore the fundamental question arises concerning whether these anomalous properties should be mainly attributed (or not) to the characteristic quasiperiodic order of QCs structure. In this regard, several experimental evidences strongly suggest that the nature of the chemical bonding determining the local atomic arrangements would play a significant role in most physical properties of these materials. [97, 98, 99] In fact, there are several hints pointing towards the important role of chemical bonding in the emergence of the unusual physical properties of QCs, namely:

1. Transport measurements show that the structural evolution from the amorphous to the quasicrystalline state (Fig.3.11) is accompanied by a parallel evolution of the electronic transport anomalies, clearly indicating the importance of short-range effects on the emergence of several transport anomalies.

2. Transport measurements also indicate that these anomalies are more pronounced in the case of QCs, hence suggesting that the relative intensity of the anomalous behavior is significantly emphasized due to the presence of long-range quasiperiodic order.

3. Many unusual physical properties of QCs are also found in approximant phases.

4. Certain anomalous transport properties, like a high resistivity value or a negative temperature coefficient, are also observed in some crystalline alloys consisting of normal metallic elements whose structure is unrelated

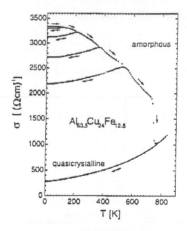

FIGURE 3.11

Temperature dependent electrical conductivity of an AlCuFe film for different annealing states for the amorphous and for the icosahedral quasicrystalline phase. (From Ref.[100]. With permission from Elsevier.)

   to the structure of QCs (as, for instance, the Heusler-type Fe_2VAl alloy) which share with them certain characteristic features in the electronic structure (i.e., a narrow pseudogap).[101]

5. Transport properties of metallic alloys with complex unit cells, having a similar number of atomic species than those of approximant phases, but not exhibiting the local isomorphism property, are typically metallic.[102]

6. Other kinds of aperiodic crystals, like incommensurately modulated phases and composites (see Section 1.3), do not show the physical anomalies observed in QCs.

According to (1)-(4) the emergence of physical anomalies in QCs should be traced back to chemical bonding effects (short-range), giving rise to some characteristic features in the electronic structure close to the Fermi level (such as the presence of a narrow pseudogap), which are generic but not specific of QCs.[104] Thus, chemical effects may ultimately become more important than quasiperiodic order effects in explaining the unusual behavior of these materials. Accordingly, crystalline approximants, which exhibit a local atomic environment very similar to their related QC alloys, appear as natural candidates to investigate the relative importance of short-range versus long-range order effects on the transport properties. This conclusion is further supported by (5) and (6), which indicate that mere structural complexity is not a sufficient condition to give rise to the emergence of anomalous transport properties in complex metallic alloys.

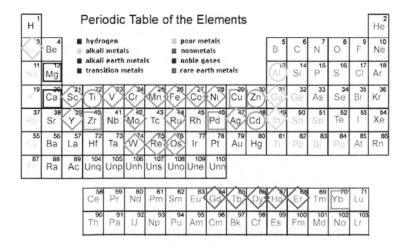

FIGURE 3.12

Chemical elements found in thermodynamically stable quasicrystal alloys. Main forming elements (Al, Ti, Zn, and Cd) are circled. The second major constituents are squared. Minor constituents are marked with a diamond.

Most atomic elements composing thermodynamically stable quasicrystalline alloys observed to date belong to the chemical family of metals, located at either alkaline, earth-alkaline, transition metals, or rare-earth groups (Fig.3.12). From this chart we see that most metallic atoms are able to participate in the formation of quasicrystalline phases under the proper stoichiometric conditions, in agreement with Cahn's conjecture quoted above. On the other hand, certain chemical trends can also be appreciated in different QC families. For instance, the minor atom constituent in the systems

$$\mathrm{Al}_{63}\mathrm{Cu}_{25}\begin{pmatrix}\mathrm{Fe}\\\mathrm{Ru}\\\mathrm{Os}\end{pmatrix}_{12} \qquad \mathrm{Al}_{70}\mathrm{Pd}_{20}\begin{pmatrix}\mathrm{Mn}\\\mathrm{Re}\end{pmatrix}_{10} \quad ,$$

belongs to the same group of the Periodic Table, hence indicating the importance of their chemical valence for the stability of the compound. This fact has been successfully exploited in order to obtain the family of stable quaternary QCs given by the formula (Fig.3.13)

$$\mathrm{Al}_{70}\mathrm{Pd}_{20}\begin{pmatrix}\mathrm{V}\\\mathrm{Cr}\\\mathrm{Mn}\\\mathrm{W}\end{pmatrix}_{5}\begin{pmatrix}\mathrm{Co}\\\mathrm{Fe}\\\mathrm{Ru}\\\mathrm{Os}\end{pmatrix}_{5} .$$

Several chemical trends are also observed in the transport properties of QCs belonging to the AlCu(Fe,RuOs) and AlPd(Mn,Re) families (Table 3.3). Thus, it is seen that increasing the atomic number of the third (incomplete d band) transition metal significantly increases the low temperature electrical resistivity of the sample as well as its temperature dependence as measured in terms of the ratio $R = \rho(4\ \mathrm{K})/\rho(300\ \mathrm{K})$.[105]

TABLE 3.3
Chemical trends in the properties of the electrical resistivity of aluminium based icosahedral QCs.

SAMPLE	$\rho(4\ \mathrm{K}) \times 10^4\ \mu\Omega\mathrm{cm}$	R
AlCuFe	1	2.2
AlCuRu	2.5	4
AlCuOs	14	4.5
AlPdMn	1	2.3
AlPdRe	120	190

This trend may be due to the relativistic contraction of the s and p states relative to the d and f states. On the one hand, this contraction lowers the orbital energies of s and p states. On the other hand, this contraction screens

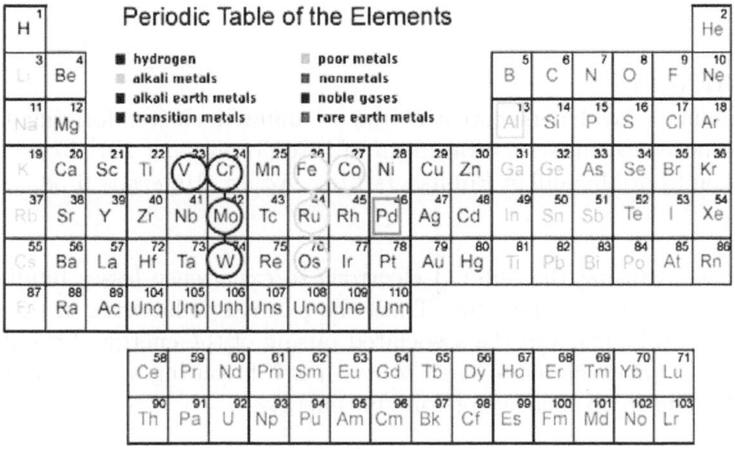

FIGURE 3.13

Chemical valence trend observed in quaternary quasicrystals associated to the AlPd(X,Y) family. [103]

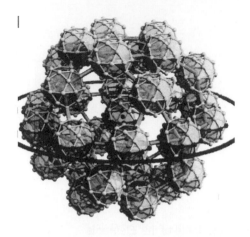

FIGURE 3.14
A three-dimensional perspective of the τ^3 inflated cluster forming the basic icosidodecahedron structural motif of i-CdYb phase. ([107] Reprinted by permission from Macmillan Publishers Ltd.: *Nature Materials* **6** 58 © 2007.)

the nucleus, causing the outer d electrons to experience lesser binding and therefore a larger spatial extent. Thus, the relativistic lowering of the energy of the s and p bands, and the associated raising of the energy of the d bands brings these bands closer to each other, hence favouring sp-d hybridization effects leading to an increase of cohesive energy.

3.2   Quasicrystals as a hierarchy of clusters

Most structural models of quasiperiodic crystals and their approximants are based on one or more characteristic structural units, commonly referred to as clusters. For instance, building units of about 50 atoms with a geometry close to that of a Mackay icosahedron have been experimentally identified in the structure of AlPdMn and AlCuFe QCs, as deduced from X-ray and neutron diffraction data.[106] More recently, the atomic structure of $YbCd_{5.7}$ was derived from a detailed x-ray diffraction analysis and described in terms of three basic building units (Fig.3.14).[107] These building units adopt well defined polyhedral shapes and can be expressed as regular arrangements of atoms in clusters, generally adopting point group icosahedral symmetries (dodecahedron, icosahedron, icosidodecahedron, triacontahedron). The role of clusters as possible structural units is supported by the fact that such clusters can be universally identified in bulk structures of most QCs, and are common also in approximants.

For instance, two families of chemically different pseudo-Mackay icosahedra (PMI) clusters (containing 51 atoms instead of the 54 atoms included in a Mackay cluster, see Fig.2.3) were identified in the icosahedral AlPdMn structure. Each PMI is made of three centrosymmetrical atomic shells: a core of nine atoms, an intermediate icosahedron of 12 atoms, and an external icosidodecahedron of 30 atoms (Fig.3.15(a)). The last two shells have practically equal radii and constitute altogether the surface of the PMI whose diameter is very close to 0.96 nm. The small inner core is a piece of a pentagonal dodecahedron whose 20 atomic sites are only partially occupied in a way which probably fluctuates from PMI to PMI within the structure. Two families of PMI can be distinguished attending to the chemical decoration of the outer shell.[108, 109, 110] The calculated atomic density of an individual PMI is 64 at/nm^3, which is close to the measured density of the bulk materials within the experimental accuracy. Starting from an individual PMI the entire self-similar atomic structure can be grown following an inflation process via successive substitutions of atoms by PMIs with proper rescaling. Here, inflation means that a subset of special points from the original structure (i.e., the PMI cluster) are found in an identical arrangement when increased by a scale factor τ^3. A planar section of the AlPdMn structure along the fivefold axis is shown in Fig.3.15(b). Note that in order to preserve the density and stoichiometry of the solid under the inflation growing scheme some overlapping is required.

On the basis that the i-AlCuFe phase is almost iso-structural to the i-AlPdMn one, with Cu (Pd) being equivalent to Fe (Mn), respectively, Gratias and co-workers proposed an unified structural model for both QCs families in terms of three basic clusters (Fig.3.16). The model is globally consistent with the chemical order obtained in previous structural investigations of these alloys.[111]

Certainly, one may consider the systematic use of geometrical clusters in structure determination as a matter of mere convenience, but both their usefulness and ubiquity naturally leads one to take a step further and speculate about the very possibility of considering QCs as molecular solids composed of actual atomic clusters arranged in a hierarchical way. Some evidences favouring the existence of clusters as stable physicochemical entities come from direct imaging techniques, such as secondary electron imaging,[112] x-ray photoelectron diffraction,[113] or STM.[114] These studies support the picture of QCs as cluster aggregates, namely, a three-dimensional quasiperiodic lattice properly decorated by atomic clusters which have the same point symmetry of the whole QC.[115] In fact, several experimental facts strongly suggest that local atomic order, on the scale of a few nanometers, plays an essential role in the emergence of the peculiar electronic properties of these materials, probably due to the formation of a number of covalent bondings among different atoms

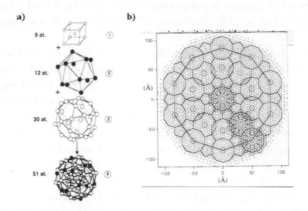

FIGURE 3.15

(a) Structure of a pseudo-Mackay icosahedron. Similar clusters can be identified in the bulk of icosahedral AlPdMn and AlCuFe, although other types of clusters can be identified as well.(b) Arrangement of pseudo-Mackay-type icosahedra showing the hierarchical, self-similar arrangement of overlapping clusters. ([108] Reprinted figures with permission from Janot C 1996 *Phys. Rev. B* **53** 181 © 1996 by the American Physical Society.)

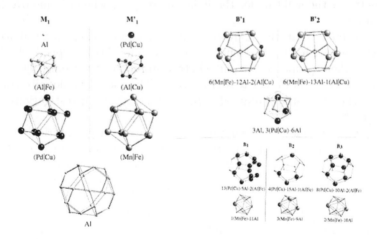

FIGURE 3.16

The main configurations of the different clusters in the unified model proposed for i-AlCuFe and i-AlPdMn. ([111] Reprinted figures with permission from Quiquandon M and Gratias D 2007 *Phys. Rev. B* **74** 214205 © 2007 by the American Physical Society.)

grouped in clusters.[22, 116]

Then, a major question in the field is whether clusters are physically significant, chemically stable entities, or simply geometrical coincidences.[117] Now, when thinking of clusters as physical entities, rather than convenient geometrical tiles, one should properly address the following issues, which are the focus of intensive current research:

- What is the number and structure of the different atomic clusters which are compatible with the chemistry of the system?

- What is the nature of the chemical bonding among the atoms belonging to a given cluster, as well as among different clusters themselves? For example, it may occur that a given cluster may act as a chemically stable structure when isolated, but it progressively loses its identity when assembled to form a solid, due to strong interactions with close neighbors. Then, along with the stability of clusters we should also consider those aspects related to their reactivity.

- What is the more appropriate packing rule (including possible overlappings) between clusters at different hierarchical stages?

Another fundamental question related to the possible existence of clusters in QCs has to do with the quasicrystalline growth process itself. In fact, how do QCs grow? So far it is not clearly known what specific qualities (if any) systems must possess in order to form quasiperiodic instead of periodic crystals. In both cases, they form via nucleation and growth, starting from a microscopic nucleus, which spontaneously arises from the solid phase, and spreads outward, converting the system from liquid to solid. A fundamental puzzle in QC physics is to understand how the growth phase nucleation, occurring at a short-range scale, can lead to a structure with long-range quasiperiodicity. In this regard, QCs cannot grow like periodic crystals, where the nucleus surface acts as a template for copying a unit cell via local interactions. Rather, QCs require specialized growth rules that dictate their formation. Using molecular simulations it has been shown that the aperiodic growth of dodecagonal QCs is controlled by the ability of the growing QC nucleus to incorporate kinetically trapped atoms into the solid phase with minimal subsequent rearrangement. This process occurs through the assimilation of stable icosahedral clusters by the growing QC.[118] In this way, the probability of formation of a highly complex structure from its elements is increased by means of the successive inclusion of nested cluster units.[119]

Cleavage and annealing experiments have also been interpreted as proof for the existence of clusters in QCs with a high mechanical stability,[120] though it seems that most mechanical properties reported for QCs (with the exception of a brittle-ductile transition at elevated temperatures) can be properly accounted for in terms of usual processes (including the additional phasonic freedom degree).

3.3   Electronic structure of icosahedral quasicrystals

3.3.1   Fermi level pseudogap

It was pointed out by William Hume-Rothery (1899-1968) in 1926 that certain metallic compounds with closely related structures but apparently unrelated stoichiometries exhibit the same ratio of number of valence atoms to number of atoms (the so-called e/a ratio).[121] For example, the isostructural phases with compositions CuZn, Cu_2Al, and Cu_5Sn share the value $e/a = 1.5$, if one considers the valence values Cu = 1, Zn = 2, Al = 3, and Sn = 4. Other alloys which may be placed in this class are CuBe, AgZn, AgCd, AgMg, Ag_3Al, and AuZn. A more striking example is provided by the so-called γ-alloys, the principal representatives of which are Cu_5Zn_8, Cu_9Al_4, $Cu_{31}Sn_8$, and Fe_5Zn_{21}. These crystals are cubic, with 52 atoms arranged within icosahedral clusters in the unit cell. Adopting the value Fe = 0 for the valence value of iron atoms, all these alloys share the value $e/a = 21/13 = 1.615384....$ Note the presence of Fibonacci numbers in the e/a ratio, which, according to the asymptotic limit given by Eq.(2.5), suggests $e/a \simeq \tau$ (1.6180...) for these alloys.

Hume-Rothery rule is explained as resulting from a perturbation of the energy of the valence electrons by their diffraction by the crystal lattice. In fact, it was pointed out by Marcel Louis Brillouin (1854-1948) that the energy distribution is perturbed when an electron has such a wave length $(\lambda = h/\sqrt{2mE})$ and direction as to permit Bragg reflection from an important crystallographic plane.[122] The perturbation is of such nature as to stabilize electrons' energy with energy just equal or less than that corresponding to Bragg reflection and to destabilize electrons with a larger energy. Hence, special stability would be expected for metals with just the right number of electrons. This number is proportional to the volume of a polyhedron in reciprocal space (the so-called Brillouin-Jones zone), corresponding to the crystallographic planes giving rise to the perturbation. For instance, the corresponding polyhedron for the γ-alloys is bounded by twelve {330} and twenty-four {411} planes (as derived from x-ray diffraction data) and contains 22.5 electrons per 13 atoms,[123] a figure close to the expected ratio $e/a = 21/13$.

Although QCs have a dense reciprocal space, only a few diffraction peaks have very strong intensities. The Hume-Rothery criterion can then be applied to QCs by introducing a pseudo-Brillouin zone with the help of the most intense diffraction spots.[124, 125] Due to their great symmetry, in the case of icosahedral QCs this zone is quite close to spherical shape, so that the diffraction condition can be expressed in the form

$$K_{hkl} = 2k_F, \qquad (3.9)$$

where K_{hkl} is the reciprocal vector of the considered diffraction plane, $k_F =$

$\sqrt[3]{3\pi^2 n}$ is the radius of the Fermi sphere, and n is the electron number per unit volume. Eq.(3.9) has been successfully used to explain the stability of i-QCs containing elements with a full d-band, like $Al_{56}Li_{33}Cu_{11}$ ($e/a = 2.12$), $Zn_{43}Mg_{37}Ga_{20}$ ($e/a = 2.2$), $Zn_{60}Mg_{30}(RE)_{10}$ ($e/a = 2.1$), or $Zn_{80}Sc_{15}Mg_5$ ($e/a = 2.15$), by adopting the valence values $Li = 1$, $Mg = 2$, $Sc = 3$, $Ga = 3$, and $RE = 3$. In all these samples the redistribution of electronic states due to the Fermi sphere-pseudo Brillouin zone interaction gives rise to a significant reduction of the density of states (pseudogap) close to the Fermi energy. [98, 104, 126]

For alloys containing a small concentration of a transition element one can properly extend the Hume-Rothery mechanism by assuming that transition atoms take electrons from the conduction band, hence adopting a negative effective valence.[127] Nevertheless, the increase of electrostatic energy due to a transfer of several electrons on one atom is unrealistic in metallic alloys. Subsequent studies based on the Linear-Muffin-Tin-Orbital method[128] provided a more suitable physical picture to account for the apparent negative valence. According to this view this effect arises from a combined effect of strong hybridization between the sp states and the transition metal d orbitals along with the diffraction of sp states by Bragg planes. As a consequence, there is an increase of the sp component of the DOS below the Fermi energy as compared to the free electron DOS, but contrarily to the d orbitals these additional states are delocalized and do not lead to a strong electrostatic energy.[28] Thus, in QCs bearing transition metal atoms, such as AlCu(Fe,Ru,Os) or AlPd(Mn,Re), the presence of hybridization effects between sp aluminum states and 3d transition metal states enhances the (structure related) Fermi surface-Brillouin zone diffraction effect, further deepening the pseudogap close to the Fermi level. In fact, the role of sp-d hybridization in both cohesion energy and transport properties has been demonstrated for a series of QCs belonging to the AlCu(Fe,Ru) and AlPd(Mn,Re) families.[129, 130, 131, 132, 133]

The precise value of the effective valence of transition metals generally depends on the approach used and the considered sample. Some representatives values are listed in Table 3.4 for the sake of illustration.

TABLE 3.4
Effective valence values for some typical transition metals.

	Cr	Mn	Fe	Co	Ni
Raynor	−4.66	−3.66	−2.66	−1.61	−0.71
LMTO[28]	−3.2	−2.7(−2.0)	−2.5	−1.3(−0.9)	−1.0

Making use of Raynor's values (along with $Pd = 0$) one obtains $e/a = 1.75$ and $e/a = 1.73$ for $Al_{65}Cu_{20}Fe_{15}$ and $Al_{70}Pd_{20}Mn_{10}$, respectively. By comparing with the electron per atom ratio obtained in the case of stable QCs not containing transition metal atoms (i.e., $e/a = 2.1 - 2.2$), we conclude that they belong to different Hume-Rothery families. In fact, on the basis that QCs and their approximants share similar electron concentrations the possibility of obtaining new quasicrystalline compounds via pseudogap electronic tuning has rendered promising results in the CaAuIn and MgCuGa systems.[134] The binary i-Cd(Yb,Ca) family, which is composed of divalent atoms, has $e/a = 2.0$, a value which lies close to that of the full d band representatives. Notwithstanding this, the role played by hybridization effects in the stability of the i-Cd(Yb,Ca) phase is significantly larger than that coming from the Fermi-surface-Brillouin zone mechanism in this binary QC.[135, 136] In this case the orbitals involved in the hybridization process come from occupied Cd-5p and unoccupied Yb-5d (or Ca-3d) orbitals, which highlights the importance of chemical bonding aspects in these quasicrystalline compounds. In fact, the influence of sp-d hybridization on the electronic structure of different Al-Mn alloys has been recently studied by photoelectron spectroscopy, and it has been confirmed that these hybridization effects alone can produce a pseudo-gap, even in the absence of Hume-Rothery mechanism.[137]

In summary, two main features can be observed in the DOS close to the Fermi energy in high quality, thermodynamically stable QCs containing transition metal atoms: a structurally induced broad minimum (~ 1 eV width) due to the Hume-Rothery mechanism and a narrow and sharply confined dip (~ 0.1 eV width) due to hybridization effects involving the transition metal bands. [138] The physical existence of the electronic pseudogap has received strong experimental support during the last decade, as indicated by measurements of the specific heat capacity, [139] photoemission, [140] soft x-ray spectroscopies,[141, 142] magnetic susceptibility, and nuclear magnetic resonance probes.[143]

3.3.2 Fine spectral features

As we have previously commented, the presence of a pseudogap is a generic feature of QCs, but it is not a specific one, since certain periodic crystals can also exhibit a substantial depletion of the electronic DOS close to the Fermi level.[144] In this regard, the possible existence of a spiky structure in the electronic DOS over an energy scale of about 10 meV, obtained in self-consistent *ab initio* calculations dealing with several suitable quasicrystalline approximants,[145] was considered as a promising characteristic feature of quasiperiodic crystals DOS. In this sense, it was argued that these peaks may stem from the structural quasiperiodicity of the substrate due to cluster aggregation,[108] or d-orbital resonance effects.[146]

Experimental investigation of AlCuFe quasicrystalline films by scanning tunneling spectroscopy at low temperatures gave evidence for a narrow, symmetric gap of about 60 meV wide located around the Fermi level.[147] A subsequent STM investigation of better resolution on AlCuFe and AlPdRe qua-

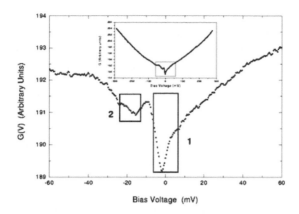

FIGURE 3.17

The differential conductance for the $Al_{63}Cu_{25}Fe_{12}$-Al tunnel junction at a temperature of $T = 2$ K at two different energy scales: ± 60 meV (main frame) and ± 300 meV (inset). Data file courtesy of R. Escudero. (Adapted from ref.[151]. Reprinted figure with permission from Maciá E 2004 *Phys. Rev. B* **69** 132201 © 2004 by the American Physical Society.)

sicrystalline ribbons confirmed the presence of a pseudogap of about 50 meV wide, but did not show evidence for finer structures in the DOS over the energy region extending about 0.5 eV from the Fermi level.[148] The existence of a sharp DOS valley of about 20 meV at the Fermi level in both quasicrystalline and approximant phases has been confirmed by nuclear magnetic resonance studies, which probe the bulk properties of the considered samples.[149] All these observations indicate that the dip centered at the pseudogap is not a surface feature and that both its width and depth are sample dependent. The dependence of the pseudogap structure with the temperature was also investigated by means of tunnelling and point contact spectroscopy, and it was reported that the width of the broad pseudogap remains essentially unmodified as the temperature is increased from 4 K to 77 K. On the contrary, the dip feature centered at the Fermi level exhibits a significant modification, deepening and narrowing progressively as the temperature is decreased.[150]

In Fig.3.17 we show low temperature tunneling spectroscopy measurements corresponding to the quasicrystalline sample i $Al_{63}Cu_{25}Fe_{12}$.[150] These measurements reveal a broad pseudogap extending over an energy scale of about 0.6 eV (shown in the inset) along with some fine structure close to the Fermi level (labeled 1 and 2 in the main frame). The broad pseudogap stems from the Fermi surface pseudo-Brillouin zone interaction, while the dips may be respectively related to hybridization effects between d-Fe states and sp-states (feature labeled 1 in Fig.3.17) and d-orbital resonance effects (feature labeled 2 in Fig.3.17). Nevertheless, the possible existence of the spiky component of the DOS is still awaiting for a definitive experimental confirmation.[152, 153] In fact, difficulties in the experimental investigation of fine structure in the DOS arise from the requirement of a high energy resolution, as the peaks and gaps to be observed are only a few meV wide. Thus, as we have previously mentioned, both high resolution photoemission and tunneling spectroscopies have failed to detect the theoretically predicted dense distribution of spiky features around the Fermi level. Several reasons have been invoked in order to explain these unsuccessful results. Among them the existence of some residual disorder present even in samples of high structural quality has been invoked as a plausible agent to smear out the finer details of the DOS.[154] It has also been argued that photoemission and STM techniques probe the near surface layers, so that sharp features close to the pseudogap could be removed by subtle structural deviations near the surface from that of the bulk, as those reported for annealed QC surfaces.[155]

On the other hand, detailed analysis of higher-resolution, extensive ab-initio calculations of several QC approximants suggests that a significant contribution to the spiky DOS component may probably stem from numerical artifacts,[156] hence explaining the absence of experimental evidences. Notwithstanding this, recent tunnelling spectroscopy measurements performed in icosahedral QCs at low temperature (5.3 K) have provided additional experimental support for the existence of a large number of energetically localized features close to the Fermi level in the electronic structure of the 5-fold surface of an i-AlPdMn sample at certain local regions.[157]

3.4   Phenomenological transport models

3.4.1   Spectral conductivity models

An important open question in the field regards whether the purported anomalies in the transport properties observed in high-quality quasicrystals can be satisfactorily accounted for by merely invoking band structure effects or, conversely, they must be traced back to the critical nature of the electronic states. At this stage, it seems quite reasonable that the proper answer should likely require a proper combination of both kinds of effects.

In fact, on the one hand, certain experimental facts, such as the relative insensitivity of the specific heat electronic term γ to thermal annealing as compared to the strong dependence of the electrical conductivity, suggest that the low values of residual conductivity $\sigma(0)$ cannot be satisfactorily explained by solely invoking the existence of the pseudogap. This conclusion

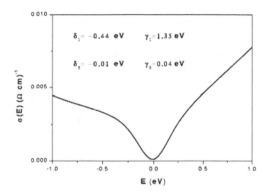

FIGURE 3.18

Spectral conductivity curve in the energy interval ±1 eV around the Fermi level as obtained from Eq.(3.10) for the electronic model parameter values γ_i and δ_i indicated in the frame.

is further stressed by the unrelated variations of σ_{4K} and γ among different AlPdRe samples.[15] On the other hand, it has been suggested that when the energy spacing between the electronic bands in the vicinity of Fermi level becomes very small, as it occurs in the case of quasicrystalline approximants, the transport may turn out to be anomalous because tunneling occurs between different bands, causing the instability of the wave packet coherence.[159]

In order to make a meaningful comparison between band structure calculations and experimental measurements one should take into account possible phason, finite lifetime, and temperature broadening effects. In so doing, it is observed that most finer details in the DOS are significantly smeared out and only the most conspicuous peaks remain in the vicinity of the Fermi level at room temperature.[101] These considerations convey one to reduce the number of main spectral features necessary to capture the most relevant physics of the transport processes. To this end, it is useful to consider the spectral conductivity function, $\sigma(E)$, defined as the $T \longrightarrow 0$ conductivity with the Fermi level at energy E. Generally speaking the conductivity spectrum should take into account both the DOS structure and the diffusivity, $D(E)$, of the electronic states, according to the relationship $\sigma(E) \propto N(E)D(E)$. Thus, although it may be tempting to assume that the $\sigma(E)$ function should closely resemble the overall structure of the DOS, it has been shown that dips in the $\sigma(E)$ curve can correspond to peaks in the DOS at certain energies.[160, 161, 162, 163] This behavior is likely to be related to the peculiar nature of critical electronic states close to the Fermi level.[163, 164, 165, 166]

Two fruitful results have been reported regarding the main features of the spectral conductivity function in QCs. On the one hand, it has been shown that the main qualitative features of the $\sigma(T)$, $S(T)$, and $R_H(T)$ curves can be accounted for by considering an *asymmetric* spectral conductivity function characterized by a broad minimum exhibiting a pronounced dip within it, hence encompassing the transport properties of both amorphous phases and QCs within a unified scheme.[22] On the other hand, a series of *ab initio* studies have shown that the electronic structure of both QCs and approximant phases belonging to the AlCu(Fe,Ru) and AlPd(Mn,Re) icosahedral families can be satisfactorily described in terms of a spectral resistivity, $\rho(E) = \sigma^{-1}(E)$, exhibiting two basic spectral features close to the Fermi level, namely, a wide and a narrow Lorentzian peak, according to the expression,[160, 161, 162, 167]

$$\sigma(E) = \bar{\sigma} \left\{ \frac{\gamma_1}{(E - \delta_1)^2 + \gamma_1^2} + \frac{\alpha\gamma_2}{(E - \delta_2)^2 + \gamma_2^2} \right\}^{-1}, \qquad (3.10)$$

where the wide Lorentzian peak is related to the Hume-Rothery mechanism and the narrow Lorentzian peak is related to sp-d hybridization effects. Quite remarkably, this model is able to properly fit the experimental $\sigma(T)$ and $S(T)$ curves in a broad temperature range. This model includes six parameters, determining the Lorentzian's heights ($\bar{\sigma}/\gamma_i$) and widths ($\sim \gamma_i$), their positions with respect to the Fermi level, δ_i, and their relative weight in the overall structure, $\alpha > 0$. The parameter $\bar{\sigma}$ is a scale factor measured in $(\Omega\text{cm eV})^{-1}$ units. Suitable values for these electronic model parameters can be obtained by properly combining *ab initio* calculations of approximant phases with experimental transport data of icosahedral samples within a phenomenological approach.[151, 168, 169, 170] In Fig.3.18 the overall behavior of the $\sigma(E)$ curve is shown for a suitable choice of the model parameters. By comparing this figure with Fig.3.17 we see that Eq.(3.10) properly captures the main spectral features of realistic samples.

3.4.2   Transport coefficients

From the knowledge of the spectral conductivity function the temperature-dependent transport coefficients can be obtained by means of the Kubo-Greenwood version of the linear response theory.[171, 172, 173] The central information quantities are the kinetic coefficients

$$\mathcal{L}_{ij}(T) = (-1)^{i+j} \int \sigma(E) \, (E - \mu)^{i+j-2} \left(-\frac{\partial f}{\partial E} \right) dE, \qquad (3.11)$$

where $f(E, \mu, T)$ is the Fermi-Dirac distribution function, E is the electron energy, and μ is the chemical potential. In this formulation all the microscopic details of the system are included in the $\sigma(E)$ function. From the knowledge of the kinetic coefficients one obtains the electrical conductivity

$$\sigma(T) = \mathcal{L}_{11}(T), \tag{3.12}$$

the thermoelectric power,

$$S(T) = \frac{1}{|e|T} \frac{\mathcal{L}_{12}(T)}{\sigma(T)}, \tag{3.13}$$

the electronic thermal conductivity,

$$\kappa_e(T) = \frac{1}{e^2 T} \mathcal{L}_{22}(T) - T\sigma(T) S(T)^2, \tag{3.14}$$

and the Lorenz function

$$L(T) \equiv \frac{\kappa_e(T)}{T\sigma(T)} \tag{3.15}$$

in a unified way. As a first approximation one generally assumes $\mu(T) \approx E_F$. Then, by expressing Eqs.(3.12-3.15) in terms of the scaled variable $x \equiv (E - \mu)\beta$, where $\beta \equiv (k_B T)^{-1}$, the transport coefficients can be rewritten as[62, 66, 174, 175]

$$\sigma(T) = \frac{J_0}{4}, \tag{3.16}$$

$$S(T) = -\frac{k_B}{|e|} \frac{J_1}{J_0}, \tag{3.17}$$

$$\kappa_e(T) = \frac{k_B^2 T}{4e^2} \left(J_2 - \frac{J_1^2}{J_0} \right), \tag{3.18}$$

$$L(T) = \left(\frac{k_B}{eJ_0} \right)^2 \begin{vmatrix} J_0 & J_1 \\ J_1 & J_2 \end{vmatrix}, \tag{3.19}$$

in terms of the reduced kinetic coefficients

$$J_n(T) = \int x^n \sigma(x) \operatorname{sech}^2(x/2) dx . \tag{3.20}$$

These kinetic coefficients, in turn, can be expressed in the form

$$J_0 c_0^{-1} = \frac{4\pi^2}{3} \beta^{-2} + a_3 \beta^{-1} H_1 + a_4 H_0 + 4a_0, \tag{3.21}$$

$$J_1 c_0^{-1} = \frac{4\pi^2}{3} a_1 \beta^{-1} + a_5 H_1 + a_3 \beta (4 - q_0 H_0),$$

$$J_2 c_0^{-1} = \frac{28\pi^4}{15} \beta^{-2} + a_6 \beta H_1 + a_5 (4 - q_0 H_0) \beta^2 + \frac{4\pi^2}{3} a_0,$$

where $c_0 \equiv \bar{\sigma} (\gamma_1 + \alpha\gamma_2)^{-1}$, and the coefficients a_i were defined in Ref.[62]. We have introduced the auxiliary integrals

$$H_k(\beta) \equiv \int_{-\infty}^{\infty} \frac{x^k}{\beta^{-2} x^2 - 2\beta^{-1} q_1 x + q_0} \operatorname{sech}^2(x/2) dx, \tag{3.22}$$

where $q_0 \equiv \varepsilon \varepsilon_1^2 \varepsilon_2^2 (\gamma_1 + \alpha \gamma_2)^{-1}$, $q_1 = (\gamma_1 \delta_2 + \alpha \delta_1 \gamma_2)(\gamma_1 + \alpha \gamma_2)^{-1}$, $\varepsilon_i^2 \equiv \gamma_i^2 + \delta_i^2$, and $\varepsilon \equiv \gamma_1 \varepsilon_1^{-2} + \alpha \gamma_2 \varepsilon_2^{-2}$. By inspecting Eq.(3.22) we realize that the auxiliary integral H_1 identically vanishes in the case $q_1 = 0$, due to the odd parity of the integrand. In that case, taking into account the Fourier transform relationship

$$\frac{1}{x^2 + a^2} = \frac{1}{2a} \int\limits_{-\infty}^{\infty} e^{-a\omega} e^{i\omega x} d\omega, \tag{3.23}$$

the auxiliary integral H_0 can be properly rearranged in the form

$$H_0(\beta) \equiv \frac{\beta^2}{2a} \int\limits_{-\infty}^{\infty} e^{-a\omega} d\omega \int\limits_{-\infty}^{\infty} e^{i\omega x} \operatorname{sech}^2\left(\frac{x}{2}\right) dx, \tag{3.24}$$

where $a^2 \equiv q_0 \beta^2$. Now, the second integral in Eq.(3.24) is just the Fourier transform of the function $4\pi\omega \operatorname{cosech}(\pi\omega)$, so that one finally obtains [66]

$$H_0 \equiv \frac{2\pi\beta^2}{a} \int\limits_{-\infty}^{\infty} e^{-a\omega} \omega \operatorname{cosech}(\pi\omega) d\omega = 4q_0^{-1} \tilde{\beta} \varsigma_H (2, 1/2 + \tilde{\beta}), \tag{3.25}$$

where $\tilde{\beta} \equiv \sqrt{q_0}\beta/2\pi$ is a scaled variable and $\varsigma_H(s,a) \equiv \sum_{k=0}^{\infty}(k+a)^{-s}$ is the Hurwitz Zeta function, which reduces to the Riemann Zeta function in the case $a = 1$.[176] Making use of these analytical expressions Eq.(3.21) can be rearranged in the matrix form

$$\begin{pmatrix} J_0 \\ J_1 \\ J_2 \end{pmatrix} = \frac{4\pi^2 c_0}{3} \begin{pmatrix} \frac{3}{\pi^2} \tilde{J}_{00} & 0 & 1 \\ 0 & \tilde{J}_{11} & 0 \\ \tilde{J}_{20} & 0 & \frac{7\pi^2}{5} \end{pmatrix} \begin{pmatrix} 1 \\ \beta^{-1} \\ \beta^{-2} \end{pmatrix}, \tag{3.26}$$

where $\tilde{J}_{00} \equiv a_0 + a_4 q_0^{-1} \tilde{\beta} \varsigma_H$, $\tilde{J}_{11} \equiv a_1 + 12 a_3 q_0^{-1} f(\tilde{\beta})$, $\tilde{J}_{20} \equiv a_0 + 12 a_4 q_0^{-1} f(\tilde{\beta})$, with $f(\tilde{\beta}) \equiv \tilde{\beta}^2(1 - \tilde{\beta} \varsigma_H)$. In this way, under the assumption that q_1 is negligible in Eq.(3.22), one obtains closed analytical expressions for the different transport coefficients. It turns out that this assumption is a reasonable one for several QCs of interest. In fact, as we will see in Section 3.4.3, the values $q_1 = -0.025$ eV, $q_1 = -0.015$ eV, and $q_1 = -8.8 \times 10^{-5}$ eV are respectively obtained for AlMnSi approximant phases,[170] i-AlCuFe QCs,[151] and i-AlPdRe QCs.[177] We notice that the smaller q_1 value corresponds to higher structural quality QCs whereas the largest one is obtained for an approximant crystal. Accordingly, we can confidently assume the limiting behavior $q_1 \to 0$ properly applies to *ideal* QCs.

In the more realistic case $q_1 \neq 0$ we can obtain useful information by expanding Eq.(3.22) in Taylor series around the Fermi level to get

$$H_0 \simeq \frac{4}{q_0} \left(1 + \frac{\pi^2}{3} \frac{4q_1^2 - q_0}{q_0^2} \beta^{-2} \right),$$
(3.27)

$$H_1 \simeq \frac{8\pi^2 q_1 \beta^{-1}}{3q_0^2} \left(1 + \frac{14\pi^2}{5} \frac{2q_1^2 - q_0}{q_0^2} \beta^{-2} \right).$$

In this way, one obtains approximate analytical expressions for the electrical conductivity and Seebeck coefficient curves, [178]

$$\sigma(T) = \sigma(0)[1 + bT^2 \Lambda(T)],$$
(3.28)

with

$$\Lambda(T) = \xi_2 + \xi_4 bT^2 + \xi_6 b^2 T^4,$$
(3.29)

and

$$S(T) = -2|e|\mathcal{L}_0 T \frac{\xi_1 + \xi_3 bT^2}{1 + \xi_2 bT^2 + \xi_4 b^2 T^4},$$
(3.30)

where $b \equiv e^2 \mathcal{L}_0$, $\mathcal{L}_0 = \pi^2 k_B^2 / 3e^2 = 2.44 \times 10^{-8}$ V^{2}K^{-2} is the Lorenz number. These expressions are valid in the low temperature regime, up to about $\sim 50 - 100$ K.[62] The coefficients ξ_n can be explicitly expressed in terms of the electronic model parameters and contain detailed information about the electronic structure of the sample. For instance, the first order phenomenological coefficients are defined in terms of the electronic model parameters as [178]

$$\xi_1 \equiv -\frac{\gamma_1 \delta_1 \varepsilon_2^4 + \alpha \delta_2 \varepsilon_1^4}{\varepsilon \varepsilon_1^4 \varepsilon_2^4},$$
(3.31)

$$\xi_2 \equiv \frac{\gamma_1 \varepsilon_2^6 \left(\varepsilon_1^2 - 4\delta_1^2 \right) + \alpha \varepsilon_1^6 \left(\varepsilon_2^2 - 4\delta_2^2 \right)}{\varepsilon \varepsilon_1^6 \varepsilon_2^6} + 4\xi_1^2,$$
(3.32)

and can be related to the topology of the spectral conductivity function $\sigma(E)$ by means of the following expressions,

$$\xi_1 = \frac{1}{2} \left(\frac{d \ln \sigma(E)}{dE} \right)_{E_F},$$
(3.33)

and

$$\xi_2 = 2\xi_1^2 + \frac{1}{2} \left(\frac{d^2 \ln \sigma(E)}{dE^2} \right)_{E_F}.$$
(3.34)

Thus, from the knowledge of the phenomenological coefficients ξ_1 and ξ_2 we can obtain suitable information concerning the slope and curvature of the DOS close to E_F.

According to Eq.(3.28), the electrical conductivity temperature dependence can be expressed as a *product* involving two different contributions. The first

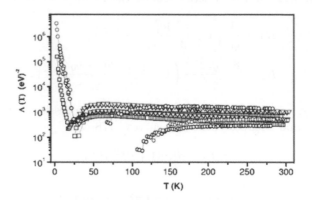

FIGURE 3.19

Diagram comparing the temperature dependences of the $\Lambda(T)$ function defined by Eq. (3.29) for different quasicrystalline samples belonging to the AlCuFe ($\square$), AlCuRu (∇), and AlPdMn ($\bigcirc$) families. Solid lines are a guide for the eye. ([178] Reprinted figure with permission from Maciá E 2002 *Phys. Rev. B* **66** 174203 © 2002 by the American Physical Society.)

one is given by the $\sigma(0)$ factor and describes the residual conductivity of the sample. This term will be the one responsible for the overall low conductivity values observed in these materials. The second contribution is given by the function $1 + bT^2\Lambda(T)$ and describes the temperature dependence of the electrical conductivity as the temperature is increased. It is worth noting that by identifying $\Delta\sigma(T) \equiv \sigma(0)\Lambda(T)bT^2$, Eq.(3.28) essentially reduces to the empirically proposed inverse Matthiessen rule given by Eq.(3.1). Therefore, the second term in Eq.(3.1) can be regarded as a product involving a *universal* parabolic function, bT^2, modulated by the sample dependent factor, $\sigma(0)\Lambda(T)$. The $\Lambda(T)$ contribution can be straightforwardly determined from experimental data. For the sake of illustration, in Fig.3.19 the temperature dependence of the $\Lambda(T)$ term corresponding to the samples shown in Fig.3.7 is plotted in a semilog plot. Quite remarkably, the temperature dependence of the $\Lambda(T)$ function exhibits a nearly universal behavior at high enough temperatures, as expected (see Section 3.1.2.2).

As we have mentioned in Section 3.1.2.5, in the low temperature regime the thermoelectric power of QCs belonging to the i-AlCu(Fe,Ru,Os) and i-AlPd(Mn,Re) families exhibits a linear dependence with T. At temperatures above $\sim 50-100$ K, however, the $S(T)$ curve clearly deviates from the linear behavior, exhibiting pronounced curvatures. This behavior can be readily described by means of Eq.(3.30). In fact, in the low temperature limit Eq.(3.30) reduces to the linear form

$$S(T \to 0) = -2|e|\mathcal{L}_0\xi_1 T \equiv m_0 T. \tag{3.35}$$

The sign of the slope m_0 is determined by the sign of the parameter ξ_1 which, in turn, depends on the electronic structure of the sample according to Eq.(3.33). Therefore, Eq.(3.35) reduces to the well-known Mott's formula $S = -|e|\mathcal{L}_0(d\ln\sigma(E)/dE)_{E=\mu}$ in the low temperature limit. It then follows that Mott's formula will properly describe the thermoelectric power of QCs as far as the remaining coefficients ξ_2, ξ_3, and ξ_4 in Eq.(3.30) are negligible as compared to ξ_1. Since these coefficients are multiplied by the temperature dependent factors bT^2 and b^2T^4, respectively, it is clear that the range of validity of Mott's formula will be strongly dependent on the electronic structure of the sample.

3.4.3 Application examples

In this Section we will illustrate the phenomenological framework introduced in the previous one by relating the main topological features of the experimental $\sigma(T)$ and $S(T)$ curves to certain characteristic features of the electronic structure of the samples. The key point of this approach relies on the analytical coefficients ξ_n, which can be regarded as phenomenological parameters containing information about the electronic structure of the sample. Since the values of the ξ_n coefficients can be *also* determined from the analysis of the experimental transport curves, one can obtain useful information about the spectral conductivity function $\sigma(E)$ from the topological features present in these curves. The first step consists in determining the values of the ξ_n coefficients from suitable fits to the experimentally obtained transport curves. The next step will be then to determine the electronic model parameters γ_i, δ_i, and α from the obtained ξ_n values making use of previously derived analytical formulae. Due to the involved nature of the analytical expressions relating the phenomenological coefficients to the model parameters, this is a rather cumbersome task. Fortunately, even the partial knowledge of some phenomenological coefficients suffices to gain some physical insight onto certain relevant features of the electronic spectrum of the sample, as we will see in the following examples.

3.4.3.1 Icosahedral quasicrystals

Let us consider the quasicrystalline sample i $Al_{63}Cu_{25}Fe_{12}$ whose differential conductance curve measured at 2 K is shown in Fig.3.17. Since some spectral features are apparent in this tunneling spectroscopy measurement one may be tempted to extract the model parameters defining the spectral conductivity function $\sigma(E)$ directly from them. From a fitting analysis of the data shown in Fig.3.17 one gets $\gamma_1 = 587 \pm 1$ meV, $\delta_1 = -5.2 \pm 0.5$ meV, $\delta_2 = -16.1 \pm 0.5$

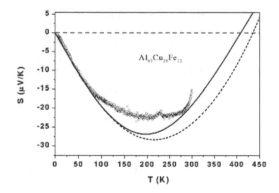

FIGURE 3.20

Comparison between the temperature dependence of the Seebeck coefficient of $Al_{63}Cu_{25}Fe_{12}$ quasicrystal in the temperature range 1-300 K as determined from experiment (open circles) and the analytical expression given by Eq.(3.30) The solid line is obtained by explicitly considering the chemical potential temperature dependence.([151] Reprinted figure with permission from Maciá E 2004 *Phys. Rev. B* **69** 132201 © 2004 by the American Physical Society.)

meV, and $\gamma_2 = 7 \pm 1$ meV. However, this γ_2 value, describing the electronic structure at very low temperatures, cannot be straightforwardly used in order to describe the transport coefficients at significantly higher temperatures. In fact, *ab initio* calculations showed that narrowest spectral features halfwidth values progressively broaden as the temperature increases.[162] Indeed, a thermal broadening of the dip around the Fermi level has been experimentally reported for different quasicrystalline samples.[150]

To circumvent this shortcoming it is convenient to derive the width of the narrowest spectral feature by taking into account physical information contained in the thermopower curve, which is extremely sensitive to the finer details in the electronic structure. From a linear fit to the thermopower data shown in Fig.3.20 (in the temperature range 6 − 70 K) one obtains $m_0 = -0.219 \pm 0.002 \ \mu VK^{-2}$, and making use of Eq.(3.35) we have $\xi_1^{exp} = +4.49 \pm 0.03$ $(eV)^{-1}$. Plugging the obtained $\gamma_1, \delta_1, \delta_2$, and ξ_1^{exp} values into Eq.(3.31) we get $\gamma_2 = 55 \pm 1$ meV. This value is about eight times larger than the value originally obtained from tunneling measurements, hence confirming the importance of thermal broadening effects. The relative weight of both Lorentzian peaks is then determined from the expression (derived from

Eq.(3.31))

$$\alpha = -\frac{\gamma_1}{\gamma_2}\left(\frac{\varepsilon_2}{\varepsilon_1}\right)^4 \frac{\delta_1 + \xi_1^{\text{exp}}\varepsilon_1^2}{\delta_2 + \xi_1^{\text{exp}}\varepsilon_2^2}, \qquad (3.36)$$

which yields $\alpha = 1.07 \pm 0.03$. Finally, the conductivity scale parameter $\bar{\sigma}$ can be determined from the knowledge of the experimental residual conductivity value $\sigma(0)$ as[178]

$$\bar{\sigma} = \frac{\sigma(0)}{\pi}(\gamma_1\varepsilon_1^{-2} + \alpha\gamma_2\varepsilon_2^{-2}), \quad (\Omega^{-1}\text{cm}^{-1}\text{eV}^{-1}). \qquad (3.37)$$

Taking the low temperature electrical conductivity value $\sigma(0) = 188$ $(\Omega\text{cm})^{-1}$ measured for an i-$Al_{63}Cu_{25}Fe_{12}$ sample at 4.2 K,[179] one gets $\bar{\sigma} = 1180 \pm 90$ $(\Omega\text{cmeV})^{-1}$. In Fig.3.20 we compare the experimental thermopower curve (open circles) with the analytical expressions given by Eq.(3.30) (dashed line). At low temperatures $S(T)$ follows a linear behavior up to about $T_1 \simeq 70$ K. At higher temperatures the thermopower progressively deviates from linearity, showing a broad minimum. Finally, as the temperature is further increased the $S(T)$ curve steadily increases, suggesting the probable existence of a crossing point where the thermoelectric power will change its sign, though experimental data do not allow for an accurate estimate of this crossing temperature.

By comparing Figs.3.17 and 3.20, we can gain some physical insight on the relationship between the electronic structure and transport properties. First, we note that the deviation from the linear behavior starts when the thermal window reaches a half-width of about $\Delta E \simeq k_B T_1 = 6.0$ meV. This value is close to the spectral peak position $\delta_1 = -5.2$ meV, hence suggesting that as far as the thermal window remains located *inside* the pseudogap's dip feature (box 1 in Fig.3.17), the thermopower exhibits a metallic-like behavior. Then, as temperature increases and charge carriers located at the little bump between both dip features start to play a more significant role in the transport properties, the $S(T)$ curve progressively deviates from linear behavior, attaining a broad minimum at $T_2 = 216$ K (determined from a 4th degree polynomial fitting). The thermal window half-width for this temperature is $\Delta E \simeq k_B T_2 = 16.9$ meV. This value is very close to the spectral resistivity peak position $\delta_2 = -16.1$ meV, hence suggesting that the minimum of the thermopower occurs when the charge carriers located *within* the second dip spectral feature (box 2 in Fig.3.17) are playing a major role in the transport properties. Afterwards, as the temperature is further increased and the states belonging to the broad pseudogap component begin to contribute significantly to the transport, the thermoelectric curve progressively rises towards positive values. Accordingly, one observes a progressive transition from metallic-like to semiconductor-like thermopower signatures as the Fermi level shifts through both spectral features due to a progressive temperature increase.[39, 145, 180]

This example illustrates the potential of this approach in order to gain information about the electronic structure of quasicrystalline samples from the study of the experimental $S(T)$ curves over a broad temperature range.

One reasonably expects that a sharper view about the main electronic features of the considered QC samples would ultimately emerge from a combined study of the different transport coefficients, $\sigma(T)$, $S(T)$ and $R_H(T)$, over different temperature ranges. The physical information gained in this way may help to clarify the possible existence of finer details in the electronic structure of quasicrystalline samples, like the much debated spiky features, which remain a fundamental open question in the science of QCs.

3.4.3.2   Quasicrystalline approximants

This phenomenological approach can straightforwardly be extended to other systems whose electronic structure around the Fermi level is characterized by two main peaks separated by a well defined pseudogap centered at the Fermi level. This includes the broad class of icosahedral quasicrystalline approximants. As a suitable sample we consider the $Al_{82.6-x}Mn_{17.4}Si_x$ ($x = 9$) α-phase,[181] which is a well documented representative of the 1/1-cubic approximants class. This approximant exhibits a sign reversal in the thermoelectric power with increasing temperature (a feature which cannot be accounted for in terms of the Mott formula usually employed to study metallic alloys), in close analogy with the behavior observed in some high quality QCs. The main goal of this example study is to gain some insight into those physical properties intrinsically related to local order effects as compared to those related to the characteristic quasiperiodic order of QCs. To this end, we first determine the phenomenological coefficients values from a combined fitting analysis of different experimental transport curves. Then we derive the approximant crystal spectral conductivity function and compare it with that corresponding to the i-QC sample studied in Section 3.4.3.1.

In Fig.3.21 we show the temperature dependence of the electrical conductivity for the $Al_{73.6}Mn_{17.4}Si_9$ cubic approximant. The curve exhibits a typical metallic behavior up to ~ 100 K, where the conductivity attains a minimum and then it progressively increases as the temperature is further increased. The $\sigma(T)$ curves of several QCs also exhibit a similar behavior in the low temperature regime (Fig.3.7).

In Fig.3.22 we show the temperature dependence of the thermoelectric power for the same approximant phase. The thermopower shows a remarkable nonlinear behavior, exhibiting a broad minimum at about $T_1 = 160$ K, and changes its sign twice at about $T_0 = 50$ K and 260 K, respectively. This anomalous behavior resembles that observed for several icosahedral QCs.[37, 38, 39, 40, 41]

From the knowledge of the complete set of phenomenological parameters one can derive the corresponding electronic model parameters following the algebraic procedure described in Ref.[170]. In Fig.3.23 we compare the spectral conductivity functions corresponding to the $Al_{73.6}Mn_{17.4}Si_9$ cubic approximant and the AlCuFe QC studied in the previous Section. By inspecting this figure we see that the spectral conductivity of the quasicrystalline phase is

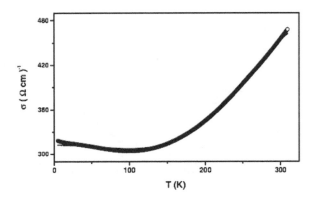

FIGURE 3.21

Electrical conductivity as a function of temperature for the $Al_{73.6}Mn_{17.4}Si_9$ cubic approximant (open circles). The solid line corresponds to the best fit curve $\sigma(T) = \sigma_0(1 + BT^2 + CT^4 + DT^6)$ with $\sigma_0 = 312.6 \pm 0.2$ $(\Omega cm)^{-1}$, $B = (-3.50 \pm 0.08) \times 10^{-6}$ K^{-2}, $C = (1.91 \pm 0.02) \times 10^{-10}$ K^{-4}, $D = (-1.07 \pm 0.02) \times 10^{-15}$ K^{-6}, with a correlation coefficient $r = 0.9824$.([170] Reprinted figure with permission from Maciá E, Takeuchi T, and Otagiri T 2005 *Phys. Rev. B* **72** 174208 © 2005 by the American Physical Society.)

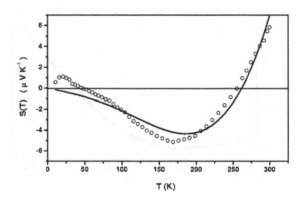

FIGURE 3.22

Thermoelectric power as a function of temperature for $Al_{73.6}Mn_{17.4}Si_9$ cubic approximant (open circles). The solid line corresponds to the best fit curve given by $S(T) = -0.0488T(a + fT^2 + gT^4)/(1 + BT^2 + CT^4 + DT^6)$ with $a = 0.29 \pm 0.05$ (eV)$^{-1}$, $f = (6 \pm 2) \times 10^{-5}$ K^{-2}, and $g = (-1.1 \pm 0.3) \times 10^{-9}$ K^{-4}, with Pearson $\chi^2 = 0.562$. ([170] Reprinted figure with permission from Maciá E, Takeuchi T, and Otagiri T 2005 *Phys. Rev. B* **72** 174208 © 2005 by the American Physical Society.)

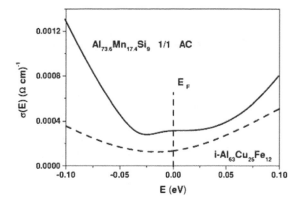

FIGURE 3.23
Spectral conductivity function derived from the electronic model parameters obtained for the $Al_{73.6}Mn_{17.4}Si_9$ cubic approximant (solid line) and an $Al_{63}Cu_{25}Fe_{12}$ icosahedral QC (dashed line). ([170] Reprinted figure with permission from Maciá E, Takeuchi T, and Otagiri T 2005 *Phys. Rev. B* **72** 174208 © 2005 by the American Physical Society.)

both deeper and broader than that corresponding to the approximant phase, thus indicating a less effective Hume-Rothery mechanism for the approximant crystal. On the other hand, the presence of a well defined spectral feature at about -0.03 eV may be indicative of hybridization effects likely related to bond formation in the approximant sample. Accordingly, these results support the view that short-range chemical effects are playing a significant role in the stabilization of approximant phases.[126]

3.4.3.3 Complex metallic alloys

As a final example let us consider alloys exhibiting complex unit cells, composed of many (10^2-10^3) atoms, but which are not quasicrystalline approximants (see Section 2.5.3). In some cases these alloys also exhibit unusual physical properties, presumably related to their structural complexity. For instance, the electrical resistivity of the ξ' phase of the Al-Pd-Mn alloys system shows an almost negligible temperature dependence between 4 and 300 K (Fig.3.24) [182] While weakly temperature dependent resistivities are not uncommon for both amorphous alloys and bulk metallic glasses lacking long-range ordered crystalline lattices,[104] the temperature-independent resistivity of ξ'-Al-Pd-Mn was observed on monocrystalline samples of good lattice perfection and structural homogeneity.

The corresponding thermopower curves are displayed in Fig.3.25. Their

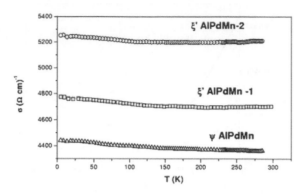

FIGURE 3.24
Electrical conductivity of AlPdMn complex alloys as a function of temperature. Solid curves are best fits obtained by a simultaneous analysis of the conductivity and thermopower data.

values are small and show a rather smooth behavior with several changes of the slope within the investigated temperature range. Following the procedure described in previous sections one obtains the spectral conductivity functions shown in Fig.3.26. In this figure the $\sigma(E)$ curves for two ξ'-AlPdMn samples are compared to those corresponding to the $Al_{63}Cu_{25}Fe_{12}$ icosahedral quasicrystal and the $Al_{73.6}Mn_{17.4}Si_9$ 1/1 cubic approximant previously studied (Fig.3.23). We observe that the spectral conductivity curves of these phases are deeper at Fermi energy and steeper in the wings, indicating the existence of a pseudogap in both the icosahedral and approximant compounds. The absence of a pseudogap in the case of ξ'-AlPdMn samples indicates that the Hume-Rothery mechanism is therefore less effective and the electrical conductivity is consequently higher. The $\sigma(E)$ curves of the ξ'-AlPdMn samples are relatively flat as compared to those corresponding to $Al_{63}Cu_{25}Fe_{12}$ and $Al_{73.6}Mn_{17.4}Si_9$ compounds. Thus, the origin of the almost temperature-independent electrical conductivity of the ξ'-AlPdMn complex alloys can be then traced back to the specific form of the spectral conductivity, which exhibits very weak variation over the energy scale of several meV around the Fermi level. In contrast to the i-AlPdMn phases, ξ'-AlPdMn complex metallic alloys do not exhibit a pseudogap at the Fermi level in the spectral conductivity. Yet, they show some fine structure that yields observable effects in the temperature-dependent thermoelectric power curves. These electronic structure related effects highlight the difference between ξ'-AlPdMn phase and conventional free-electron alloys.

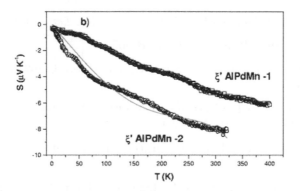

FIGURE 3.25

Thermoelectric power of AlPdMn complex alloys as a function of temperature. Solid curves are best fits obtained by a simultaneous analysis of the conductivity and thermopower data. (From ref.[169]. With permission from IOP Publishing Ltd.)

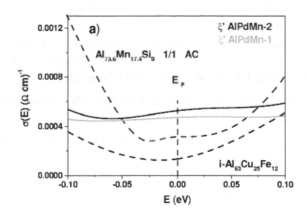

FIGURE 3.26

Comparison among the spectral conductivity functions corresponding to quasicrystals, approximant phases, and complex metallic alloys. (From ref.[169]. With permission from IOP Publishing Ltd.)

References

[1] Dubois J M 2005 *Useful Quasicrystals* (World Scientific, Singapore)

[2] Levine D and Steinhardt P J 1984 *Phys. Rev. Lett.* **53** 2477

[3] Pauling L 1987 *Phys. Rev. Lett.* **58** 365

[4] Roche S, Trambly de Laissardière G, and Mayou D 1997 *J. Math. Phys.* **38** 1794

[5] Wälti Ch, Felder E, Chernikov M A, Ott H R, de Boissieu M, and Janot C 1998 *Phys. Rev. B* **57** 10504

[6] Biggs B D, Poon S J, and Munirathnam N R 1990 *Phys. Rev. Lett.* **65** 2700

[7] Mayou D, Berger C, Cyrot-Lackmann F, Klein T, and Lanco P 1993 *Phys. Rev. Lett.* **70** 3915

[8] Rapp Ö 1999 *Physical Properties of Quasicrystals*, Ed. Stadnik Z M, Springer Series in Solid-State Physics **126** (Springer-Verlag, Berlin) p 127

[9] Mayou D 2000 in *Quasicrystals Current Topics*, Eds. Belin-Ferré E, Berger C, Quiquandon M, and Sadoc A (World Scientific, London) p 445.

[10] Grimm U and Schreiber M 2002 *Quasicrystals: An Introduction to Structure, Physical Properties, and Applications*, Eds. Suck J B, Schreiber M, and Häussler P (Springer, Berlin) p 49

[11] Maciá E, Dubois J M, and Thiel P A 2002 *Quasicrystals* entry in *Ullmann's Encyclopedia of Industrial Chemistry*, Sixth Edition, 2002 January Release on CD-ROM; Wiley-VCH, Winheim

[12] Rosenbaum R, Lin S T, and Su T I 2003 *J. Phys. Condens. Matter* **15** 4169

[13] Delahaye J, Berger C, and Fourcaudot G 2003 *J. Phys. Condens. Matter* **15** 8753

[14] Lay Y Y, Jan J C, Chion J W, Tsai H M, Pong W F, Tsai M H, Pi T W, Lee J F, Ma C I, Tseng K L, Wang C R, and Lin S T 2003 *Appl. Phys. Lett.* **82** 2035

[15] Pierce F S, Guo Q, and Poon S J 1994 *Phys. Rev. Lett.* **73** 2220

[16] Tamura R, Waseda A, Kimura K, and Ino H 1994 *Phys. Rev. B* **50** 9640

[17] Kirihara K and Kimura K 2000 *Science and Technology Ad. Materials* **1** 227

[18] Carlsson A 1991 *Nature* **353** 353

[19] Shu-yuan L, Xue-mei W, Dian-lin Z, He L X, and Kuo K X 1990 *Phys. Rev. B* **41** 9625; Dian-lin Z, Shao-chun C, Yun-ping W, and Xue-mei W 1991 *Phys. Rev. Lett.* **66** 2778; Xue-mei W, Dian-lin Z, and Chen L F 1993 *Phys. Rev. B* **48** 10542; Shuyuan L, Guohong L, and Dianlin Z 1996 *Phys. Rev. Lett.* **77** 1998

[20] Edagawa K, Chernikov M A, Bianchi A D, Felder E, Gubler U, and Ott H R 1996 *Phys. Rev. Lett.* **77** 1071

[21] Quivy A, Quiquandon M, Calvayrac Y, Faudot F, Gratias D, Berger C, Brand R A, Simonet V, and Hippert F 1996 *J. Phys. Condens. Matter* **8** 4223

[22] Häussler P, Nowak H, and Haberken R 2000 *Mater. Sci. Eng.* **294-296** 283; Roth C, Schwalbe G, Knöfler R, Zavaliche F, Madel O, Haberkern R, and Häussler P 1999 *J. Non-Cryst. Solids* **252** 869

[23] Lalla N P, Tiwari R S, and Srivastava O N 1995 *J. Phys. Condens. Matter* **7** 2409

[24] Tritt T M, Pope A L, Chernikov M, Feuerbacher M, Legault S, Gagnon R, and Strom-Olsen J 1999 *Mat. Res. Soc. Symp. Proc.* **553** 489; Pope A L, Tritt T M, Chernikov M, Feuerbacher M, Legault S, Gagnon R, and Strom-Olsen J 1999 *ibid.* **545** 413

[25] Pierce F S, Poon S J, and Guo Q 1993 *Science* **261** 737

[26] Delahaye J and Berger C 2001 *Phys. Rev. B* **64** 094203

[27] Klein T and Symko O G 1994 *Phys. Rev. Lett.* **73** 2248

[28] Trambly de Laissardière G, Nguyen-Manh D, and Mayou D 2005 *Prog. Materials Sci.* **50** 679

[29] Homes C C, Timusk T, Wu X, Altounian Z, Sahnoune A and Ström-Olsen J O 1991 *Phys. Rev. Lett.* **67** 2694

[30] Degiorgi L, Chernikov M A, Beeli C and Ott H R 1993 *Solid State Commun.* **87** 721

[31] Wu X, Homes C C, Burkov S E, Timusk T, Pierce F S, Poon S J, Cooper S L and Karlow M A 1993 *J. Phys. Condens. Matter* **5** 5975

[32] Bianchi A D, Bommeli F, Chernikov, Gubler U, Degiorgi L and Ott H R 1997 *Phys. Rev. B* **55** 5730

[33] Demange V, Milandri A, de Weerd M C, Machizaud F, Joandel G and Dubois J M 2002 *Phys. Rev. B* **65** 144205

[34] Burkov S E, Timusk T and Ashcroft N W 1992 *J. Phys. Condens. Matter* **4** 9447

[35] Mayou D 2000 *Phys. Rev. Lett.* **85** 1290

[36] Basov D N, Timusk T, Barakat F, Greedan J, and Grushko B 1994 *Phys. Rev. Lett.* **72** 1937

[37] Pierce F S, Poon S J, and Biggs B D 1993 *Phys. Rev. Lett.* **70** 3919

[38] Biggs B D, Li Y, and Poon S J 1991 *Phys. Rev. B* **43** 8747

[39] Pierce F S, Bancel P A, Biggs B D, Guo Q, and Poon S J 1993 *Phys. Rev. B* **47** 5670

[40] Haberken R, Fritsch G, and Härting M 1993 *Appl. Phys. A* **57** 431

[41] Bilušic A, Pavuna D, and Smontara A 2001 *Vacuum* **61** 345; Bilušic A, Smontara A, Lasjaunias J C, Ivkov J, and Calvayrac Y 2000 *Mater. Sci. Eng. A* **294-296** 711

[42] Giannò K, Sologubenko A V, Chernikov M A, Ott H R, Fisher I R, and Canfield P C 2000 *Mater. Sci. Eng. A* **294-296** 715

[43] Pope A L, Tritt T M, Gagnon R, and Strom-Olsen J 2001 *Appl. Phys. Lett.* **79** 2345

[44] Ashcroft N W and Mermin N D 1976 *Solid State Physics* (Saunders College Publ., Cornell) p 255

[45] Klein T, Gozlan A, Berger C, Cyrot-Lackmann F, Calvayrac Y, and QuivyA 1990 *Europhys. Lett.* **13** 129

[46] Pierce F S, Bancel P A, Biggs B D, Guo Q, and Poon S J 1993 *Phys. Rev. B* **47** 5670

[47] Haberken R, Lindqvist P, and Fritsch G 1993 *J. Non-Cryst. Solids* **153&154** 303

[48] Lindqvist P, Berger C, Klein T, Lanco P, Cyrot-Lackmann F, and Calvayrac Y 1993 *Phys. Rev. B* **48** 630

[49] Tamura R, Waseda A, Kimura K, and Ino H 1994 *Phys. Rev. B* **50** 9640

[50] Giraud F, Grenet T, Berger C, Lindqvist P, Gignoux C, and Fourcaudot G 1996 *Czech. J. Phys.* **46** 2709

[51] Bilušic A, Bešlić I, Lasjaunias J C, and Smontara A 1999 *Fizika A* **8** 183

[52] Lin C R, Lin S T, Wang C R, Chou S L, Horng H E, Cheng J M, Yao Y D, and Lai S C 1997 *J. Phys. Condens. Matter* **9** 1509

[53] Chernikov M A, Paschen S, Felder E, Vorburger P, Ruzicka B, Degiorgi L, Ott H R, Fischer I R, and Canfield P C 2000 *Phys. Rev. B* **62** 262

[54] Haberken R, Khedhri K, Madel C, and Häussler P 2000 *Mater. Sci. Eng. A* **294-296** 475

[55] Dubois J M, Kang S S, Archembault P, and Colleret B 1993 *J. Mat. Res.* **8** 38

[56] Chernikov M A, Bianchi A, and Ott H R 1995 *Phys. Rev. B* **51** 153; Kalugin P A, Chernikov M A, Bianchi A, and Ott H R 1996 *ibid.* **53** 14 145

[57] Dubois J M 2001 *J. Phys. Condens. Matter* **13** 7753

[58] Naumis G G, Salazar F, and Wang C 2005 *Phil. Mag.* **332** 141

[59] Maciá E 2003 *J. Appl. Phys.* **93** 1014

[60] Mahan G D and Bartkowiak M 1999 *Appl. Phys. Lett.* **74** 953

[61] Vekilov Yu K, Isaev E I, and Johansson B 2006 *Phys. Lett. A* **352** 524

[62] Landauro C V, Maciá E, and Solbrig H 2003 *Phys. Rev. B* **67** 184206

[63] Perrot A and Dubois J M 1993 *Ann. Chim. Fr.* **18** 501

[64] Bihar Z, Bilusic A, Lukatela J, Smontara A, Leglic P, McGuiness P J, Dolinsek J, Jaglicic A, Jamovec J, Demange V, and Dubois J M 2006 *J. Alloys Compd.* **407** 65

[65] Giannò K, Sologubenko A V, Chernikov M A, Ott H R, Fisher I R, and Canfield P C 2000 *Phys. Rev. B* **62** 292

[66] Maciá E and Rodríguez-Oliveros R 2007 *Phys. Rev. B* **75** 104210

[67] Hippert F, Simonet V, and Trambly de Laissardiere G 2000 in *Quasicrystals: Current Topics*, Eds. Belin-Ferré E, Berger C, Quiquandon M, and Sadoc A (World Scientific, Singapore) p 475

[68] Lanco P, Klein T, Berger C, Cyrot-Lackmann F, Fourcaudot G, and Sulpice A 1992 *Europhys. Lett.* **18** 227

[69] Matsuo S, Nakano H, Ishimasa T, and Fukano Y 1989 *J. Phys.: Condens. Matter* **1** 6893

[70] Hippert F, Audier M, Préjean J J, Sulpice A, Lhotel E, Simonet V, Calvayrac Y 2003 *Phys. Rev. B* **68** 134402

[71] Trambly de Laissardiere G and Mayou D 2000 *Phys. Rev. Lett.* **85** 3273

[72] Sato T J, Takakura H, and Tsai A P 2000 *Phys. Rev. B* **61** 476

[73] Lifshitz R 1997 *Rev. Mod. Phys.* **69** 1181

[74] Charrier B, Ouladdiaf B, and Schmitt D 1997 *Phys. Rev. Lett.* **78** 4637

[75] Reisser R and Kronmüller H 1994 *J. Mag. Mats.* **131** 00

[76] Islam Z, Fisher I R, Zarestky J, Canfield P C, Stassis C, and Goldman A I 1998 *Phys. Rev. B* **57** R11047

[77] Hattori Y, Fukamichi K, Suzuki K, Aruga-Katori H, and Goto T 1995 *J. Phys. Condens. Matter* **7** 4193

[78] Mancinelli C, Jenks C J, Thiel P A, and Gellman A J 2003 *J. Mater. Res.* **18** 1447

[79] He G, Muser M H, and Robbins M O 1999 *Science* **284** 1650

[80] Rajasekaran E, Zeng X C, and Diestler D J 1997 *Micro/Nanotribology and Its Applications* Ed. Bhushan B (Kluwer Academic Publishers, Dordrecht, The Netherlands)

[81] Braun O M and Kivshar Y S 2004 *The Frenkel-Kontorova Model: Concepts. Methods and Applications*, (Springer-Verlag, Berlin); Braun O M and Kivshar Y S 1998 *Phys. Rep.* **306** 1

[82] Shaub Th, Bürgler D E, Güntherodt H J, Suk J B, and Audier M 1995 *Appl. Phys. A* **61** 491

[83] Unal B, Lograsso T A, Ross A, Jenks C J, and Thiel P A 2005 *Phys. Rev. B* **71** 165411; Ledieu J, Cox E J, MacGrath R, Richardson N V, Chen Q, Fournée V, Lograsso T A, Ross A R, Caspersen K J, Unal B, Evans J W, and Thiel P A 2005 *Surf. Sci.* **583** 4

[84] Ledieu J, Hoeft J T, Reid D E, Smerdon J A, Diehl R D, Lograsso T A, Ross A R, and McGrath R 2004 *Phys. Rev. Lett.* **92** 135507

[85] Cai T C, Ledieu J, MacGrath R, Fournée V, Lograsso T A, Ross A R, and Thiel P A 2003 *Surf. Sci.* **526** 115

[86] Braun O M, Vanossi A, and Tosatti E 2005 *Phys. Rev. Lett.* **95** 026102

[87] Park J Y, Ogletree D F, Salmeron M, Ribeiro R A, Canfield P C, Jenks C J, and Thiel P A 2005 *Science* **309** 1354; *Phys. Rev. B* **72** 220201(R)

[88] Jenks C J and Thiel P A 1998 *Langmuir* **14** 1392

[89] Thiel P A 2004 *Prog. Surf. Sci.* **75** 69; 191

[90] Belin-Ferré E and Dubois J M 2006 *Int. J. Mat. Res.* **97** 1

[91] Jenks C J and Thiel P A 1998 *J. Mol. Catal. A Chem.* **131** 301; Jenks C J, Lograsso T A, and Thiel P A 1998 *J. Am. Chem. Soc.* **120** 12668

[92] Yoshimura M and Tsai A P 2002 *J. Alloys Compd.* **342** 451; Tanabe T, Kameoka S, and Tsai A P 2006 *Catal. Today* **111** 153

[93] Sadoc A, Majzoub E H, Huette W T, and Kelton K F 2003 *J. Alloys Compd.* **356-7** 96

[94] Phung Ngoc B, Geantet C, Aouine M, Bergeret G, Raffy S, and Marlin S 2008 *Int. J. Hidrogen Energy* **33** 1000

[95] Hargittan I 1997 *The Chemical Intelligencer* October p 25-48

[96] Pauling L 1954 *The Nature of the Chemical Bond* (Cornell University Press, Ithaca, 1960)

[97] Tamura R, Asao T, Tamura M, and Takeuchi S 2001 *Mat. Res. Symp. Proc.* **643** K13.3.1; Tamura R, Asao T, and Takeuchi S 2001 *Mater. Trans.* **42** 928

[98] Sato H, Takeuchi T, and Mizutani U 2001 *Phys. Rev. B* **64** 094207; Takeuchi T, Onogi T, Banno E, and Mizutani U 2001 *Mater. Trans.* **42** 933

[99] Kirihara K, Nagata T, Kimura K, Kato K, Takata M, Nishibori E, and Sakata M 2003 *Phys. Rev. B* **68** 014205

[100] Haberken R, Khedhri K, Madel C, and Häussler P 2000 *Mater. Sci. Eng.* **294-296** 475

[101] Stadnik Z M 1998 in *Physical Properties of Quasicrystals*, Ed. Stadnik Z M, Springer Series in Solid-State Physics **126** (Springer-Verlag, Berlin)

[102] Smontara A, Smiljanic I, Bilusic A, Jaglicic Z, Klanjsek M, Roitsch S, Dolinsek J, and Feuerbacher M 2007 *J. Alloys Compd.* **430** 29

[103] Tsai A P 1999 *Physical Properties of Quasicrystals*, Ed.Stadnik Z M, Springer Series in Solid-State Physics **126** (Springer-Verlag, Berlin) p 5

[104] Mizutani U 2001 *Introduction to the Electron Theory of Metals* (Cambridge University Press, Cambridge, 2001)

[105] Grenet T 2000 in *Quasicrystals Current Topics*, Eds. Belin-Ferré E, Berger C, Quiquandon M, and Sadoc A (World Scientific, Singapore)

[106] Boudard M and de Boissieu M 1999 in *Physical Properties of Quasicrystals*, Ed. Z. M. Stadnik, Springer Series in Solid-State Sciences **126** (Springer, Berlin)

[107] Takakura H, Pay-Gómez C, Yamamoto A, de Boissieu M, and Tsai A P 2007 *Nature Materials* **6** 58

[108] Janot C and de Boissieu M 1994 *Phys. Rev. Lett.* **72** 1674; C. Janot 1996 *Phys. Rev. B* **53** 181

[109] Janot C 1997 *J. Phys. Condens. Matter* **9** 1493

[110] Janot C and Dubois J M 2002 in *Quasicrystals: An Introduction to Structure, Physical Properties and Applications*, Eds. Suck J B, Shreiber M, and Haüssler P, Springer Series in Materials Science **55** (Springer, Berlin) p 183

[111] Quiquandon M and Gratias D 2007 *Phys. Rev. B* **74** 214205

[112] Zurkirsh M et al. 1996 *Phil. Mag. Lett.* **73** 107

[113] Naumovic D et al. 1997 in *New Horizons in Quasicrystals*, Eds. Goldman A I, Sordelet D J, Thiel P A, and Dubois J M (World Scientific, Singapore) p 86

[114] Ebert P et al. 1996 *Phys. Rev. Lett.* **77** 3827

[115] Abe E, Yan Y, and Pennycook S J 2004 *Nature Materials* **3** 759

[116] Kirihara K, Nagata T, Kimura K, Kato K, Takata M, Nishibori E, and Sakata M 2003 *Phys. Rev. B* **68** 014205

[117] Thiel P A 2002 *J. Alloys Compd.* **342** 477

[118] Keys A S and Glotzer S C 2007 *Phys. Rev. Lett.* **99** 235503

[119] Bernal J D 1960 *The Scale of Structural Units in Biopoesis* in *Aspects on the Origin of Life*, Ed. Florkin M (Pergamon Press, Oxford)

[120] Ebert P 2004 *Prog. Surf. Sci.* **75** 109

[121] Hume-Rothery W 1926 *J. Inst. Metals* **35** 295

[122] Brillouin L 1930 *Comp. Rend.* **191** 198

[123] Jones H 1934 *Proc. Roy. Soc. London* **A144** 225; **A147** 396

[124] Friedel J 1988 *Helv. Phys. Acta* 61 538; Friedel J and Dénoyer F. 1987 *C. R. Acad. Sci. Paris* **305** 171

[125] Massalski T B and Mizutani U 1978 *Prog. Mater. Sci.* **22** 151

[126] Asahi R, Sato H, Takeuchi T, and Mizutani U 2005 *Phys. Rev. B* **71** 165103

[127] Raynor G V 1949 *Prog. Met.* **1** 1

[128] Andersen O K 1975 *Phys. Rev. B* **12** 3060

[129] Trambly de Laissardière G, Mayou D, and Nguyen Manh D 1993 *Europhys. Lett.* **21** 25

[130] Trambly de Laissardière G, Nguyen Manh D, Magaud L, Julien J P, Cyrot-Lackmann F, and Mayou D 1995 *Phys. Rev. B* **52** 7920

[131] Mizutani U 2001 *Mater. Trans.* **42** 901; Mizutani U, Takeuchi T, Banno E, Fourneé V, Takata M, and Sato H 2001 *Mat. Res. Soc. Symp. Proc.* **643** K13.1.1

[132] Krajčí M and Hafner J 2002 *J. Phys. Condens. Matter* **14** 1865

[133] Fournee V, Belin-Ferré E, Pecheur P, Tobola J, Dankhazi Z, Sadoc A, and Müller H 2002 *J. Phys. Condens. Matter* **14** 87

[134] Lin Q and Corbett J D 2007 *J. Am. Chem. Soc.* **129** 6789; 2005 *ibid.* **127** 12786

[135] Ishii Y and Fujiwara T 2001 *Phys. Rev. Lett.* **87** 206408

[136] Tamura R, Takeuchi T, Aoki C, Takeuchi S, Kiss T, Yokoya T, and Shin S 2004 *Phys. Rev. Lett.* **92** 146402

[137] Shukla A K, Biswas C, Dhaka R S, Das S C, Krüger P, and Barman S R 2008 *Phys. Rev. B* **77** 195103

[138] Fujiwara T and Yokokawa T 1991 *Phys. Rev. Lett.* **66** 333

[139] Klein T, Berger C, Mayou D, and Cyrot-Lackmann F 1991 *Phys. Rev. Lett.* **66**, 2907; Pierce F S, Bancel P A, Biggs B D, Guo Q, and Poon S J 1993 *Phys. Rev. B* **47** 5670; Chernikov M A, Bianchi A, Felder E, Gubler U, and Ott H R 1996 *Europhys. Lett.* **35** 431

[140] Mori M, Matsuo S, Ishimasa T, Matsuura T, Kamiya K, Inokuchi H, and Matsukawa T 1991 *J. Phys. Condens. Matter* **3** 767

[141] Belin E, Dankhazi Z, Sadoc A, Calvayrac A, Klein T, and Dubois J M 1992 *J. Phys. Condens. Matter* **4** 4459

[142] Belin E 2004 *J. Non-Cryst. Solids* **334&335** 323

[143] Shastri A, Borsa F, Goldman A I, Shield J E, and Torgeson D R 1993 *J. Non-Cryst. Solids* **153&154** 347; 1994 *Phys. Rev. B* **50** 15 651

[144] Mizutani U, Takeuchi T, and Sato H 2004 *J. Non-Cryst. Solids* **334&335** 331

[145] Fujiwara T, Yamamoto S, and Trambly de Laissardiére G 1993 *Phys. Rev. Lett.* **71** 4166; Trambly de Laissardiére G and Fujiwara T 1994 *Phys. Rev. B* **50** 5999; 1994 *ibid.* **50**, 9843

[146] Trambly de Laissardiére G and Mayou D 1997 *Phys. Rev. B* **55** 2890; Trambly de Laissardiére G, Roche S and Mayou D 1997 *Mater. Sci. Eng. A* **226-228** 986

[147] Klein T, Symko O G, Davydov D N, and Jansen A G M 1995 *Phys. Rev. Lett.* **74** 3656

[148] Davydov D N, Mayou D, Berger C, Gignoux C, Neumann A, Jansen A G M and Wyder P 1996 *Phys. Rev. Lett.* **77** 3173

[149] Tang X P, Hill E A, Wonnell S K, Poon S J, and Wu Y 1997 *Phys. Rev. Lett.* **79** 1070

[150] Escudero R, Lasjaunias J C, Calvayrac Y, and Boudard M 1999 *J. Phys. Condens. Matter* **11** 383

[151] Maciá E 2004 *Phys. Rev. B* **69** 132201

[152] Zhang G W, Stadnik Z M, Tsai A P, and Inoue A 1994 *Phys. Rev. B* **50** 6696; Shastri A, Baker D B, Conradi M S, Borsa F, and Torgeson D R 1995 *ibid.* **52** 12 681

[153] Lindqvist P, Lanco P, Berger C, Jansen A G M, and Cyrot-Lackmann F 1995 *Phys. Rev. B* **51** 4796

[154] Stadnik Z M, Pordie D, Garnier M, Baer Y, Tsai A P, Inoue A, Edagawa K, and Takeuchi S 1996 *Phys. Rev. Lett.* **77** 1777; Stadnik Z M, Pordie D, Garnier M, Baer Y, Tsai A P, Inoue A, Edagawa K, Takeuchi S, and Buschow K H J 1997 *Phys. Rev. B* **55** 10938

[155] Ebert Ph, Feuerbacher M, Tamura N, Wollgarten M, and Urban K 1996 *Phys. Rev. Lett.* **77** 3827

[156] Zijlstra E S and Janssen T 2000 *Mater. Sci. Eng. A* **294-296** 886

[157] Widmer R, Gröning O, Ruffieux P, and Gröning P 2006 *Phil. Mag.* **86** 781

[158] Guohong L, Haifeng H, Yunping W, Li L, Shanlin L, Xiunian J, and Dianlin Z 1999 *Phys. Rev. Lett.* **82** 1229

[159] Roche S and Fujiwara T 1998 *Phys. Rev. B* **58** 11338

[160] Solbrig H and Landauro C V 2000 *Physica B* **292** 47

[161] Landauro C V and Solbrig H 2000 *Mater. Sci. Eng. A* **294-296** 600

[162] Landauro C V and Solbrig H 2001 *Physica B* **301** 267

[163] Roche S and Mayou D 1997 *Phys. Rev. Lett.* **79** 2518

[164] Maciá E and Domínguez-Adame F 2000 *Electrons, Phonons, and Excitons in Low Dimensional Aperiodic Systems* (Editorial Complutense, Madrid)

[165] Thiel P A and Dubois J M 2000 *Nature (London)* **406** 570

[166] Maciá E and Domínguez-Adame F 1996 *Phys. Rev. Lett.* **76** 2957

[167] Solbrig H and Landauro C V 2002 *Adv. Solid State Phys.* **42** 151

[168] Maciá E 2003 *J. Appl. Phys.* **93** 1014

[169] Maciá E and Dolinšek J 2007 *J. Phys.: Condens. Matter* **19** 176212

[170] Maciá E, Takeuchi T, and Otagiri T 2005 *Phys. Rev. B* **72** 174208

[171] Chester G V and Thellung A 1961 *Proc. Phys. Soc. London* **77** 1005

[172] Greenwood D A 1958 *Proc. Phys. Soc. London* **71** 585

[173] Kubo R 1957 *J. Phys. Soc. Jpn.* **12** 570; Kubo R, Yokota M, and Nakajima S 1957 *ibid.* **12** 1203

[174] Maciá E 2000 *Appl. Phys. Lett.* **77** 3045

[175] Maciá E 2002 *Appl. Phys. Lett.* **81** 88

[176] Apostol T M 1995 *Introduction to Analytic Number Theory* (Springer-Verlag, New York)

[177] Maciá E 2004 *Phys. Rev. B* **69** 184202

[178] Maciá E 2002 *Phys. Rev. B* **66** 174203

[179] Sahnoune A, Ström-Olsen J O, and Zaluska A 1992 *Phys. Rev. B* **46** 10 629

[180] Fujiwara T 1993 *J. Non-Cryst. Solids* **156-158** 865

[181] Takeuchi T, Otagiri T, Sakagami H, Kondo T, and Mizutani U 2004 *Mat. Res. Soc. Symp. Proc. 2003* **805** 105

[182] Dolinšek J, Jeglič P, McGuiness P J, Jagličić Z, Bilušić A, Bihar Ž, Smontara A, Landauro C V, Feuerbacher M, Grushko B, and Urban K 2005 *Phys. Rev. B* **72** 064208

4

Aperiodically layered materials

4.1 A novel design: Fibonacci superlattices

The notion of quasiperiodic crystal was a major epistemological breakthrough in condensed matter science. It provides a natural extension of the classical crystal notion when the periodic order of the atoms is replaced by a quasiperiodic one. It was then natural that the notion of quasiperiodic order was readily applied to other possible building blocks in the arrangements of matter. Thus, shortly after the pioneering works by Shechtman and Steinhardt, the notion of quasiperiodic order was extended from the atomic scale proper of metallic alloys to the submicrometer scale typical of semiconductor heterostructures (see Section 1.4). In fact, on 21 October 1985, Roberto Merlin and co-workers reported in *Physical Review Letters* the first realization of a semiconductor-based quasiperiodic superlattice.[1] The sample, grown by molecular beam epitaxy, consisted of two basic building blocks arranged according to the Fibonacci sequence (Fig.4.1). These building blocks were each composed of a bilayer of AlAs and GaAs and satisfied the thickness ratio $d_A/d_B \simeq 1.595$, a value relatively close to the Golden mean value $\tau = 1.618....$ The sample consisted of $F_{14} = 377$ bilayers and had a total thickness of ~ 1.85 μm (nominal value $L = d_A F_{13} + d_B F_{12} = 1.91$ μm).

The room-temperature x-ray diffraction pattern showed a significant number of peaks superimposed to the main satellite reflections of the GaAs layers (inset Fig.4.2). These peaks occur in a geometric progression with τ as a common ratio, according to the expression

$$q_{m_1,m_2} = \frac{2\pi}{\Lambda} m_1 \tau^{m_2},\tag{4.1}$$

where $\Lambda \equiv \tau d_A + d_B$ is an average of the relative thicknesses of blocks A and B, and m_i are integers. This expression indicates that the peaks can be labeled as a series of multiples of the Golden mean, hence highlighting the self-similar arrangement of the diffraction peaks (we recall that for a periodic superlattice the distance between successive peaks is approximately constant). In fact, one can easily check that Eq.(4.1) satisfies the relationship $q_{m_1,m_2+1} = q_{m_1,m_2} + q_{m_1,m_2-1}$. Making use of Eq.(4.1) the x-ray results were also used to obtain an experimental value for the Golden mean: $\tau_{\exp} = 1.630 \pm 0.015$.

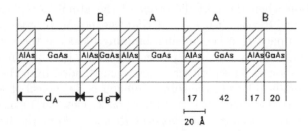

FIGURE 4.1

Schematic arrangement of GaAs and AlAs bilayers deposited following the Fibonacci sequence ABAAB.... The AlAs layers (electronic barriers) have all the same (nominal) thicknesses (17 Å), whereas the GaAs layers (electronic wells) take on two different values (42 and 20 Å). (Values taken from ref.[2].)

The Fourier transform of the ideal Fibonacci sequence consists of a dense set of diffraction peaks given by the wave vectors,[3]

$$q_{n_1,n_2} = \frac{2\pi}{\Lambda}(n_1 + n_2\tau),$$ (4.2)

where n_i are integers. Keeping expression (2.3) in mind one realizes that Eq.(4.2) reduces to Eq.(4.1) when n_1 and n_2 are successive Fibonacci numbers. In actual experiments the entire diffractogram profile is dominated by peaks which can be properly labeled in terms on integers belonging to the set $\{1,2,3\}$, which are Fibonacci numbers themselves. This fact indicates that Eq.(4.1) and (4.2) are essentially equivalent in practice.

According to these expressions one expects that every increase in experimental resolution will reveal new peaks in what was previously unresolved background. This characteristic feature was nicely demonstrated by high-resolution x-ray diffraction (synchrotron) studies, clearly demonstrating the presence of two kinds of order (namely, periodic and quasiperiodic) coexisting in the *same* sample at *different length scales* (Fig.4.2).[4] This novel feature distinguishes quasiperiodic superlattices from usual periodic ones, opening promising avenues for new materials design.

In fact, since different physical phenomena have their own relevant physical scales, by properly matching the characteristic length scales of elementary excitations propagating through the system, one can exploit the physical properties related to the quasiperiodic order we have introduced in the system. To this end, one should consider the possible role of structural imperfections which are inevitably introduced during the growth process. Quite interestingly, detailed structural characterization studies have shown that disorder does not seriously disrupt the overall coherence of the quasiperiodic sequence, so that most physical properties related to quasiperiodic order are robust enough. This property prompted the interest in the potential applications of quasiperiodic layered structures composed of different materials arranged according to different kinds of aperiodic sequences.[2]

4.2 General aperiodic heterostructures

4.2.1 Substitution sequences and matrices

The rapid progress achieved in growth technologies, like molecular beam epitaxy, magnetron sputtering, or vacuum deposition, has made it possible to grow artificial structures with different aperiodic modulations of chemical composition along the growth direction. For the sake of illustration in Table 4.1 we list some representatives among the plethora of aperiodic heterostructures grown during the last two decades.

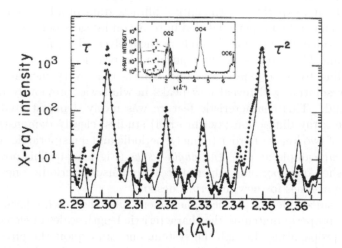

FIGURE 4.2

High-resolution diffraction profile of the GaAs/AlAs Merlin's Fibonacci superlattice. The dots represent synchrotron x-ray data and the line is the calculated Fourier intensity. A low-resolution scan showing the overall appearance of the x-ray scattering is plotted in the inset. The shaded region indicates the range of the high-resolution scan shown in the main figure. For relatively great reciprocal space vectors one observes a series of evenly spaced reflections, corresponding to the (002), (004), and (006) GaAs planes, while for smaller reciprocal space vectors one obtains a dense set of peaks which can be properly labeled by successive powers of τ progression.([4] Reprinted figure with permission from Todd J, Merlin R, Clarke R, Mohanty K M and Axe J D 1986 *Phys. Rev. Lett.* **57** 1157 © 1986 by the American Physical Society.)

TABLE 4.1
Aperiodic superlattices based on semiconductor materials have been mainly grown by molecular beam epitaxy (MBE), most metallic multilayers have been grown by magnetron sputtering (MS), and dielectric multilayers have been grown by vacuum deposition (VD), lithographic techniques (L), or electric poling (EP).

MATERIALS	SEQUENCE	GROWTH	REF
GaAs/AlAs	Fibonacci	MBE	1985 [1]
GaAs/AlAs	Thue-Morse	MBE	1991 [5]
GaAs/Ga$_{1-x}$Al$_x$As	Fibonacci	MBE	1987 [6]
Si/Ge$_x$Si$_{1-x}$	Fibonacci	MBE	1987 [7]
GaAs/Ga$_{0.5}$Al$_{0.5}$As/AlAs	Ternary self-similar	MBE	1990 [8]
Nb/Cu	Fibonacci	MS	1986 [9]
Nb/Ta	Fibonacci	MBE	1988 [10]
Ta/Al	Fibonacci	MS	1991 [11]
Ta/Al	Ternary Fibonacci	MS	1992 [12]
Mo/V	Precious means	MS	1990 [13]
W/Ti	Fibonacci	MS	1996 [14]
LiNbO$_3$	Fibonacci	L	1989 [15]
SiO$_2$/TiO$_2$	Fibonacci	VD	1994 [16]
LiTaO$_3$	Fibonacci	EP	1997 [17]
PbS/CdS	Fibonacci/TM	VD	1997 [18]
Na$_3$AlF$_6$/ZnS	Cantor	VD	2002 [19]
porous Si	Thuc-Morse		2007 [20]

A common feature of all these systems is that the ordering of layers can be specified in terms of the so-called substitution sequences. A substitution sequence is formally defined by its action on an alphabet $\mathcal{A} = \{A, B, C...\}$, which consists of certain number of letters. In actual realizations each letter will correspond to a different type of building block in the heterostructure. The substitution rule starts by replacing each letter by a finite word, as it is illustrated in Table 4.2. The corresponding aperiodic sequence is then obtained by iterating the substitution rule g starting from a given letter of the set $\mathcal{A}$ in order to obtain an aperiodic string of letters. For instance, the Fibonacci sequence is obtained from the continued process $A \rightarrow AB \rightarrow ABA \rightarrow ABAAB \rightarrow ABAABABA \rightarrow ABAABABAABAAB \rightarrow ....$

TABLE 4.2

Substitution rules most widely considered in the study of self-similar layered systems, where n and m are positive integers.

SEQUENCE	SET $\mathcal{A}$	SUBSTITUTION RULE
Fibonacci	$\{A, B\}$	$g(A) = AB \quad g(B) = A$
Silver mean	$\{A, B\}$	$g(A) = AAB \quad g(B) = A$
Bronze mean	$\{A, B\}$	$g(A) = AAAB \quad g(B) = A$
Precious means	$\{A, B\}$	$g(A) = A^n B \quad g(B) = A$
Copper mean	$\{A, B\}$	$g(A) = ABB \quad g(B) = A$
Nickel mean	$\{A, B\}$	$g(A) = ABBB \quad g(B) = A$
Metallic means	$\{A, B\}$	$g(A) = AB^n \quad g(B) = A$
Mixed means	$\{A, B\}$	$g(A) = A^n B^m \quad g(B) = A$
Thue-Morse	$\{A, B\}$	$g(A) = AB \quad g(B) = BA$
Period-doubling	$\{A, B\}$	$g(A) = AB \quad g(B) = AA$
ternary Fibonacci	$\{A, B, C\}$	$g(A) = AC \quad g(B) = A$ $g(C) = B$
Rudin-Shapiro	$\{A, B, C, D\}$	$g(A) = AB \; g(B) = AC$ $g(C) = DB \; g(D) = DC$
paper folding	$\{A, B, C, D\}$	$g(A) = AB \; g(B) = CB$ $g(C) = AD \; g(D) = CD$

Other popular sequence was first introduced by Axel Thue in 1906,[21] and then rediscovered by Morse in 1921.[22] The so-called Thue-Morse sequence has been extensively studied in the mathematical literature as the prototype of a sequence generated by substitution. In this case the continued process reads $A \rightarrow AB \rightarrow ABBA \rightarrow ABBABAAB \rightarrow ABBABAABBAABABBA \rightarrow ....$ The number of letters in this sequence increases geometrically, $N = 2^n$, where n indicates the iteration order. In the infinite limit the relative frequency of both kinds of letters in the sequence takes the same value, i.e., $v_A = v_B = 1/2$. This result contrasts with that corresponding to the Fibonacci sequence, where $v_A = \tau^{-1}$, and $v_B = \tau^{-2}$. Another important difference is that in the

Fibonacci sequence B letters always appear isolated, whereas in Thue-Morse sequence both dimers AA and BB appear alike.

The introduction of the so-called period-doubling sequence originated in the theory of dynamical systems. It describes the behavior of any system at the accumulation point of a period-doubling cascade.[23] For this sequence the continued process reads $A \to AB \to ABAA \to ABAAABAB \to$ .... The number of letters in this sequence increases as $N = 2^n$ (like the Thue-Morse sequence) but letters B appear always isolated (like the Fibonacci sequence) and their relative frequency in the infinite limit is $v_A = 2/3$, and $v_B = 1/3$.

To each substitution rule we can associate a substitution matrix as follows

$$\mathbf{S} = \begin{pmatrix} n_A[g(A)] & n_A[g(B)] & n_A[g(C)] & ... \\ n_B[g(A)] & n_B[g(B)] & n_B[g(C)] & ... \\ n_C[g(A)] & n_C[g(B)] & n_C[g(C)] & ... \\ ... & ... & ... & ... \end{pmatrix}, \tag{4.3}$$

where $n_i[g(j)]$ indicates the number of times a given letter i appears in the substitution rule $g(j)$, irrespective of the order in which these letters occur. The dimension of the substitution matrix is then determined by the number of different letters included in the basic alphabet $\mathcal{A}$. For the sake of illustration, the substitution matrices corresponding to the aperiodic sequences shown in Table 4.2 are listed in second column of Table 4.3.

It can be checked that all these matrices are primitive, that is, all entries of $\mathbf{S}^N$ are strictly positive integers (i.e., $s_{ij} \neq 0$), for some $N \geq 1$. This condition guarantees that, (i) the word resulting from the successive application of the corresponding substitution sequence is self-similar in the $N \to \infty$ limit, and (ii) the substitution matrix eigenvalue with larger modulus (sometimes referred to as Frobenius eigenvalue), say μ_+, is real, positive and larger than one.[24] The components of the Frobenius eigenvector, once normalized, read

$$v_A = \frac{\mu_+ - n_B[g(B)]}{\mu_+ + n_B[g(A)] - n_B[g(B)]} \qquad v_B = \frac{n_B[g(A)]}{\mu_+ + n_B[g(A)] - n_B[g(B)]} \tag{4.4}$$

for binary sequences (analogous expressions are obtained for sequences containing three or more letters) and they respectively indicate the frequencies of letters A and B in the infinite sequence $N \to \infty$ limit. For instance, making use of the data included in Table 4.3 we readily obtain the well-known results $\boldsymbol{v} = (\tau^{-1}, \tau^{-2})$, $\boldsymbol{v} = (1/2, 1/2)$, and $\boldsymbol{v} = (2/3, 1/3)$, for the Fibonacci, Thue-Morse, and period-doubling sequences, respectively.

An alternative way of constructing self-similar aperiodic sequences relies on a successive concatenation process. This process starts with an appropriate set of short length words, say $S_0 = A$ and $S_1 = AB$, and then proceeds by successively applying a concatenation rule. For instance, in the case $S_n = S_{n-1} * S_{n-2}$, where $*$ denotes the concatenation operation, we get $S_2 = S_1 * S_0 = ABA$, $S_3 = S_2 * S_1 = ABAAB$, $S_4 = S_3 * S_2 = ABAABABA$, and so on. It is realized that the sequence obtained in this way corresponds to the

TABLE 4.3
Substitution matrices and their related eigenvalues for the aperiodic sequences listed
in Table 4.2. Those sequences satisfying the Pisot property are explicitly indicated
(bolded names).

SEQUENCE	S	detS	trS	EIGENVALUES
Fibonacci	$\begin{pmatrix} 1 & 1 \\ 1 & 0 \end{pmatrix}$	-1	1	$\mu_\pm = \frac{1 \pm \sqrt{5}}{2}$
Silver mean	$\begin{pmatrix} 2 & 1 \\ 1 & 0 \end{pmatrix}$	-1	2	$\mu_\pm = 1 \pm \sqrt{2}$
Bronze mean	$\begin{pmatrix} 3 & 1 \\ 1 & 0 \end{pmatrix}$	-1	3	$\mu_\pm = \frac{3 \pm \sqrt{13}}{2}$
Precious means	$\begin{pmatrix} n & 1 \\ 1 & 0 \end{pmatrix}$	-1	n	$\mu_\pm = \frac{n \pm \sqrt{n^2+4}}{2}$
Copper mean	$\begin{pmatrix} 1 & 1 \\ 2 & 0 \end{pmatrix}$	-2	1	$\mu_+ = 2, \ \mu_- = -1$
Nickel mean	$\begin{pmatrix} 1 & 1 \\ 3 & 0 \end{pmatrix}$	-3	1	$\mu_\pm = \frac{1 \pm \sqrt{13}}{2}$
Metallic means	$\begin{pmatrix} 1 & 1 \\ n & 0 \end{pmatrix}$	$-n$	1	$\mu_\pm = \frac{1 \pm \sqrt{1+4n}}{2}$
Mixed means	$\begin{pmatrix} n & 1 \\ m & 0 \end{pmatrix}$	$-m$	n	$\mu_\pm = \frac{n \pm \sqrt{n^2+4m}}{2}$
Thue-Morse	$\begin{pmatrix} 1 & 1 \\ 1 & 1 \end{pmatrix}$	0	2	$\mu_+ = 2 \ \mu_- = 0$
Period-doubling	$\begin{pmatrix} 1 & 2 \\ 1 & 0 \end{pmatrix}$	-2	1	$\mu_+ = 2, \ \mu_- = -1$
ternary Fibonacci	$\begin{pmatrix} 1 & 1 & 0 \\ 0 & 0 & 1 \\ 1 & 0 & 0 \end{pmatrix}$	1	1	$\mu_+ \simeq 1.466 \quad \lvert \mu_{2,3} \rvert \simeq 0.826$
Rudin-Shapiro	$\begin{pmatrix} 1 & 1 & 0 & 0 \\ 1 & 0 & 1 & 0 \\ 0 & 1 & 0 & 1 \\ 0 & 0 & 1 & 1 \end{pmatrix}$	0	2	$\mu_+ = 2, \ \mu_2 = 0, \ \mu_{3,4} = \pm\sqrt{2}$
paper folding	$\begin{pmatrix} 1 & 0 & 1 & 0 \\ 1 & 1 & 0 & 0 \\ 0 & 1 & 0 & 1 \\ 0 & 0 & 1 & 1 \end{pmatrix}$	0	3	$\mu_+ = 2, \ \mu_1 = 1, \ \mu_{3,4} = 0$

Fibonacci sequence. In a similar way, the ternary Fibonacci sequence listed in Table 4.2 can be obtained from the concatenation rule $S_n = S_{n-1} * S_{n-3}$, starting with $S_0 = A$, $S_1 = AB$, and $S_2 = ACB$. The construction rule of the Thue-Morse sequence is somewhat more involved. It is defined by the concatenation rule $S_n = S_{n-1} * S_{n-1}^+$, where S_{n-1}^+ is the complement of S_{n-1}, which is obtained from the substitution rule $A^+ = B$ and $B^+ = A$. This concatenation rule guarantees the presence of a mirror symmetry plane at the concatenation point for the odd order sequences, which also form a palindrome, that is a word which reads backwards the same as forwards (e.g., $ABBABAABBAABABBA$ for the $n = 5$ Thue-Morse sequence).

4.2.2 Diffracting heterostructures

By analyzing the spectrum of a given substitution matrix one can obtain relevant information about the nature of the order present in the related sequence. In fact, according to the Bombieri-Taylor theorem,[25] if the spectrum of the substitution matrix $\mathbf{S}$ contains a Pisot-Vijayaraghavan number, then the lattice is quasiperiodic; otherwise it is not. A Pisot-Vijayaraghavan number (also called a Pisot number for short) is a positive algebraic number (i.e., a number which is obtained from the solution of an algebraic equation) greater than one, all of whose conjugate elements (the other solutions of the algebraic equation) have absolute value less than unity. The smallest and second smallest Pisot numbers are respectively given by the positive root of the equations $x^3 - x - 1 = 0$ (i.e., $x_+ \simeq 1.3247$) and $x^4 - x^3 - 1 = 0$ (i.e., $x_+ \simeq 1.3803$). The golden mean, satisfying the algebraic equation $x^2 - x - 1 = 0$, is another instance of Pisot number.

We have seen that the presence of a quasiperiodic distribution of scattering centers is a sufficient condition for efficient diffraction, so that this mathematical property has deep physical implications. Quite remarkably, given a multilayered structure whose layers are arranged according to a certain substitution sequence, the study of the Pisot property allows one to relate some basic features of its diffraction spectra to the algebraic properties of the corresponding substitution matrix. As an illustrative example let us consider the case of the so-called precious mean sequences, defined by the substitution rule $A \to A^n B$, $B \to A$ (Table 4.2). Among the members of this family we find the golden mean (Fibonacci), silver mean or bronze mean, among others. Their Frobenius eigenvalues read $(n \pm \sqrt{n^2 + 4})/2$ (Table 4.3). Since $\sqrt{n^2 + 4} > n$ we conclude that all the representatives of this family satisfy the Pisot property, so that the corresponding layered structures are quasiperiodic. Physically this means that their Fourier spectrum can be expressed as a finite sum of weighted δ-Dirac functions (i.e., Bragg peaks).[24] In fact, several experimental studies[13] have confirmed that quasiperiodic heterostructures satisfying the Pisot property exhibit discrete diffraction peaks which can be properly labeled in terms of the expression

$$q_{m_1,m_2} = \frac{2\pi}{\Lambda_0} m_1 \mu_+^{m_2}. \tag{4.5}$$

This expression properly generalizes Eq.(4.1), where the Frobenius eigenvalue μ_+ now plays the role of the golden mean in Eq.(4.1), and Λ_0 is a suitable superlattice average wavelength. On the contrary, heterostructures based on the metallic mean sequence, defined by the substitution rule $A \rightarrow AB^n$, $B \rightarrow A$, do not exhibit sharp diffraction peaks.[26] Among the members of this family we find copper and nickel means (Table 4.2) and their Frobenius eigenvalues read $(1 \pm \sqrt{1+4n})/2$ (Table 4.3). We realize that none of the representatives of this family satisfies the Pisot property for $n \geq 2$, so that the corresponding sequences are not quasiperiodic. It is interesting to note that the metallic and precious mean sequences are complementary among them, in the sense that the invariants associated to their substitution matrices are related through the relationship $\det \mathbf{S} \leftrightarrow - \text{tr} \mathbf{S}$ (see Table 4.3).

On the basis of these algebraic complementary behaviors one can understand some properties of the substitution sequences referred to as mixed means in Tables 4.2 and 4.3. In fact, although these sequences were originally thought of as possible generalizations of Fibonacci one,[27] it was subsequently realized that they do not satisfy the Pisot property, so that they exhibit poor diffraction spectra. The reason for labeling them as *mixed* means stems from the fact that their substitution matrix can be properly decomposed as a sum of the substitution matrices corresponding to the precious means, the metallic means, and the Fibonacci sequence, according to the expression

$$\mathbf{S} = \begin{pmatrix} n & 1 \\ m & 0 \end{pmatrix} \equiv \begin{pmatrix} n & 1 \\ 1 & 0 \end{pmatrix} + \begin{pmatrix} 1 & 1 \\ m & 0 \end{pmatrix} - \begin{pmatrix} 1 & 1 \\ 1 & 0 \end{pmatrix}. \tag{4.6}$$

Accordingly, the term *generalized* Fibonacci sequence should be properly restricted to sequences exhibiting the same basic properties as in the Fibonacci case, namely, inflation symmetry and a Fourier spectrum consisting of Bragg peaks. Quite remarkably, the sequences generated from the substitution rules $A \rightarrow B^{n-1}AB$, $B \rightarrow B^{n-1}A$, nicely fit those requirements. In fact, one can readily check that their related substitution matrices,

$$\mathbf{S} = \begin{pmatrix} 1 & 1 \\ n & n-1 \end{pmatrix}, \tag{4.7}$$

have the same algebraic invariants than those of precious means (see Table 4.3). In particular, both kinds of substitution matrices share the same characteristic polynomial, and can thus be considered as formally equivalent from the viewpoint of the Pisot property.[28]

It has been experimentally confirmed that superlattices composed of more than two building blocks arranged according to substitution sequences satisfying the Pisot property also diffract. For instance, ternary Fibonacci multilayers composed of three kinds of Ta/Al bilayers building blocks exhibit a series of Bragg peaks which can be labeled in terms of the expression,[12]

$$q_{n_1,n_2,n_3} = \frac{2\pi}{D}(n_1\mu_+^2 + n_3\mu_+ + n_2), \tag{4.8}$$

where $D = \mu_+^2 d_A + \mu_+ d_C + d_B$, and the value of μ_+ is given in Table 4.3.

By inspecting Table 4.3 we see that all the diffracting superlattices we have considered so far not only satisfy the Pisot property but also have $|\det \mathbf{S}| = 1$. It turns out that the Pisot nature of a substitution sequence provides a first criterion which demarcates those sequences possessing Bragg peaks from the rest. The unimodular condition $|\det \mathbf{S}| = 1$ is a second criterion to distinguish between strictly quasiperiodic sequences and limit-quasiperiodic ones.[24] In the first case, the diffraction spectrum consists in Bragg peaks supported by a Fourier module with rank equal to the letters of the alphabet set $\mathcal{A}$. In the limit-quasiperiodic case one also finds a discrete component of its diffraction spectrum consisting of Bragg peaks. However, this component is supported by a Fourier module with a countably infinity of generators over the integers (i.e., the reciprocal space has infinite dimensions).

An interesting exceptional case is provided by the Thue-Morse sequence which satisfies the Pisot property (see Table 4.3), but exhibits a singular continuous Fourier intensity.[29] This peculiar behavior can be explained by the fact that this sequence contains palindromes of arbitrary length,[30, 31] along with mirror symmetry planes hierarchically distributed through the chain.[32] Note also that the substitution matrix associated to the Thue-Morse sequence (see Table 4.3) coincides with that corresponding to the *periodic* sequence $A \to AB$, $B \to AB$, hence suggesting one may expect some periodic-like features in its physical properties. In fact, high-resolution diffraction studies of Thue-Morse heterostructures (having a finite size) have shown the presence of relatively broad diffraction peaks which can be indexed in terms of a set of integer couples according to the expression

$$q_{m_1,m_2} = q_0\frac{2m_1 + 1}{3 \times 2^{m_2}}, \tag{4.9}$$

which gives the accurate locations of most of the *main* diffraction peaks.[5] Accordingly, *finite* realizations of Thue-Morse lattices exhibit an *essentially discrete* diffraction pattern, and according to the definition of aperiodic crystal given by the IUCr, Thue-Morse heterostructures can be properly regarded as aperiodic crystal representatives, albeit they can not be regarded as quasiperiodic ones.

4.2.3 Cantor-like heterostructures

A fundamentally different type of multilayered structure can be constructed in terms of the fractal sets introduced in Section 1.5. For instance, one may consider the triadic Cantor set, which is obtained through the repetition of a

FIGURE 4.3

XTEM micrograph of the sixth generation of a triadic Cantor superlattice grown by magnetron sputtering (only the first five generations are resolved in the image). The segments and gaps correspond to nanometer sized ($d = 1.4$ nm) layers of amorphous Ge and Si, respectively. The superlattice contains $3^6 = 729$ layers in total (nominal thickness of 1020.6 nm). ([33] Reprinted figure with permission from Järrendahl K, Dulea M, Birch J, and Sundgren J E 1995 *Phys. Rev. B* **51** 7621 ⓒ 1995 by the American Physical Society.)

simple rule: divide any given segment into three equal parts, then eliminate the central one, and continue this process. Though this is a usual way of obtaining a Cantor set, it is by no means the only one. More general Cantor sets can be generated by iterating the operation consisting in the division of a segment in $s = 2r - 1$ equal parts ($r > 2$) and the removal of $r - 1$ of its pieces. The resulting structures are self-similar and have a fractal dimension $D = \ln r / \ln s$. Alternatively, a Cantor structure can be obtained by successively applying the substitution rules $A \rightarrow ABA...BA$ (containing r A's) and $B \rightarrow BBB...B$ (containing s B's). Thus, the triadic Cantor set is obtained from the inflation process $A \rightarrow ABA$ and $B \rightarrow BBB$. Making use of this procedure different kinds of Cantor heterostructures have been grown by alternatively depositing two different materials, say A and B, playing the role of segments and gaps, respectively (Fig.4.3).[33, 34] The obtained structure can be regarded as a finite order approximant of a mathematical Cantor set.

We note that the resulting segments and gaps are mutually commensurate by construction. Therefore, from a structural viewpoint quasiperiodic lattices and fractal lattices belong to two different classes of aperiodic systems, since quasiperiodic structures are composed of building blocks exhibiting two (or more) incommensurate periods, while fractal lattices are not. In fact, one can readily check that the substitution matrix corresponding to a general Cantor superlattice

$$\mathbf{S} = \begin{pmatrix} r & 0 \\ s-r & s \end{pmatrix} \qquad (4.10)$$

has the eigenvalues $\mu_+ = s > 1$ and $\mu_- = r > 1$, so that none of the related sequences satisfies the Pisot property. The lack of quasiperiodic order, however, does not prevent diffraction at all, since the self-similar arrangement of layers is perfectly able to do the job. In fact, diffractograms corresponding to the Cantor superlattice shown in Fig.4.3 showed a series of relatively broad peaks which can be indexed in terms of three integers, n_1, n_2, and n_3, according to the expression[33]

$$q_{n_1,n_2,n_3} = \frac{2\pi}{d} \frac{n_1}{\mu_+^{n_2}(\mu_+^{n_3} - 1)}. \qquad (4.11)$$

The Fourier transform of a Cantor superlattice of order n can be conveniently split in a contribution coming from the progressively thicker B blocks, describing the gaps, (B_n), plus a term due to the A layers ensemble contribution (A_n), as follows [33, 35]

$$F_n(q) = \left(1 - \frac{f_A}{f_B}\right) A_n(q) + B_n(q), \qquad (4.12)$$

where $f_{A,B}$ is the Fourier transform of layer A or B, respectively. Now, as the order of the Cantor heterostructure is increased, the ensemble of A layers progressively defines the characteristic self-similar pattern encoded in the inflation rule, meanwhile B layers simply stack together to form progressively thicker and homogeneous blocks. Accordingly, in the thermodynamic limit the only contribution due to the B system is a δ peak at $q = 2\pi/d$, where d is the width of the B layers, and the main information regarding the fine peak structure is contained in the $A_n(q)$ function, which exhibits conspicuous self-similar features.[33, 35] Therefore, when comparing the structural properties of quasiperiodic and fractal lattices one realizes that the self-similarity of the underlying structure (present in both kinds of lattices) is more readily reflected in the Fourier spectra of fractal lattices than quasiperiodic ones.

At this point, some reflection on the fractality and self-similarity notions is in order. Let us consider an initial segment and we take on it two points generated by a *random* algorithm, then we discard the fraction of the segment defined by these two points and continue applying the same procedure successively. In the thermodynamic limit the resulting structure is certainly a Cantor set, but it is not self-similar at all.[36, 37] This interesting remark clearly illustrates the fact that Cantor-like and self-similar features are not necessarily equivalent notions.

This result is particularly relevant when considering the energy spectra of both quasiperiodic and fractal structures. As we will see in Chapter 5, in both cases we have highly fragmented spectra, exhibiting a well-defined, hierarchical band splitting as the system's size is increased. Thus, both kinds

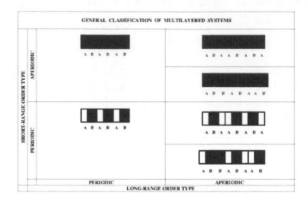

FIGURE 4.4

Classification of multilayered systems attending to the nature of the order they exhibit at different scales. (From ref.[40] Reprinted figure from Maciá E 2006 *Rep. Prog. Phys.* **69** 397 with permission from IOP Publishing Ltd.)

of structures possess energy spectra supported by Cantor sets in the thermodynamic limit. Nevertheless, the energy spectra of fractal systems usually exhibit strict self-similarity, characterized by the existence of a single scaling factor along the spectrum. On the contrary, the scaling properties of the energy spectrum of aperiodic lattices must be described using the formalism of multifractal geometry, since in this case a broad distribution of scaling factors must be considered even in the simplest models.[38, 39] Accordingly, only in the case of fractal systems a direct correlation between the geometry of the structure and the related energy spectra has been clearly observed, as for example in terms of the scalability and sequential splitting of light transmission spectra through Cantor dielectric multilayers,[19, 34] as we will discuss in Chapters 7 and 8.

4.2.4   A classification scheme

The possibility of growing devices based on an aperiodic stacking of different layers introduces an additional degree of freedom, related to the presence of two different kinds of order in the same sample at different length scales. This characteristic property is illustrated in Fig.4.4, where we present a classification of multilayered systems attending to the kind of order present at different scales.

On the bottom panels we have systems based on usual periodic crystalline layers which are stacked either periodically or aperiodically. Both classes of multilayered structures have been grown and fully characterized during the last two decades. On the contrary, the structures presented on the top panels are still awaiting for a definitive experimental realization. In fact, they are based on quasicrystalline layers piled up either periodically or aperiodically. Since the epitaxial growth of quasicrystalline thin films over both crystalline and quasicrystalline substrates is a topic under intensive current research,[41, 42, 43, 44, 45] we can confidently expect that some representatives of these classes may be obtained in the near future. In addition, one may think also of hybrid multilayered structures where both crystalline and quasicrystalline layers are stacked either periodically or aperiodically. Finally, one may consider multilayers being composed of amorphous slabs (hence lacking long-range order on the atomic scale) which are either periodically or aperiodically stacked. These systems may be regarded as hybrid order systems in which both disorder and structural order coexist at different scales. A physical realization of such multilayers is provided by Cantor superlattices considered in Section 4.2.3 (Fig. 4.3).

4.3 Signatures of quasiperiodicity

In section 3.1 we saw that quasiperiodic crystals exhibit a number of anomalous transport properties, partially related to the presence of a quasiperiodic order in the underlying atomic structure. Is there a similar set of anomalous physical properties in aperiodically layered systems?

Earlier studies focused on metallic multilayers in which one has the potential to study the effects of aperiodicity on a wide variety of problems which may also lead to some insight into the behavior of three-dimensional quasicrystalline alloys. In this sense, metallic multilayers can be used as model systems in which it is possible to systematically study one-dimensional aperiodic order by varying a broad set of design parameters: the chemical nature of layer constituents, the layer dimensions, the type of aperiodic arrangement of layers, and the total number of layers.

Metallic multilayers were originally proposed by Ivan Schuller as artificial structures where the constituent layers have different crystal structures. Therefore, one may find large differences in the lattice parameters of metallic multilayers, at variance with semiconductor based superlattices made of similar materials exhibiting very close lattice parameters.[46] Within the broad spectrum of possible metallic multilayers a lot of works were originally devoted to the study of periodically alternating systems of the form $\cdots S-N-S-N \cdots$, where S and N denote superconducting and normal-metal (or lower critical temperature superconducting layers), respectively. This class of artificially prepared superconducting materials was of interest due to the presence of strong coupling between neighbor layers (proximity effect). As a consequence of the proximity of the lower critical temperature layers, the multilayer critical temperature, T_c, is reduced from the high-T_c layers bulk value. Another interesting physical magnitude to study is the parallel upper critical field $H_{c\parallel}$, which provides a direct probe of the interplay between the superconducting coherence length in the growth direction $\xi_\perp$, and the scales associated with the artificially imposed multilayer modulation. Thus, a dimensional crossover from anisotropic three-dimensional behavior to two-dimensional behavior was observed in periodic Nb/Cu multilayers when $\xi_\perp$ is of the order of the copper layer thickness.[47]

Following the first report of Fibonacci superlattices (FSLs) by Merlin and co-workers, a series of Mo ($T_c = 4.82$ K) and V ($T_c = 0.62$ K) Fibonacci multilayers were grown by magnetron-sputtering.[48] The dependence of the superconducting transition temperature was studied as a function of the characteristic wavelength Λ defined in Eqs.(4.1) and (4.2). It was observed that T_c slowly increased as Λ increased, approaching the bulk Mo value, but this effect can be interpreted in terms of proximity effects, as it was done in the periodic multilayers case, so that no evidence of quasiperiodic related effects could be appreciated. On the other hand, the study of the parallel field behavior was also inconclusive as to whether some reported anomalies were due (or not) to quasiperiodicity effects.[48] Subsequent studies on Nb ($T_c = 9.3$ K) and Ta ($T_c = 4.4$ K) Fibonacci multilayers reported on the presence of two upturns in the parallel upper critical field $H_{c\parallel}$ with decreasing temperature. These upturns were associated with dimensional crossover occurring twice as the superconductor coherence length in the growth direction sample the two characteristic length scales $2d_A$ and d_A present in the Fibonacci multilayer.[10] At first sight this result may be interpreted as a first evidence of quasiperiodicity effects in a physical magnitude. However, some reflection indicates that a similar double crossover should also appear in periodic $S - N$ multilayers having two different S-layer thicknesses of appropriate size relative to $\xi_\perp$. In fact, the presence of AA layers in Fibonacci sequence is, in essence, a short-range feature which is shared with a lot of both periodic and aperiodic sequences.

A clearer evidence about the role of quasiperiodic order in the physical properties of semiconductor based aperiodic multilayers was provided by the study of the energy-level structure when a magnetic field is applied parallel to the layers. In this field configuration, carriers move in cyclotron orbits of radius $\sqrt{\hbar/eB}$, in the multilayer growth direction. Therefore, a magnetic field can be considered as a very appropriate tool to study vertical motion of charge carriers through an aperiodically modulated potential profile, provided that the potential barriers are thin and low enough to allow an efficient tunneling of the carriers through them.[49] The energy-level structure of GaAs/GaAlAs Fibonacci superlattices was thus probed measuring the luminescence intensity as a function of the excitation energy (luminescence excitation spectroscopy) and the obtained results were compared with similar experiments in periodic superlattices. In this way, it was verified that the shape of the excitation spectra properly reflected the shape of the DOS and that the one-dimensional quasiperiodic distribution of the barriers along the growth direction affects the carrier motion perpendicular to the layers very differently from that of parallel layers. In fact, the luminescence excitation spectra of Fibonacci superlattices in magnetic fields perpendicular and parallel to the superlattice layers are strongly anisotropic. In particular, only the parallel-field spectra show well resolved Landau levels which, in addition, show a characteristic self-similar behavior as a function of the intensity of the applied parallel field.[49]

The observation of quasiperiodicity related effects in the luminescence experiments is directly related to the fact that the radius of the cyclotron orbits, which span the range from 5.47 nm ($B_\parallel = 22$ T) to 9.07 nm ($B_\parallel = 8$ T), can extend over a significant portion of the superlattice, which is composed of GaAlAs barriers of width 1.12 nm separated by GaAs wells of width 1.69 nm. On the contrary, Fibonacci metallic multilayers were not well suited to this end, because the superconducting coherence length in the considered samples ($\xi_\perp \simeq 80 - 130$ Å) was much larger than the normal metal thickness ($15 - 30$ Å), so that the superconducting layers are effectively coupled and the possible effects due to quasiperiodicity are negligible.

Guided by this experience the next experiments were aimed at detecting quasiperiodicity signatures in the physical properties of quasiperiodic heterostructures by focusing on the study of the electronic structure of semiconductor based superlattices. Two different types of quasiperiodic modulation were considered in GaAs/GaAlAs FSLs: one chose the GaAs to have the same width but the GaAlAs barriers could have either different widths or different heights. Photoluminescence excitation spectroscopy measurements showed that, in the first case, the electronic states are localized in relatively short portions of the heterostructure, whereas in the latter case some states may delocalize through the structure.[50] In this way, FSLs with different type of quasiperiodic modulation exhibited different electronic properties, somewhat intermediate between those of periodic superlattices (completely delocalized states) and random ones (exponentially localized states). Accordingly, one should expect that related transport properties should also exhibit a similar behavior.

The first application of optical spectroscopy to the investigation of perpendicular transport properties in FSLs was reported by Yamaguchi and co-workers.[51] The sample used in their work consisted of two GaAs/AlAs FSLs and an enlarged well, which is inserted between them (Fig.4.5). A portion of the carriers generated in the FSL by photoexcitation recombine radiatively or non-radiatively in the superlattice, and the other carriers flow into the enlarged well. Thus, the enlarged well acts as a sink for the carriers, which subsequently undergo recombination. Consequently, two emission peaks are observed in the photoluminescence spectrum. Comparison of the photoluminescence spectra between the enlarged well and the superlattice emission provides information about the perpendicular transport of the carriers through

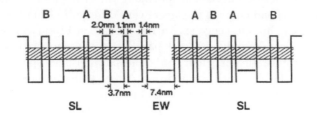

FIGURE 4.5

Sketch of the conduction band potential profiles for the Fibonacci superlattice
(SL) and the enlarged well (EW) inserted at the center. The energy level in
the EW and the miniband in the superlattice are also shown. (From ref.[51].
With permission from Elsevier.)

the superlattice.

In addition, the carrier transport was also investigated by measuring time
behavior of the superlattice and enlarged well peaks of the emission spectra.
From these measurements (in the picosecond time-scale) the time dependence
of carrier transport was calculated as well. By comparing the photoluminescence spectra corresponding to three different superlattices (random, periodic,
and quasiperiodic) several conclusions were drawn:

- the obtained spectra are quite different from one another, so that this
  technique can be successfully used to discriminate the type of order (i.e.,
  periodic, quasiperiodic, and random) present in a given heterostructure;

- the experimental results reasonably fit with the calculated ones. In
  particular, an overall trifurcation pattern can be observed in the FSL,
  with clear indication of additional fragmentation in at least one of the
  main subbands (Fig.4.6);

- the localization degree of the carriers is largest for the random superlattice, intermediate for the Fibonacci system, and smallest for the periodic
  superlattice (even if one takes into account a common localization effect
  present in all the systems due to the monolayer width fluctuations);

- at intermediate temperatures ($T = 50$ K) the phonon assisted carrier
  transport is faster in Fibonacci and periodic systems than in the random

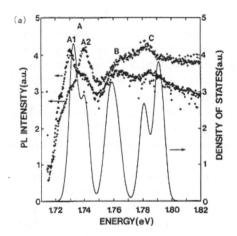

FIGURE 4.6

Photoluminescence spectrum of the superlattice emission (solid circles) and the enlarged well emission (crosses) on the Fibonacci system at 4 K. There appear three main peaks labeled A, B, and C. The peak A has substructure designated A1 and A2. The solid line is the density of states calculated by using the transfer matrix technique. (From ref.[51]. With permission from Elsevier.)

system, clearly suggesting that the electronic wavefunctions are more extended in the former than in the latter.

As we will see in more detail in Chapter 5 spatial quasiperiodicity leads to a hierarchical fragmentation pattern of electronic energy spectra in general Fibonacci systems. This pattern can be probed by experiments in which an electric field applied parallel to the superlattice quantum well layers induces a change in the charge transmission, ΔT, through exciton lifetime broadening or exciton ionization. The detected electromodulated signal is directly related to the change in absorption due to the applied electric field. Measurements of the differential transmission, $\Delta T/T$, provide a way of determining the change in the absorption coefficient of the excitonic transition, which is correlated with the transition oscillator strength. Using this electromodulation spectroscopy technique the first generations of the electronic spectrum hierarchical fragmentation pattern were observed in different kinds of GaAs/GaAlAs Fibonacci superlattices.[52] The hierarchical structure of the electronic spectrum, composed of three main bands with some inner structure, was also reported from normal-incidence infrared reflectance experiments in a GaAs/GaAlAs Fibonacci superlattice.[53] Nonetheless, these observations lacked resolution enough to clearly appreciate the prefractal structure of the energy spectra.

In fact, in order to fully appreciate specific features of quasiperiodic systems, arising from the fractal nature of their energy spectra and the critical nature of their states, the study of classical waves has a number of advantages over the study of quantum elementary excitations since, in this case, the presence of electron-phonon, electron-electron, or spin-orbit interactions makes difficult the analysis of data. Thus, a number of experimental studies dealing with surface[54] and ultrasonic waves[15] in Fibonacci structures have been reported, confirming that characteristic self-similar features in the transmission spectra are observable when the long-range aperiodic modulation is established at the micrometer range. However, the clearest demonstration of self-similar features in the physical properties of aperiodic systems was obtained by measuring the optical transmission of quasiperiodic and fractal stacks of dielectric layers.

The appealing possibility of probing the degree of localization of electromagnetic waves propagating through an optical multilayer constructed following the Fibonacci sequence was originally proposed by Mahito Kohmoto, Bill Sutherland, and Kazumoto Iguchi in 1987, who analysed the rich fractal structure of the transmission coefficient as a function of the wavelength of light at normal incidence.[55] Their theoretical proposal spurred the interest for possible optical applications[56] as well as for new theoretical aspects of light transmission in aperiodic media.[58, 59] In a subsequent work, the optical transmission of a Fibonacci dielectric multilayer composed of stacks of SiO_2 and TiO_2 thin films was reported.[16] Relative film thicknesses were chosen such that the phase shift for normally incident light was the same, i.e.,

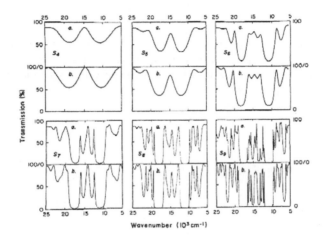

FIGURE 4.7

Optical transmission as a function of the wave number for Fibonacci dielectric multilayers composed of a different number of layers, ranging from $N = 5$ (S_4) to $N = 55$ (S_9). The indices of refraction of the A (silicon dioxide) and B (titanium dioxide) layers at 700 nm are $n_A = 1.45$ and $n_B = 2.30$. Relative film thicknesses were $d_A = 121.0$ nm and $d_B = 76.4$ nm. The dielectric stacks are sandwiched between 6.5 mm thick fused silica substrates. Curves a. are experimental results and b. theoretical calculations. ([57] Reprinted figure with permission from Gellermann W, Kohmoto M, Sutherland B and Taylor P C 1994 *Phys. Rev. Lett.* **72** 633 © 1994 by the American Physical Society.)

$n_A d_A = n_B d_B$. Furthermore, the individual layers were taken as quarter-wave layers, since previous theoretical results indicated that quasiperiodicity effects are enhanced in this case. In Fig.4.7 experimental and theoretical results are reported for a series of Fibonacci dielectric multilayers of increasing size. As the number of layers composing the structure is progressively increased one clearly appreciates a progressive fragmentation of the transmission spectra, which follows the trifurcation hierarchical pattern predicted by theory. In addition, the self-similarity of spectra around the central wavelength is evident by comparing the S_6 and S_9 (or S_5 and S_8) spectra. Similar results have been recently reported from the experimental study of a one-dimensional waveguide serial structure made of segments and loops of coaxial cable arranged according to the Fibonacci sequence.[60] In this way, the study of classical waves propagation through aperiodically modulated substrates has become a very active research field in materials science in recent years, to which we will come back in Chapters 7 and 8.

References

[1] Merlin R, Bajema K, Clarke R, Juang F Y, and Bhattacharya P K 1985 *Phys. Rev. Lett.* **55** 1768

[2] Merlin R and Clarke R 1990 U.S. Patent No. 4.955.692

[3] Levine D and Steinhardt P J 1984 *Phys. Rev. Lett.* **53** 2477

[4] Todd J, Merlin R, Clarke R, Mohanty K M and Axe J D 1986 *Phys. Rev. Lett.* **57** 1157

[5] Axel F and Terauchi H 1991 *Phys. Rev. Lett.* **66** 2223; Kolář M 1994 *Phys. Rev. Lett.* **73** 1307; Axel F and Terauchi H 1994 *Phys. Rev. Lett.* **73** 1308

[6] Lockwood D J, MacDonald A H, Aers G C, Dharma-wardana M W C, Devine R L S and Moore W T 1987 *Phys. Rev. B* **36** 9286

[7] Dharma-wardana M W C, MacDonald A H, Lockwood D J, Baribeau J M and Houghton D C 1987 *Phys. Rev. Lett.* **58** 1761

[8] Terauchi H, Kamigaki K, Okutani T, Nishihata Y, Kasatani H, Kasano H, Sakaue K, Kato H and Sano N 1990 *J. Phys. Soc. Jpn.* **59** 405

[9] Hu A, Tien C, Li X J, Wang Y H and Feng D 1986 *Physics Letters A* **119** 313

[10] Cohn J L, Lin J J, Lamelas F J, He H, Clarke R and Uher C 1988 *Phys. Rev. B* **38** 2326

[11] Peng R W, Hu A and Jiang S S 1991 *Appl. Phys. Lett.* **59** 2512; Jiang S S, Zou J, Cockayen D J H, Sikorski A, Hu A and Peng R W 1992 *Phil. Mag. B* **66** 229

[12] Peng R W, Hu A, Jiang S S, Zhang C S and Feng D 1992 *Phys. Rev. B* **46** 7816

[13] Birch J, Severin M, Wahlström U, Yamamoto Y, Radnoczi G, Riklund R and Wallenberg L R 1990 *Phys. Rev. B* **41** 10398

[14] Pan F M, Jin G J, Wu X L, Feng J W, Hu A and Jiang S S 1996 *J. Appl. Phys.* **80** 4063

[15] Zhu Y Y, Ming N B and Jiang W H 1989 *Phys. Rev. B* **40** 8536

[16] Hattori T, Tsurumachi N, Kawato S, and Nakatsuka H 1994 *Phys. Rev. B* **50** 4220

[17] Zhu S N, Zhu Y Y, Qin Y Q, Wang H F, Ge C Z, and Ming N B 1997 *Phys. Rev. Lett.* **78** 2752

[18] Musikhin S F, Il'in V I, Rabizo O V, Nakueva L G and Yudintseva T V 1997 *Semiconductors* **31** 46

[19] Lavrinenko A V, Zhukovsky S V, Sandomirski K S, and Gaponenko S V 2002 *Phys. Rev. E* **65** 036621

[20] Moretti L, Rea I, De Stefano L, and Rendina I 2007 *Appl. Phys. Lett.* **90** 191112

[21] Thue A 1906 *Norske Vididensk. Selsk. Skr. I.* **7** 1

[22] Morse M 1921 *Trans. Am. Math. Soc.* **22** 84

[23] Collet P and Eckmann J P 1980 *Iterated Maps on the Interval as Dynamical Systems* (Birkhäuser, Boston)

[24] Luck J M, Godrèche C, Janner A, and Janssen T 1993 *J. Phys. A: Math. Gen.* **26**, 1951

[25] Bombieri E and Taylor J E 1987 *Contemp. Math.* **64**, 241; Godrèche C, Luck J M and Vallet F 1987 *J. Phys. A* **20**, 4483

[26] Severin M and Riklund R 1989 *J. Phys. Condens. Matter* **1** 5607

[27] Gumbs G and Ali M K 1988 *Phys. Rev. Lett.* **60** 1081

[28] Fu X, Liu Y, Zhou P and Sritrakool W 1997 *Phys. Rev. B* **55** 2882

[29] Godrèche C and Luck J M 1992 *Phys. Rev. B* **45** 176

[30] Hof A, Knill O and Simon B 1995 *Comm. Math. Phys.* **174** 149

[31] Baake M 1999 *Lett. Math. Phys.* **49** 217

[32] Barache D and Luck J M 1994 *Phys. Rev. B* **49** 15004

[33] Järrendahl K, Dulea M, Birch J, and Sundgren J E 1995 *Phys. Rev. B* **51** 7621

[34] Zhukovsky S V, Lavrinenko A V, and Gaponenko S V 2004 *Europhys. Lett.* **66** 455

[35] Hamburger-Lidar D A 1996 *Phys. Rev. E* **54** 354

[36] García-Moliner F 2005 *Microelectron. J.* **36** 870

[37] Pérez-Álvarez R and García-Moliner F 2001 *Some Contemporary Problems in Condensed Matter Physics* Eds. Vlaev S J and Gaggero M (Nova Science Publishers, New York)

[38] Maciá E and Domínguez-Adame F 2000 *Electrons, Phonons, and Excitons in Low Dimensional Aperiodic Systems* (Editorial Complutense, Madrid)

[39] Pérez-Álvarez R, García-Moliner F, and Velasco V R 2001 *J. Phys.: Condens. Matter* **13** 3689

[40] Maciá E 2006 *Rep. Prog. Phys.* **69** 397

[41] Smith A R, Chao K J, Niu Q, and Shih C K 1996 *Science* **273** 226

[42] Ledieu J, Hoeft J T, Reid D E, Smerdon J A, Diehl R D, Ferralis N, Lograsso T A, Ross A R, and McGrath R 2005 *Phys. Rev. B* **72** 035420

[43] Sharma H R, Shimoda M, Ross A R, Lograsso T A, and Tsai A P 2005 *Phys. Rev. B* **72** 045428

[44] Curtarolo S, Setyawan W, Ferralis N, Diehl R D, and Cole M W 2005 *Phys. Rev. Lett.* **95** 136104

[45] Fournée V, Sharma H R, Shimoda M, Tsai A P, Unal B, Ross A R, Lograsso T A, and Thiel P A 2005 *Phys. Rev. Lett.* **95** 155504

[46] Schuller I K 1980 *Phys. Rev. Lett.* **44** 1597

[47] Chun C S L, Zheng G C, Vicent J L and Schuller I K 1984 *Phys. Rev. B* **29** 4915

[48] Karkut M G, Triscone J-M, Ariosa D, and Fischer Ø 1986 *Phys. Rev. B* **34** 4390

[49] Toet D, Potemski M, Wang Y Y, Maan J C, Tapfter L, and Ploog K 1991 *Phys. Rev. Lett.* **66** 2128

[50] Laruelle F and Etienne B 1988 *Phys. Rev. B* **37** 4816

[51] Yamaguchi A A, Saiki T, Tada T, Ninomiya T, Misawa K, and Kobayashi T 1990 *Solid State Commun.* **75** 955

[52] Dinu M, Nolte D D and Melloch M R 1997 *Phys. Rev. B* **56** 1987

[53] Munzar D, Bočáek L, Humlíček J and Ploog K 1994 *J. Phys.: Condens. Matter* **6** 4107

[54] Kono K, Nakada S, Narahara Y, and Ootuka Y 1991 *J. Phys. Soc. Jpn.* **60** 368

[55] Kohmoto M, Sutherland B, and Iguchi K 1987 *Phys. Rev. Lett.* **58** 2436

[56] Schwartz C 1988 *Appl. Opt.* **27** 1232

[57] Gellermann W, Kohmoto M, Sutherland B, and Taylor P C 1994 *Phys. Rev. Lett.* **72** 633

[58] Dulea M, Severin M, and Riklund R 1990 *Phys. Rev. B* **42** 3680

[59] Latgé A and Claro F 1992 *Opt. Commun.* **94** 389

[60] El Boudouti E H, El Hassouani Y, Aynaou H, Djafari-Rouhani B, Akjouj A, and Velasco V R 2007 *J. Phys.: Condens. Matter* **19** 246217

5

One-dimensional quasiperiodic models

5.1 Effective one-dimensional systems

Broadly speaking, an obvious motivation for the recourse to one-dimensional (1D) models in Solid State Physics is the complexity of the full-fledged problem. In the particular case of quasicrystalline matter this general motivation is further strengthened by the lack of translational symmetry, though the presence of a well-defined long-range orientational order in the system also prevents a naive application of procedures specifically developed for the study of random structures in this case. On the other hand, we can also invoke more fundamental reasons supporting the use of one-dimensional models as a first approximation to the study of realistic quasiperiodic systems. In fact, as we will discuss in the following sections, most characteristic features of quasiperiodic systems, like the fractal structures of their energy spectra and related eigenstates, can be explained in terms of resonant coupling effects in the light of Conway's theorem. Therefore, the physical mechanisms at work are not so dependent on the dimension of the system, but are mainly determined by the self-similarity of the underlying structure. Consequently, in order to ascertain whether some of the purported anomalies in the transport properties of quasicrystals are directly related to the kind of order present in their structure (a basic point for any general theory of quasicrystalline matter), the recourse to one-dimensional models can be considered as a promising starting point, for such models encompass, in the simplest possible manner, most of the novel physics attributable to the quasiperiodic order.

Thus, shortly after the discovery of QCs, the study of phonon propagation through aperiodic linear chains was considered as a first approximation to understand quasiperiodic effects in quasicrystals. Earlier models were based on a harmonic chain composed of two kinds of masses, m_A and m_B, which are arranged aperiodically, and two kinds of springs, K_{AA} and $K_{AB} = K_{BA}$, whose value will generally depend on the type of joined atoms (Fig. 5.1). According to this description different kinds of models can be considered. In the on-site model one assumes all the spring constants to be equal at every site of the chain (Fig. 5.1b). In the transfer model, all the masses are assumed to be identical instead (Fig. 5.1a). From a physical point of view, one expects that the nature of the chemical bonding between the different

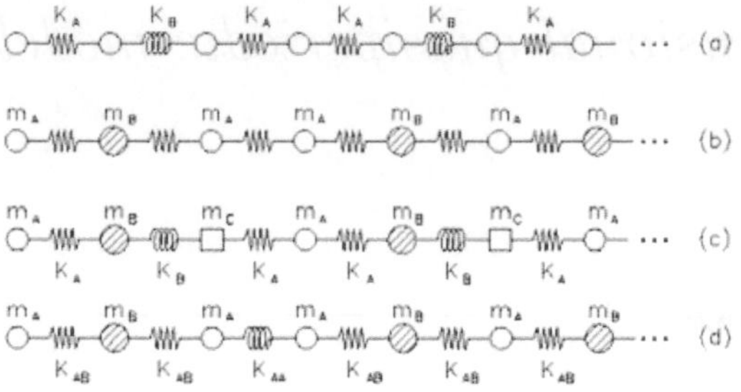

FIGURE 5.1

Sketch illustrating different models considered in the literature: a) binary transfer model, b) binary on-site model, c) ternary mixed model [13], and d) Fibonacci binary mixed model.

atoms (and thereof the value of the spring constant representing the bond) will depend on the nature of the involved atoms. In this case, the aperiodic distribution of masses in the system *induces* a (generally different) aperiodic distribution of spring constants in the chain (Fig. 5.1c-d). Therefore, in most physical situations of interest, one must consider the so-called mixed models. Earlier studies focused on the study of either transfer[1, 2, 3] or on-site models. [4, 5] Subsequently, more general models were considered.[6, 7, 8, 9, 10, 11] Ternary models, defined in terms of two different spring constants and three different kinds of masses aperiodically distributed have been also considered (Fig. 5.1c).[12, 13] Since most stable QCs found to date are ternary alloys this class of models makes a closer connection with the chemical complexity of these materials.[14] More general models, considering three different types of bonds connecting five different types of masses have also been considered in the literature.[15]

A standard way to study the acoustic and thermal properties of a lattice is to consider a nearest-neighbor harmonic chain given by the Lagrangian

$$L = \frac{1}{2} \sum_{n=1}^{N} m_n \dot{\eta}_n^2 - \frac{1}{2} \sum_{n=1}^{N-1} K_{n,n+1} (\eta_n - \eta_{n+1})^2, \qquad (5.1)$$

where η_n is the displacement of the nth atom from its equilibrium position; m_n, with $n = A, B$, is the corresponding mass, $K_{n,n\pm1}$ denotes the strength of the harmonic coupling between neighbor atoms, and N is the number of particles in the system. The dynamical equation for the normal modes $\eta_n =$

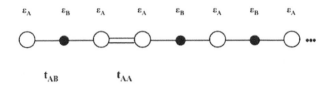

FIGURE 5.2
Tight-binding chain model describing the electron dynamics in terms of on-site energies ε_n and transfer integrals $t_{n,n\pm1}$.

$u_n e^{i\omega t}$ then reads

$$(K_{n,n-1} + K_{n,n+1} - m_n\omega^2)u_n - K_{n,n-1}u_{n-1} - K_{n,n+1}u_{n+1} = 0, \qquad (5.2)$$

where ω is the vibration frequency. The main properties of the frequency spectra of different kinds of aperiodic lattices are fully addressed in Section 5.4.

In a similar way, within the electron independent approximation, the problem of the electron dynamics moving through a mixed Fibonacci chain can be described in terms of a tight-binding model, where the on-site energies ε_n (accounting for the atomic potentials) are arranged according to the Fibonacci sequence, and the transfer integrals $t_{n,n\pm1}$ (accounting for the hopping of the electron between neighboring atoms) are also arranged aperiodically (Fig.5.2). The corresponding Schrödinger equation takes the form

$$(E - \varepsilon_n)\psi_n - t_{n,n-1}\psi_{n-1} - t_{n,n+1}\psi_{n+1} = 0, \qquad (5.3)$$

where E is the electron energy and ψ_n is the wave function amplitude at site n. By comparing Eqs.(5.2) and (5.3) we see that they are formally analogous, hence indicating that the electron and phonon problems will share some common properties, which we will exploit in the following sections.

Another interesting physical system which can be approximately described in terms of an effective one-dimensional description is provided by superlattices and multilayered systems introduced in Chapter 4. We recall that in order to grow a superlattice we must define two distinct building blocks, say

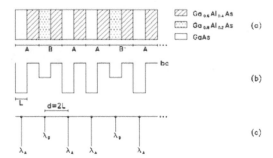

FIGURE 5.3
Sketch showing the key features of a continuous model describing a general
Fibonacci superlattice in terms of Dirac function potentials.

A and B, and order them according to either a periodic pattern or a given
aperiodic sequence. Each building block can be composed of one or more
layers of different materials and can have arbitrary thickness. In the case of
a chemically modulated structure one takes equally spaced blocks defining a
periodic array of period d and introduce a quasiperiodic modulation by means
of an appropriate choice of their chemical composition, as it is illustrated in
Fig.5.3(a) for a AlGaAs/GaAs semiconductor-based Fibonacci superlattice.
In principle one may consider a number of possible potential profiles: square
barriers, V-shaped, sawtooth, parabolic. The square barrier conduction band
profile is shown in Fig.5.3(b) for the sake of illustration. More generally, one
can approximate the interaction of an electron with the underlying substrate
in terms of an arbitrarily sharply peaked potential approaching the δ-function
limit (zero width and constant area), as depicted in Fig.5.3(c).

In this way, one should consider the following Schrödinger equation

$$\left[-\frac{\hbar^2}{2m}\frac{d^2}{dx^2} - \sum_n \lambda_n \delta(x - nd) \right] \psi(x) = E\psi(x), \qquad (5.4)$$

where the potential strength $\lambda_n > 0$ (attractive potentials) takes on two values, λ_A and λ_B, arranged according to the Fibonacci sequence. Expressing the electron wave function as a linear combination of atomic orbitals
$\psi(x) = \sum_n c_n \phi_n(x - nd)$, where $\phi_n(x - nd) = \sqrt{\lambda_n}\exp(-\lambda_n|x - nd|)$ is
the normalized eigenfunction of a delta function placed at $x = nd$, and ne-
glecting the overlap involving three different centres, one obtains the following

FIGURE 5.4

Sketch of a Fibonacci trans-polyacetylene chain modeled in terms of spherical δ-functions.

tight-binding equation [16, 17]

$$(E - \epsilon_n)c_n = t_{n,n+1}c_{n+1} + t_{n,n-1}c_{n-1}, \tag{5.5}$$

with $\epsilon_n = -\lambda_n^2/2$, and

$$t_{n,n\pm1} = -\sqrt{\lambda_n^3 \lambda_{n\pm1}} \exp(-\lambda_{n\pm1}d). \tag{5.6}$$

Another illustrative example of approximate one-dimensional system is provided by an array of CH units roughly representing the trans-polyacetylene chain (Fig.5.4). There are two parameters that can be varied in a transfer model, namely, the values of the short, R_S, and long, R_L, bond lengths between neighboring units.

It has been shown that the electronic band structure of quasi-1D polymers, such as trans-polyacetylene, can be quite accurately described in terms of δ-functions of the form

$$V(r) = \frac{1}{r^2}\delta(r - R). \tag{5.7}$$

Physically this potential describes a highly localized force field which vanishes everywhere except on a spherical shell of radius R around the lattice site. In that case the Schrödinger equation can be expressed as [18]

$$(\kappa - \kappa_n)\psi_n - \frac{e^{-\kappa R_{n,n+1}}}{R_{n,n+1}}\psi_{n+1} - \frac{e^{-\kappa R_{n,n-1}}}{R_{n,n-1}}\psi_{n-1} = 0, \tag{5.8}$$

where $\kappa = \sqrt{-2E}$, $\kappa_n = \sqrt{2|E_n|}$, is the on-site energy of the CH units, with $E < 0$ the electronic energy, and $R_{n,n\pm1}$ measures the effective bond lengths between neighboring CH units. We realize that Eq.(5.8) is formally analogous to the previously considered tight-binding Eqs.(5.2), and (5.3), respectively describing the phonon and electron dynamics through a chain, or Eq.(5.5) describing the electron propagation through a superlattice along the growth direction, by simply defining $t_{n,n\pm1} \equiv e^{-\kappa R_{n,n\pm1}}/R_{n,n\pm1}$. Note that these effective hopping integrals explicitly depend on the electron energy and they decrease exponentially with the distance between nearest-neighbors, adopting the functional form typical of s-like orbitals. Many other physical systems of interest can be described in terms of tight-binding equations similar to those introduced in the previously considered examples. In this way, some important results have been obtained regarding the dynamics of elementary excitations, like electrons, phonons, excitons, polaritons, spin waves, plasmons, or magnons propagating through different classes of aperiodic lattices. [19, 20, 21] We will discuss the main features of the obtained results in the following sections.

5.2   Classification schemes based on spectral properties

A key question in any general theory of aperiodic systems regards the relationship between their atomic topological order, determined by a given aperiodic density function, and the physical properties stemming from their structure. At the time being a general theory describing such a relationship is still lacking, though most relevant physical properties can be reasonably analyzed in terms of appropriate model Hamiltonians. Most rigorous, mathematical results in the field have been derived from the study of nearest-neighbor, tight-binding models described in terms of the Hamiltonian

$$(H\psi)_n = t\psi_{n+1} + t\psi_{n-1} + \lambda V_n\psi_n, \quad n \in Z, \tag{5.9}$$

where $\lambda > 0$ measures the strength potential, and the potential sequence V_n is generated according to some aperiodic substitution sequence (see Table 4.2 in Section 4.2.1), and then assigning a potential value to each letter of the string, i.e., $A \to V_A$ and $B \to V_B$. From a mathematical point of view these models belong to the class of almost periodic Schrödinger operators, which display unusual spectral properties.

According to Lebesgue's decomposition theorem, the energy (or frequency) spectrum of any measure in $\mathbb{R}^n$ can be uniquely decomposed in terms of just three kinds of spectral measures (and mixtures of them), namely: pure-point (μ_P), absolutely continuous (μ_{AC}), and singularly continuous (μ_{SC}) spectra, in the form

$$\mu = \mu_P \cup \mu_{AC} \cup \mu_{SC}. \tag{5.10}$$

Suitable examples of physical systems containing both the pure-point and/or the absolutely continuous components in their energy spectra are well known, the hydrogenic atom being a paradigmatic instance. On the other hand, the absence of actual physical systems exhibiting the singular continuous component relegated this measure as a merely mathematical issue. From this perspective, the discovery of quasicrystals, followed by the fabrication of Fibonacci semiconductor heterostructures, bridged the long standing gap between the theory of spectral operators in Hilbert spaces and the condensed matter theory, spurring a considerable amount of work from both mathematician and physicist communities.[22, 23, 24] In fact, the energy spectrum of most self-similar systems considered to date seems to be a purely singular continuous one, which is supported on a Cantor set of zero Lebesgue measure. Thus, the spectrum exhibits an infinity of gaps and the total bandwidth of the allowed states vanishes in the thermodynamic limit. This property has been proven rigorously for systems based on the Fibonacci,[25, 26, 27] Thue-Morse, and period doubling sequences.[28]

A particularly relevant result obtained from the study of Eq.(5.9) is the so-called gap-labeling theorem, which provides a relationship between reciprocal space (Fourier) spectra and Hamiltonian energy spectra. In fact, this theorem relates the position of a number of gaps in the energy spectra of elementary excitations to the singularities of the Fourier transform of the substrate lattice.[29, 30, 31] It is then convenient to focus on the nature of the measure associated with the lattice Fourier transform, which is related to the main features of the diffraction pattern through the expression

$$I_N(q) = |\mathcal{F}_N(q)|^2 \,, \tag{5.11}$$

where I_N is the intensity of the diffraction peaks, N measures the system size, q denotes the scattering vector, and $\mathcal{F}_N$ is the Fourier transform of the lattice atomic or electronic densities, depending on the type of scattering process. A convenient description of the diffraction spectra in the thermodynamic limit can be made in terms of the integrated intensity function defined as

$$H(q) = \lim_{N \to \infty} \int_0^q \frac{I_N(q')}{N} dq'. \tag{5.12}$$

This function represents the normalized distribution of the diffracted intensity peaks up to a given q point in the reciprocal space. Note that a completely analogous function can be introduced in the study of the energy spectra by replacing the normalized intensity distribution by the density of electronic or vibrational states, respectively. In such case we would express Eq.(5.12) in the form

$$H(q) = \int_0^q d\mu(q'), \tag{5.13}$$

where μ describes the corresponding spectral measure. Accordingly, we can visualize the main features of the different Lebesgue measures by considering diffraction experiments as well.

In the case of both periodic and quasiperiodic crystals there are intervals along a given q axis where the intensity vanishes, so that $H(q)$ remains constant. These intervals are separated by Bragg peaks, where $H(q)$ has finite jumps. The resulting scenario can be formally written as

$$H(q) = \int_0^q \sum_n g_n \delta(q' - q_n) dq',$$
(5.14)

where the sum runs over the countable set of Bragg peaks. Accordingly, the weight of any set of diffraction points in reciprocal space equals the sum of the corresponding coefficients g_n in Eq.(5.14). A measure accomplishing this property is said to be pure-point. For example, Fibonacci lattice's Fourier transform has a purely point measure, with a countable set of peaks.

On the other hand, the Thue-Morse lattice Fourier transform is no longer composed of a countable set of points separated by well-defined intervals (it does not contain δ peaks), but it exhibits a structure similar to that of a Cantor set. More precisely, the support of its Fourier spectrum can be covered by an ensemble of open intervals with arbitrary small total length. This property characterizes a purely singular continuous measure. It should be noted that from a mathematical point of view the nature of a measure is determined by its asymptotic limit (infinite system size), so that one can observe a number of relatively broad peaks in finite systems, which progressively smear out as the system size is increased, as it has been discussed in Section 4.2.2.

As an example of the third type of primitive measure component we have the Rudin-Shapiro lattice possessing a purely absolutely continuous Fourier transform. In this case, we have a diffuse spectrum, where the contribution to $H(q)$ of any interval on a given q axis is roughly proportional to its length. By making these intervals arbitrarily small it can be proved that any single point in the spectrum has zero weight and $H(q)$ is both continuous and derivable, so that it can be formally expressed as

$$H(q) = \int_0^q f(q') dq',$$
(5.15)

where $f(q)$ is a positive function. This property defines an absolutely continuous measure and it is curiously shared with random sequences, which also have the same kind of spectral measure. Such a result suggests that the structure of the Rudin-Shapiro sequence could give rise to physical properties qualitatively similar to those usually observed in disordered systems.[32]

In order to gain additional insight on the relationship between the type of structural order present in an aperiodic solid (as determined by its lattice Fourier transform) and its related transport properties (as determined by the main features of the energy spectrum and the nature of its eigenstates) it is convenient to introduce the chart depicted in Fig.5.5. In this chart we present a classification scheme of aperiodic systems based on the nature of their diffraction spectra (in abscissas) and their energy spectra (in ordinates). In this

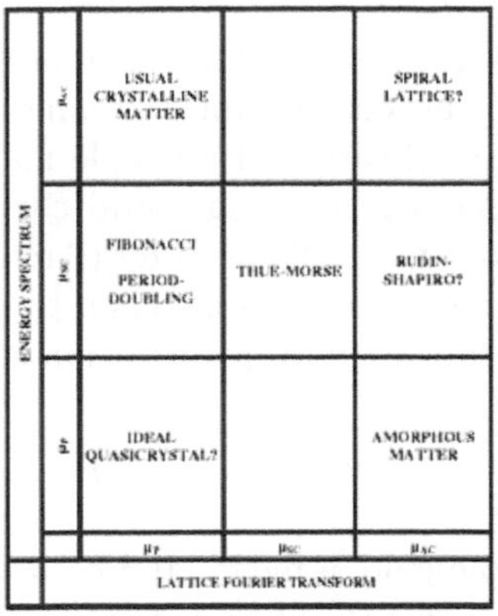

USUAL CRYSTALLINE MATTER			SPIRAL LATTICE?
FIBONACCI PERIOD-DOUBLING	THUE-MORSE		RUDIN-SHAPIRO?
IDEAL QUASICRYSTAL?			AMORPHOUS MATTER

FIGURE 5.5
Classification of aperiodic systems attending to the spectral measures of their lattice Fourier transform and their Hamiltonian spectrum energy. (From ref.[33]: Maciá E 2006 *Rep. Prog. Phys.* **69** 397. With permission from IOP Publishing Ltd.)

way, we clearly see that the simple classification scheme based on the periodic-amorphous dichotomy is replaced by a much richer one, including nine different entries. In the upper left corner we have the usual periodic crystals exhibiting pure point Fourier spectra (well defined Bragg diffraction peaks) and an absolutely continuous energy spectrum (Bloch-like wave-functions in allowed bands). In the lower right corner we have amorphous matter exhibiting an absolutely continuous Fourier spectrum (diffuse spectra) and a pure point energy spectrum (exponentially localized wave-functions). By inspecting this chart, one realizes that although Fibonacci and Thue-Morse lattices share the same kind of energy spectrum (a purely singular continuous one), they have different Fourier transforms, so that these lattices must be properly classified in separate categories.

At the time being, the nature of the energy spectrum corresponding to the Rudin-Shapiro lattice is yet an open question. Numerical studies suggested that some electronic states are localized in these lattices, in such a way that the rate of spatial decay of the wave-functions is intermediate between power and exponential laws. This means that the charge distribution is less spread than in the case of the critical states observed in Fibonacci and Thue-Morse lattices (which follow a power law), but the localization degree is still weaker than that characteristic of a random medium.[32] These results clearly illustrate that there is not a simple relation between the spectral nature of the Hamiltonian describing the dynamics of elementary excitations propagating through an aperiodic lattice and the spatial structure of the lattice potential.

Spiral lattices provide an interesting instance of perfectly ordered systems where both translational and orientational symmetries are discarded. These lattices are based on the application of a simple mathematical algorithm. In the first place we consider a generating spiral curve, which can adopt any general form in polar coordinates, like $r = a\theta$ (Archimedean), $r = a\sqrt{\theta}$ (parabolic), $r = ae^\theta$ (logarithmic), and so on. Then the spiral lattice is obtained by restricting r and θ according to a quantization condition of the form [34]

$$r = an^\tau, \qquad \theta = \phi_d n, \qquad (5.16)$$

where $n = 0, 1, 2...$, τ is the golden mean, and $\phi_d = 2\pi/\tau$ is referred to as the divergence angle, and measures the angle between adjacent radius vectors $r(n)$ and $r(n+1)$. Since the divergence angle yields irrational fractions of 2π we obtain lattices entirely lacking rotational symmetry. Accordingly, their Fourier transform does not show well-defined sharp spots, but diffuse rings similar to the electron diffraction patterns obtained from small areas of some amorphous materials.[35] The spiral lattices generated in this way exhibit arrangements analogous to those observed in many botanical structures and are characterized by self-similar inflation or deflation operations, like those observed in quasiperiodic crystals. In fact, some examples of known mineral structures, like clino-asbestos, hallosyte, or cylindrite, have been interpreted in terms of spiral lattice based structural principles.[36] To the best of our knowledge, the energy spectrum of elementary excitations propagating through spiral lattices has not been yet reported, so that we have tentatively allocated them in the upper right corner of the Fig.5.5 chart. In a similar way, we include in the left lower corner of that chart the ideal quasicrystals, i.e., those without any sort of structural defect, since theoretical arguments suggest that in a perfect, self-similar structure the presence of coherent resonance effects among the electronic states may efficiently induce their localization, leading to a perfectly isolating solid phase.[37]

5.3 Remarkable properties of singular continuous spectra

5.3.1 Fractal nature of spectra

The study of the 1D Schrödinger equation with an aperiodic potential has been the subject of a large number of works. Most of the studies devoted to the nature of spectra in aperiodic systems have been mainly focused on one-dimensional and two-dimensional systems by considering the dynamics of elementary excitations obeying the canonical equation

$$v_n\phi_n = t_{n,n-1}\phi_{n-1} + t_{n,n+1}\phi_{n+1}, \qquad (5.17)$$

along with an appropriate set of boundary conditions. In Eq.(5.17), ϕ_n is the amplitude of the elementary excitation at the nth lattice position, and v_n depends on the excitation energy (frequency), E (or ω), as well as on other characteristic physical magnitudes of the system, like atomic masses m_n, elastic constants $K_{n,n\pm1}$, or electronic binding energies ε_n (see Table 9.1). It turns out that this basic equation properly describes many systems of physical interest (see Section 5.1), which can be described in terms of tight-binding models within the single band approximation. For instance, in the case of Fibonacci superlattices (see Fig.5.3) we have seen that this corresponds to the situation in which we choose the (say GaAs) wells to have all the same width but the barriers (say GaAlAs) to have either different widths (transfer models) or different heights (on-site models).

The most relevant results reported by different authors from the study of different systems described in terms of Eq.(5.17) can be summarized as follows:

- The energy spectrum of most self-similar systems exhibits an infinity of gaps and the total bandwidth of the allowed states vanishes in the $N \to \infty$ limit. This has been proven rigorously for systems based on the Fibonacci [26, 27], Thue-Morse, and period doubling sequences;[26] and it has been conjectured to be a general property of self-similar systems.[38]

- The position of the gaps can be precisely determined through the gap labeling theorem in some definite countable set of numbers.[29, 30, 31]

- Scaling properties of the energy spectrum can be described using the formalism of multifractal geometry.[39, 40, 41]

For transfer models, a fragmentation scheme based on three main sub-bands, which successively trifurcate in turn according to the typical Cantor set scheme was reported.[42, 43] On the other hand, the energy spectrum corresponding to on-site models exhibits four main subbands, which are successively fragmented following a trifurcation pattern similar to that observed for transfer models.[4] A similar fragmentation pattern is observed in mixed models in which both diagonal and off-diagonal terms are present in the Hamiltonian.[16, 17] An illustrative example of the spectrum structure corresponding to this model class is shown in Figs.5.6 and 5.7. By inspecting these figures we clearly appreciate the following prefractal signatures:

- the spectra exhibit a highly-fragmented structure generally constituted by as many fragments as the number of elements present in the chain,

- the energy levels appear in subbands which concentrate a high number of states, and which are separated by relatively wide forbidden intervals,

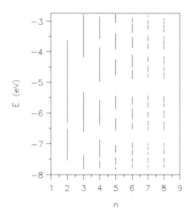

FIGURE 5.6

Fragmentation pattern of the energy spectrum of a trans-polyacetylene quasiperiodic chain. The number of allowed subbands increases as a function of
the system size expressed in terms of the Fibonacci order n as $N = F_n$.(From
ref.[44]. With permission from Elsevier.)

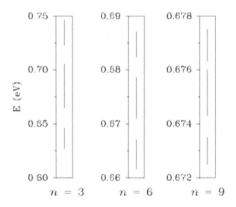

FIGURE 5.7

Self-similarity in the energy spectrum of a InAs/GaSb Fibonacci superlattice.(From ref.[45]. With permission from IOP Publishing Ltd.) The left
panel shows the whole spectrum for an order $n = 3$ superlattice, whereas the
central and right panels show a detail of the spectrum for superlattices of
orders $n = 6$ and $n = 9$, respectively.

- the degree of internal structure inside each subband depends on the total length of the chain, and the longer the chain, the finer the structure, which displays distinctive features of a self-similar distribution of levels.

Taken altogether these features provide compelling evidence about the intrinsic fractal nature of the numerically obtained spectra, which will eventually show up with mathematical accuracy in the thermodynamic limit $N \to \infty$.

5.3.2 Critical eigenstates

The notion of *critical* wave function has evolved continuously since its introduction in the study of aperiodic systems,[46] leading to a somewhat confusing situation. For instance, references to self-similar, chaotic, quasiperiodic, lattice-like, or quasilocalized wave functions can be found in the literature depending on the different criteria adopted to characterize them.[3, 47, 48, 49, 50]. Generally speaking, critical states exhibit a rather involved oscillatory behavior, displaying strong spatial fluctuations which show distinctive self-similar features in some instances (Fig.5.8). As we can see the wave function is peaked on short chain sequences but reappear far away on chain sequences showing the same lattice ordering. This is a direct consequence of the underlying lattice self-similarity and, as a consequence, the notion of an envelope function, which has been most fruitful in the study of both extended and localized states, is mathematically ill-defined in the case of critical states, and other approaches are required to properly describe them and to understand their structure.

Most interestingly, the possible existence of *extended* states in several kinds of aperiodic systems, including both quasiperiodic [7, 9, 51, 52, 53, 54] and non-quasiperiodic ones,[48, 55] has been discussed in the last years spurring the interest on the precise nature of critical wave functions and their role in the physics of aperiodic systems. From a rigorous mathematical point of view the nature of a state is uniquely determined by the measure of the spectrum to which it belongs. In this way, since it has been proven that Fibonacci lattices have purely singular continuous energy spectra [25, 26] (see Section 5.2), we must conclude that the associated electronic states cannot be, strictly speaking, extended in Bloch's sense. This result holds for other aperiodic lattices (Thue–Morse, period doubling) as well, and it may be a general property of the spectra of self-similar aperiodic systems.[38] We will discuss this issue in more detail in Section 5.5.

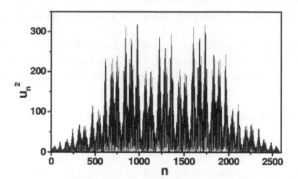

FIGURE 5.8

Squared amplitude distribution of a critical phonon normal mode in a Fibonacci lattice composed of $N = 2584$ atoms with a mass ratio $m_A/m_B = 34/21$.

5.4 Frequency spectra of general Fibonacci lattices

In order to illustrate the main properties of one-dimensional aperiodic systems let us consider, as a model example, the phonon dynamics in the general Fibonacci lattice introduced in Section 5.1 (see Fig.5.1d). A convenient way to study transport properties in one-dimensional aperiodic systems relies on the so-called transfer matrix technique (see Section 9.2). Within this approach the equation of motion given by Eq.(5.2) is cast in the form

$$\begin{pmatrix} u_{n+1} \\ u_n \end{pmatrix} = \begin{pmatrix} \frac{v_n}{K_{n,n+1}} & -\frac{K_{n,n-1}}{K_{n,n+1}} \\ 1 & 0 \end{pmatrix} \begin{pmatrix} u_n \\ u_{n-1} \end{pmatrix} \equiv \mathbf{T}_n \begin{pmatrix} u_n \\ u_{n-1} \end{pmatrix}, \qquad (5.18)$$

where $v_n \equiv K_{n,n-1} + K_{n,n+1} - m_n\omega^2$. The allowed regions of the frequency spectrum are determined from the condition

$$|\mathrm{tr} M_N(\omega)| \equiv \left| \mathrm{tr}(\prod_{n=N}^{1} \mathbf{T}_n) \right| \leq 2, \qquad (5.19)$$

where $M_N(\omega)$ is the global transfer matrix, and N is the number of atoms in the chain. The overall structure of the frequency spectrum corresponding

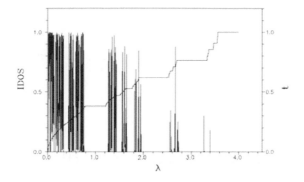

FIGURE 5.9
Overall structure of the frequency spectrum for a mixed Fibonacci lattice with
$N = 610$, $\alpha = 2$, $\gamma = 1.2$, $m_A = 1$, and $k_{AB} = 1$, as given by the integrated
density of states (IDOS, see Section 9.5.1) and the transmission coefficient, t
(see Section 9.5.2). ([10] Reprinted figure with permission from Maciá E 2000
Phys. Rev. B **61** 6645 © 2000 by the American Physical Society.)

to this system is illustrated in Figs. 5.9 and 5.10, where we show the frequency dependence of several diagnostic tools (see Section 9.5) in terms of
the parametrized frequency $\lambda \equiv m_A \omega^2 / K_{AB}$ and some representative values
of the ratios $\alpha = m_A / m_B$, and $\gamma = K_{AA} / K_{AB}$ (the overall structure of the
frequency spectrum does not significantly depend on the values adopted for
the different model parameters).[10]

As we can see, the frequency spectrum shows a splitting scheme, characterized by the presence of several main subbands separated by well-defined
gaps. Inside each main subband the fragmentation scheme follows a trifurcation pattern, in which each subband further trifurcates obeying a hierarchy
of splitting from one to three subsubbands. At low and intermediate frequencies ($0 \leq \lambda \leq 2$), the minima of the Lyapunov coefficient take significantly
low values. Conversely, starting about $\lambda = 2$ we realize that these minima
monotonically increase with λ. Since the Lyapunov coefficient is related to
the inverse of the localization length of the corresponding vibration modes
(see Section 9.5.4), these results indicate that the high frequency phonons are
more localized than the low frequency ones. This result is additionally supported by the dependence of the transmission coefficient with the frequency.
In fact, we observe that most of the low frequency phonons exhibit transmission coefficients close to unity. On the contrary, starting at about $\lambda \sim 1.5$,
we observe that, as the phonon frequency increases, the values of the corre-

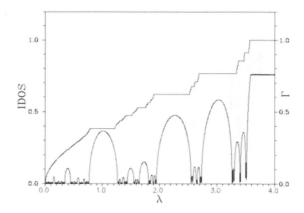

FIGURE 5.10
Overall structure of the frequency spectrum for a mixed Fibonacci lattice with
$N = 610$, $\alpha = 2$, $\gamma = 1.2$, $m_A = 1$, and $k_{AB} = 1$, as given by the integrated
density of states (IDOS) and the Lyapunov coefficient, Γ (see Section 9.5.4).
([10] Reprinted figure with permission from Maciá E 2000 *Phys. Rev. B* **61**
6645 © 2000 by the American Physical Society.)

sponding transmission coefficients progressively decrease.

We note that the presence of allowed states in the range $1 \leq \lambda \leq 2$ is directly
related to the quasiperiodic order of the chain, for the periodic binary chain
exhibits a broad gap over this frequency interval (Fig.5.11).

Additional physical insight can be gained by analyzing the spectrum within
the framework of real-space renormalization techniques (see Section 9.4). To
this end, the original Fibonacci chain is decomposed into a certain number of
minor subchains, according to a given criteria referred to as blocking scheme.
In principle, the choice of a suitable blocking scheme is arbitrary, although
in the study of quasiperiodic systems, the minimization of the information
entropy (in Shannon's sense) seems to play a relevant role in the choice of appropriate blocking schemes.[17] Guided by this criterion the original lattice is
decomposed in a series of *trimers* and *tetramers* of the form BAB and BAAB.
According to the construction rules for the Fibonacci lattice, the number of
trimers present in the chain, n_{BAB}, equals the number of isolated A atoms.
Analogously, the number of tetramers coincides with the number of AA pairs.
Then, in the thermodynamic limit we have the limits $\lim (n_{BAB}/N) = \tau^{-4}$
and $\lim (n_{BAAB}/N) = \tau^{-3}$ (see Sections 2.4.3 and 4.2.1). Now, neglecting the
trivial translation modes $\lambda_0 = 0$, each trimer, if considered as an independent
dynamical system, contributes with two different normal vibration modes,
whose respective frequencies are given by $\lambda_b = \alpha$, $\lambda_d = 2 + \alpha$, and, anal-

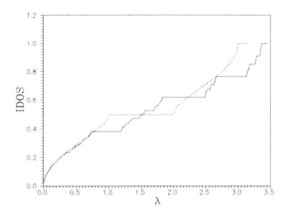

FIGURE 5.11

Comparison between the frequency spectrum of a periodic binary lattice (dashed) and a Fibonacci lattice (solid) of the same size.

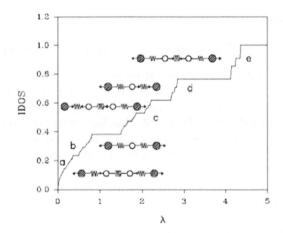

FIGURE 5.12

Correspondence between the main subbands in the frequency spectrum of a Fibonacci lattice with $N = 610$, $\alpha = 0.5$, and $\gamma = 2$, and the normal modes associated to the trimers and tetramers introduced in the renormalization scheme. ([10] Reprinted figure with permission from Maciá E 2000 *Phys. Rev. B* **61** 6645 © 2000 by the American Physical Society.)

ogously, each tetramer contributes with three different normal modes given by

$$\lambda_c = 1 + \alpha \,, \lambda_e = \gamma + \eta + \sqrt{(\eta - \gamma)^2 + 2\gamma} \,, \lambda_a = \gamma + \eta - \sqrt{(\eta - \gamma)^2 + 2\gamma}, \quad (5.20)$$

where $\eta \equiv (1 + \alpha)/2$. Hence, we have *five* basic normal modes describing the *fundamental* dynamical state of the Fibonacci lattice. If we assume that these normal modes are coupled by a resonance effect, we realize that the number and frequencies of the normal modes which appear at the first stage of the renormalization process respectively determine the number and approximate location of the main subbands in the resulting phonon spectrum.[17, 56] In this way, we can relate every main subband appearing in the frequency spectrum to a specific normal mode. Such a procedure is illustrated in Fig. 5.12, where we see that the lower frequency region of the spectrum ($\lambda < 1$) contains two main contributions: the lowest frequency contribution ($\lambda \leq 0.5$), which is related to the tetramers λ_a (contributing with τ^{-3} states) and the frequency interval $0.5 \leq \lambda < 1$, which is related to the trimer's normal mode λ_b (contributing with τ^{-4} states). Therefore, although both subbands are separated by a quite narrow gap, their origin can be traced back to the dynamics of quite different vibrating blocks in the lattice.

5.5 Nature of critical states in aperiodic systems

As we have seen, from a mathematical point of view all the states belonging to the singular continuous spectra characteristic of Fibonacci, Thue-Morse, and period-doubling aperiodic systems can be properly referred to as critical eigenstates in a broad sense. However, this fact does not necessarily imply that all these critical states behave in exactly the same way from a physical viewpoint. In fact, physically states can be classified according to their transport properties. Thus, conducting states in crystalline systems are described by periodic Bloch functions, whereas insulating systems exhibit exponentially decaying functions corresponding to localized states. Within this scheme the notion of critical states is somewhat imprecise because, generally speaking, critical states exhibit a rather involved oscillatory behavior, displaying strong spatial fluctuations at different scales.

A first step towards a better understanding of critical states was provided by the demonstration that the amplitudes of critical states in a Fibonacci lattice do not tend to zero at infinity, but are bounded below through the system.[27] This result suggests that the physical behavior of critical states might be more similar to that corresponding to extended states than to localized ones. Accordingly, the possible existence of extended critical states in several kinds of aperiodic systems was extensively discussed,[9, 7, 54] and arguments supporting the convenience of widening the very notion of extended state in aperiodic systems to include critical states which are not Bloch functions were put forward.[52]

The use of multifractal methods to analyze the electronic states in Fibonacci lattices provided conclusive evidence on the existence of different kinds of critical states, depending on their location in the highly fragmented energy spectra. Thus, while states located at the edges or the band centers of the main subbands exhibit a distinctive self-similar spatial structure, most of the remaining states do not show any specific pattern.[47] A similar situation also occurs in the phonon spectrum. In fact, when studying band structure effects in the thermal conductivity of Fibonacci quasicrystals one finds a great variety of critical normal modes,[11, 52] exhibiting quite different physical behaviors, which range from highly conducting extended states to critical states whose transmission coefficient oscillates periodically between two extreme values, depending on the system's length.[51]

The term "critical" was originally borrowed from thermodynamics, where it has usually been applied to describe a conventional phase transition where a state undergoes fluctuations in certain physical properties which are the same on all length scales. This situation is referred to as a passage through a "critical point." At the critical temperature various thermodynamic functions develop a singular behavior which is related to long-range correlations and large fluctuations. Actually, the system should appear identical on all length scales at, exactly, the critical temperature and, consequently, it would be scale invariant. All these features, characteristic of thermodynamic phase transitions, have been progressively incorporated to the study of both incommensurate and quasiperiodic systems. In fact, in the study of the Aubry model, given by the Hamiltonian Eq.(5.9) with the potential $V_n = \mu \cos(nQ + v)$, Q being an irrational number,[57] it was early realized that a metal-insulator transition appears when the potential strength takes on the threshold value $\mu = 2$, and that this process can be formally described as a phase transition affecting the electronic energy spectrum topological structure. From this point of view, the potential strength plays the same role as the temperature plays in a usual thermodynamic system.

Following a chronological order the concept of critical wave function was born in the study of the Anderson Hamiltonian which describes a regular lattice with site-diagonal disorder. This model is known to have extended states for weak disorder in 3D systems, as well as in 2D samples with a strong magnetic field. For strong disorder, on the other hand, the electronic states are localized. For 1D systems it was proved that localized states decay exponentially in space in most cases.[58] However, this exponential decay relates to the asymptotic evolution of the envelope of the wave function while the short-range behavior is characterized by strong fluctuations. The magnitude of these fluctuations seems to be related to certain physical parameters, such as the degree of disorder which, in turn, controls the appearance of the so-called mobility edges. Approaching a mobility edge, from the insulator regime, the exponential decay constant diverges, so that the wave function amplitudes can be expected to feature fluctuations on all length scales larger than the lattice spacing. This singular fact turns out to be very convenient to explain metal-insulator transitions. Actually, a localized state occupies only a minute fraction of space. On the other hand, extended states should spread homogeneously over the whole sample. Both characteristics can be accommodated at the mobility edge if one assumes a wave function with a filamentary structure like a net over the sample. Schreiber and Grussbach[59] gave an intuitive picture of a critical wave function in 3D:

> "A filamentary structure over the whole sample, like a mesh with openings on all scales or a curdled structure with lumps of all sizes, could represent an extended state which nevertheless does not fill any finite fraction of the volume."

The connection between this picture and that of a fractal object is not casual. It was early suggested by Aoki[60] that critical wave functions might display self-similar fluctuations and, in this sense, they may be characterized by some fractal dimensionality. Later on Soukoulis and Economou [61] numerically demonstrated the fractal character of certain eigenfunctions in disordered systems and characterized their amplitude behavior by a fractal dimensionality. What is more interesting, the fractal character of the wave function itself is suggested as a new method for finding mobility edges. The observation of anomalous scaling of both the moments of the probability distribution and the participation ratio near the localization threshold in the Anderson model strongly suggested that a critical wave function cannot be adequately treated as simply fractal.[62] Rather, since different moments scale in different ways, the more general concept of multifractality has to be employed, yielding a set of generalized fractal dimensions. In fact, as the computed wave function amplitudes are in general nowhere exactly zero, the dimension describing the support of the wave function coincides with the dimension of the lattice (usually the Euclidean dimension). Accordingly, the wave functions cannot be described as homogeneous fractals.[61, 63, 64, 65, 66] From these studies it is concluded that wave function amplitudes exhibit a fractal behavior not only at the mobility edge but more generally in both the insulator and metallic regimes of disordered systems up to length scales of the order of the localization length or the coherence length, respectively.

To the best of my knowledge, the first reference to "critical" wave functions in incommensurate systems was due to Ostlund and Pandit.[67] In their original treatment the term "critical" is intended to describe the existence of a wave function which is not Bloch-like extended nor localized, but somewhere in between[47, 68]

> "wave functions which are neither localized nor extended in an standard manner."

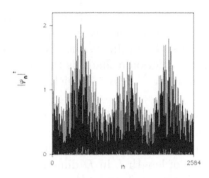

FIGURE 5.13
A representative example of the overall spatial charge distribution of a critical wavefunction in a quasiperiodic (Fibonacci) chain. ([52] Reprinted figure with permission from Maciá E and Domínguez-Adame F 1996 *Phys. Rev. Lett.* **76** 2957 © 1996 by the American Physical Society.)

At this rather fuzzy level the notion of "criticality" can be understood as follows. An extended wave function is expected to extend homogeneously over the whole sample. On the other hand, for a localized wave function at a particular site of the sample, one expects its probability density to display a single dominant maximum at, or around, this site, and its envelope function is generally believed to decay exponentially in space. On the contrary, a critical state is characterized by strong spatial fluctuations of the wave function amplitudes. This unusual behavior, consisting of an alternatively decaying and recovering of the wave function amplitudes, is illustrated in Fig.5.13.

Then, one can describe a critical state in a quasiperiodic system in the following qualitative way:[69, 70] Let us assume that a given state ψ^L spreads over a region of characteristic length L. Then, Conway's theorem implies that a similar region must exist at a distance $\leq 2L$. If L is sufficiently long, then both regions will be good candidates for a tunneling process between them, so that we might express $\psi^{2L} = z\psi^L$, where z is a damping factor roughly measuring the probability amplitude of the tunneling event. Within such a description the case $z = 0$ corresponds to strictly localized states, whereas $|z| = 1$ is the signature of extended states. For intermediate localization cases, one can write

$$\psi^L \simeq L^{-\ln|z|/\ln 2} \simeq L^{-\alpha}, \tag{5.21}$$

where the precise value of $|z|$ will be dependent on the parameters of the considered model. In this way the spatial structure of the wave function amplitudes is directly related to the topological properties of the quasiperiodic substrate. In particular, the self-similar properties of most critical wave functions can be traced back to the self-similarity of the lattice itself, through a series of hierarchical tunneling events involving the overlap of different subsystems at different length scales. Accordingly, one of the main results concerning electronic localization in quasiperiodic chains is the power law behavior of the envelope of the wave function $\psi^N \simeq N^{-\alpha}$ which characterizes most critical states.

The nature of critical states can also be related to the scaling properties of the electronic (or frequency) spectrum. In fact, one can focus on the scaling properties of the bandwidth of a series of approximants of a quasicrystal. If we consider an initial cube of length L in D dimensions, the spectrum of the infinite periodic system of unit cell L and L^D atoms will be composed by L^D bands, irrespective of the nature (periodic, disordered, or quasiperiodic) of the atoms in the considered cell. The typical bandwidth is then related to the overlap degree between two states, namely, ψ_a and ψ_b, localized in two adjacent cells, and can be qualitatively measured as $W \simeq \langle \psi_a | H | \psi_b \rangle$. For Bloch states with moduli $\psi \simeq L^{-D/2}$ and with an average hopping amplitude t from one site to another, one gets $W \simeq t/L$, as it is well known. The same argument for a disordered unit cell involves a localization length ξ, and the bandwidth is then given by $W \simeq tL^{D-1}e^{-L/\xi}$, which corresponds to a purely discrete spectrum, as expected, in the limit $L \to \infty$. In a quasiperiodic system the algebraic localization of typical wave functions, as described by Eq.(5.21), gives rise to a scaling behavior of the bandwidths of the form $W \simeq tL^{-\beta}$, where the exponent $\beta > 1$ is related to the distribution of α's.[69]

5.6 The role of critical states in transport properties

As we have seen, from a physical viewpoint, the states can be classified according to their *transport properties* which, in turn, are determined by the spatial distribution of the wave function amplitudes. An overall estimation about the influence of critical states in the transport properties of quasiperiodic systems can be inferred from expression $W \simeq tL^{-\beta}$. In fact, the mean group velocity for a critical state can be approximated as $v \simeq LW \simeq tL^{1-\beta}$. This expression indicates that the mobility of the charge carriers goes to zero as the system size grows, but this asymptotic limit is reached more slowly than it is achieved in the case of exponentially localized states, whose mobility vanishes at a pace determined by the relationship $v \simeq tL^D e^{-L/\xi}$.[69] This qualitative result provides strong support to the view of critical states as occupying an intermediate position between localized and extended states, although one may be tempted to consider them closer to the last from a physical point of view.

In this Section we will focus on a class of critical wave functions belonging to general aperiodic and fractal systems which are extended from a physical point of view. This result widens the notion of extended wave function to include electronic states which *are not* Bloch functions, and it is a relevant first step to clarify the precise manner in which the aperiodic order of these systems influences their transport properties. To this end, we will exploit an algebraic formalism which allows one to give a detailed analytical account of the transport properties of critical wave functions for certain particular values of the energy. In this way, we study the relationship between the spatial structure of critical wave functions and their transport properties showing that, although *most* critical functions exhibit rather low transmission coefficients, it is possible to find certain wave functions which are among those exhibiting higher transmission coefficients in finite aperiodic systems.

5.6.1 Extended critical states in general Fibonacci lattices

Let us consider the tight-binding model introduced in Section 5.1 (Fig.5.2) describing the electron dynamics through a mixed Fibonacci chain in terms of Eq.(5.3). Making use of suitable renormalization techniques (see Section 9.4) one can express the global transfer matrix in terms of a set of properly renormalized transfer matrices $\mathbf{R}_A$ and $\mathbf{R}_B$ which are themselves arranged according to the Fibonacci sequence. Each matrix $\mathbf{R}_\nu$ can be interpreted as an effective transfer matrix describing the propagation of the electron through the basic building blocks AB and ABA in terms of which the entire chain can be decomposed. From the explicit evaluation of such matrices (see Eqs.(9.45)) one obtains the commutator[9, 52]

$$[\mathbf{R}_A, \mathbf{R}_B] = \frac{\epsilon(1+\gamma^2) - E(1-\gamma^2)}{\gamma} \begin{pmatrix} 1 & 0 \\ E+\epsilon & -1 \end{pmatrix}, \qquad (5.22)$$

where $\gamma \equiv t_{AA}/t_{AB}$, the origin of energies is defined in such a way that $\varepsilon_A = \epsilon = -\varepsilon_B$, and the energy scale is given by $t_{AB} \equiv 1$. This commutator considerably simplifies for the two cases mostly discussed in the literature, namely the on-site ($\gamma = 1$) and transfer ($\epsilon \equiv 0$) models. The expression (5.22) shows that the on-site model is *intrinsically* non-commutative, for the commutator vanishes only in the trivial periodic case. On the contrary, in the transfer model the $\mathbf{R}$ matrices commute for the energy value $E = 0$, which corresponds to the center of the energy spectrum. Most interestingly, according to expression (5.22), there exists *always one* energy satisfying the relation

$$E_* = \epsilon \frac{1+\gamma^2}{1-\gamma^2}, \qquad (5.23)$$

for any realization of the mixed model (i.e., for any combination of $\epsilon \neq 0$ and $\gamma \neq 1$ values). For these energies the condition $[\mathbf{R}_A, \mathbf{R}_B] = \mathbf{0}$ is fulfilled

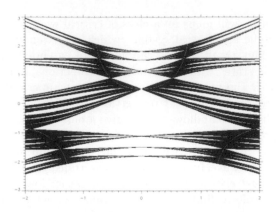

FIGURE 5.14

Phase diagram (γ in abscissas and E in ordinates) for a general Fibonacci chain with $N = 34$ and $\epsilon = 0.5$. The energies corresponding to extended wave functions are marked with a thin white line. (Courtesy of Roland Ketzmerick).

and, making use of the Cayley-Hamilton theorem (see Section 9.2) the global transfer matrix of the system, $\mathcal{M}_N(E_*) \equiv \mathbf{R}_A^{n_A}\mathbf{R}_B^{n_B}$, can be explicitly evaluated. From the knowledge of $\mathcal{M}_N(E_*)$ the condition for the considered energy value to be in the spectrum, $|\mathrm{tr}[\mathcal{M}_N(E_*)]| \leq 2$, can be readily checked and, afterwards, relevant magnitudes describing their transport properties can be explicitly determined as follows.

The global transfer matrices corresponding to the energies given by Eq.(5.23) can be expressed in the closed form

$$\mathcal{M}_N(E_*) = \begin{pmatrix} U_N & -\gamma U_{N-1} \\ \gamma^{-1}U_{N-1} & -U_{N-2} \end{pmatrix}, \tag{5.24}$$

where $U_k(x)$ are Chebyshev polynomials of the second kind (see Section 9.2) and $x \equiv \sqrt{E_*^2 - \epsilon^2}/2 \equiv \cos\phi$. From expression (5.24) we get $\mathrm{tr}[\mathcal{M}_N(E_*)] = 2\cos(N\phi)$, where we take into account the relationship $U_n - U_{n-2} = 2T_n$ between Chebyshev polynomials of the first and second kinds (see Section 9.3). Consequently, we can ensure that these energies belong to the spectrum in the quasiperiodic limit ($N \to \infty$). An illustrative example of the energy spectra of mixed Fibonacci chains for different values of the model parameters is shown in Fig.5.14. Its characteristic fragmentation scheme is clearly visible. One appreciates that the extended states given by Eq.(5.23) are located in the densest regions of the phase diagram.

The Landauer conductance (see Sec.9.5.3) can be obtained by embedding the Fibonacci lattice in an infinite periodic arrangement of identical atoms

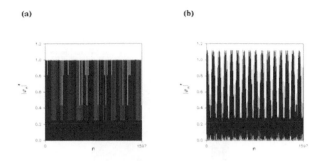

FIGURE 5.15

Electronic charge distribution in Fibonacci lattices with $N = F_{16}$ and (a) $\gamma = 2$, $\epsilon = 0.75$, $E_1 = -1.25$ and (b) $\gamma = 2$, $\epsilon = 0.5$, $E_2 = -5/6$. Their transmission coefficients are, respectively, $T(E_1) = 0.5909\ldots$ and $T(E_2) = 0.9046\ldots$. The wave function amplitudes have been calculated with the aid of the matrix formalism making use of the initial conditions $\varphi_0 = 0$ and $\varphi_1 = 1$. ([52] Reprinted figure with permission from Maciá E and Domínguez-Adame F 1996 *Phys. Rev. Lett.* **76** 2957 © 1996 by the American Physical Society.)

connected by hopping integrals $t \equiv t_{AB}$. In this way one gets[52]

$$G_N(E_*) \equiv G_0 T_N(E_*) = \frac{2e^2}{h} \frac{1}{1 + \frac{(1-\gamma^2)^2}{(4-E_*^2)\gamma^2} \sin^2(N\phi)}. \qquad (5.25)$$

Two important conclusions can be drawn from this expression. In the first place, the Landauer conductance is always bounded below for *any* lattice length, which proves the true extended character of the related states. In the second place, since the factor multiplying the sine in the denominator of expression (5.25) only vanishes in the case $\gamma = 1$, the critical states we are considering *do not* verify, in general, the transparency condition $T_N(E_*) = 1$ ($G_N(E_*) = G_0$) in the quasiperiodic limit.

For the sake of illustration, two representative examples are respectively shown in Fig.5.15. The charge distribution shown in Fig.5.15(a) corresponds to the state of energy $E_1 = -1.25$ in a Fibonacci chain with $N = F_{16} = 1597$ sites and lattice parameters $\gamma = 2$ and $\epsilon = 0.75$. The overall behavior of the wave function amplitudes clearly indicates its extended character. At this point it is worth to mention that, albeit its appearance, this wave function is non-periodic: the sequence of values taken by the wave function amplitude is arranged according to a quasiperiodic sequence. Fig 5.15(b) shows the charge

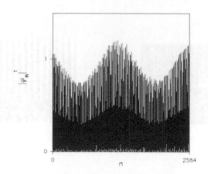

FIGURE 5.16

Electronic charge distribution corresponding to the eigenstate $E_*(k = 1193) = -13/50$, belonging to a general Fibonacci lattice with $N = F_{17} = 2584$, $\gamma = 1.5$, and $\epsilon = 0.1$. Its transmission coefficient is $T_N(E_*) = 1$.

distribution corresponding to the state of energy $E_2 = -5/6$ in a system of the same length and model parameters $\gamma = 2.0$ and $\epsilon = 0.5$. At first sight, by comparing both figures, one may be tempted to think that the transmission coefficient corresponding to the wave function plotted in Fig.5.15(a) must be higher than that corresponding to the wave function shown in Fig.5.15(b), because the charge distribution of the former along the system is more homogeneous than that corresponding to the latter. Actually, however, making use of expression (5.25), one finds $T(E_1) = 0.5909\ldots$ and $T(E_2) = 0.9046\ldots$, which is precisely the opposite case.

However, it is also possible to find states satisfying the transparency condition in *finite Fibonacci systems* whose lengths satisfy the relationship $N\phi = k\pi$, $k = 1, 2\ldots$, in Eq.(5.25) which, in turn, implies

$$E_*(k) = \pm\sqrt{\epsilon^2 + 4\cos^2(k\pi/N)}. \tag{5.26}$$

In this way, these transparent states can be classified according to a well defined scheme determined by the integer k. An illustrative example is shown in Fig.5.16.

From this plot we notice the existence of two different superimposed structures. In fact, a periodic-like, long-range oscillation with a typical wavelength of about N sites is observed to modulate a quasiperiodic series of short-range minor fluctuations of the wave function amplitude, typically spreading over a few lattice sites. This qualitative description receives a quantitative support from the study of its Fourier transform, showing two major components in

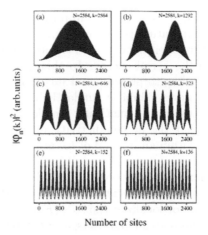

FIGURE 5.17
The wave function versus the site number for a Fibonacci lattice with $N = 2584$ sites and $E(k) = -2.5\cos(\pi/k)$ for different k values.([71] Reprinted figure with permission from Huang X and Gong C 1998 *Phys. Rev. B* **58** 739 © 1998 by the American Physical Society.)

the Fourier spectrum corresponding to the low and high frequency regions, respectively.

By combining Eqs.(5.23) and (5.26) we obtain the following relationship for the values of the model parameters satisfying the transparency condition

$$\epsilon = \pm \frac{1}{\gamma} \frac{\gamma^2}{\gamma} \cos\left(\frac{k\pi}{N}\right). \tag{5.27}$$

A particularly interesting case occurs when N is exactly divisible by k. In that case, the wave functions exhibit a remarkable spatial distribution, where the quasiperiodic component (characteristic of short spatial scales) is nicely modulated by a long scale periodic component (Fig.5.17). The periodicity of the long scale modulation is a direct consequence of the divisibility properties of Fibonacci numbers (since $N = F_n$),[71] namely that F_k is a divisor of F_{pk} for all $p > 0$.[72] Accordingly, when one chooses the model parameters in such a way they satisfy Eq.(5.27), with $N/k = p \in \mathbb{N}$, one will always get a periodically modulated critical wavefunction.

Subsequent numerical studies of the energy spectrum of mixed Fibonacci lattices have shown that a significant number of electronic states exhibiting very large transmission coefficients ($T_N(E) = 0.99999$) are located around the transparent states given by Eq.(5.23).[73] This result suggests that these critical states behave in a way quite similar to conventional extended states from a physical viewpoint, albeit they can not be rigorously described in terms of Bloch functions. To further analyze this important issue the study of the ac conductivity at zero temperature is very convenient, since it is very sensitive to the distribution nature of eigenvalues and the localization properties of the wave function close to the Fermi energy. In this way, by comparing the ac conductivities corresponding to periodic and mixed Fibonacci lattices it was concluded that both systems exhibit a similar behavior, though the value of the ac conductivity takes on systematically smaller values in the Fibonacci case, due to the fact that the ac conductivity involves the contribution of non-transparent states within an interval of $\hbar\omega$ around the Fermi level in this case.[73]

In summary, for general Fibonacci systems in which both diagonal and off-diagonal quasiperiodic order is present in their model Hamiltonian, we have critical states which are not localized (i.e., $T_N(E_*) \neq 0 \ \forall N$, when $N \to \infty$). For finite Fibonacci chains one can find *transparent* states exhibiting a physical behavior completely analogous to that corresponding to usual Bloch states in periodic systems (i.e., $T_N(E_*) = 1$) for a given choice of the model parameters, prescribed by Eq.(5.27). There exist a second class of critical states, those located close to the transparent ones, which are not strictly transparent (i.e., $T_N(E) < 1$), but exhibit transmission coefficient values very close to unity. Finally, the remaining states in the spectrum show a broad diversity of possible values of the transmission coefficient (i.e., $0 < T_N(E) \ll 1$), in agreement with the earlier view of critical states as intermediate between periodic Bloch wave functions ($T_N(E) = 1$) and Anderson localized states ($T_N(E) = 0$).

5.6.2   Critical modes tuning

The rich variety of critical states in general Fibonacci systems suggests the appealing possibility of *modulating* the transport properties of normal modes propagating through a Fibonacci lattice by properly selecting the values of the masses composing the chain (isotopic effect). To this end, we will consider the system studied in Section 5.4 and will extend the renormalization approach discussed in Section 5.6.1 in order to study the phonon dynamics as well. The commutator corresponding to the phonon problem reads (see Section 9.4)

$$[\mathbf{R}_A, \mathbf{R}_B] = \frac{\lambda}{\gamma}(2\gamma - 1 - \alpha\,[1 + \lambda(\gamma - 1)]) \begin{pmatrix} 1 & 0 \\ 2 - \alpha\lambda & -1 \end{pmatrix}. \qquad (5.28)$$

Aside from the trivial, limiting case $\lambda \to 0$, this commutator vanishes for the frequencies given by the expression

$$\lambda^* = \frac{\alpha - 2\gamma + 1}{\alpha(1 - \gamma)}. \qquad (5.29)$$

For the sake of illustration, let us consider the particular case given by the condition $K_{AA} = K_{AB}/2$. In this case ($\gamma = 1/2$), expression (5.29) reduces to $\lambda^* = 2$ *for any arbitrary choice of the masses* m_A *and* m_B. In other words, the commutation frequency becomes independent of the values assigned to the mass distribution through the Fibonacci lattice. The renormalized matrices $\mathbf{R}_A$ and $\mathbf{R}_B$ (see Eq.(9.46)) then adopt the simple form

$$\mathbf{R}_A = \begin{pmatrix} 1 & 0 \\ 2(\alpha - 2) & 1 \end{pmatrix} \qquad \mathbf{R}_B = \begin{pmatrix} -1 & 0 \\ 2(1 - \alpha) & -1 \end{pmatrix}, \qquad (5.30)$$

and the corresponding power matrices can be easily evaluated by induction, so that the transfer matrix reads

$$\mathbf{M}_N(\lambda^*) \equiv \mathbf{R}_A^{n_A} \mathbf{R}_B^{n_B} = (-1)^{n_B} \begin{pmatrix} 1 & 0 \\ 2(\alpha F_{n-2} - F_{n-1}) & 1 \end{pmatrix}. \qquad (5.31)$$

Two interesting consequences can be extracted from expression (5.31). In the first place, we realize that the frequency $\lambda^* = 2$ belongs to the spectrum regardless of the system length, since $|\mathrm{tr}[\mathbf{M}_N(\lambda^*)]| = 2$ in this particular case. In the second place, if we choose the values for the masses in such a way that their ratio satisfies the relationship $\alpha = F_{n-1}/F_{n-2}$, we get $\mathbf{M}_N(\lambda^*) = \pm I$, where I is the identity matrix. Consequently, when the parameter α is a rational approximant of the golden mean $\tau = \lim_{n \to \infty} (F_{n-1}/F_{n-2}) = (\sqrt{5} + 1)/2$, the state corresponding to the resonance frequency λ^* is a *transparent* state with $T_N(\lambda^*) = 1$. An illustrative example of this kind of state is shown in the inset of Fig. 5.18 for a lattice with $N = 2584$ and $\alpha = 1597/987$. The normal mode amplitudes have been obtained by iterating the dynamical equation (5.18) with the initial conditions $u_0 = 0$ and $u_1 = 1$. The extended nature of the state is clearly appreciated.

At this point, however, we must stress that the spatial structure of this critical normal mode is determined by two different contributions, which correspond to two separate scale lengths. Thus, although at long scales (greater than, say, 100 sites) the state shows a distinct periodic-like pattern, such an alternating pattern resolves into a series of quasiperiodic oscillations at shorter length scales. Therefore, the structure of the critical normal mode is *not* periodic, since the separation between two successive peaks takes on two different values (108 and 109 sites) which alternate in a quasiperiodic fashion.

The relationship between the spatial structure of the normal modes and their related transport properties is further explored by means of a power spectrum analysis which allows us to describe the overall structure of the normal mode as a superposition of two basic contributions involving different scale lengths. The existence of both contributions is conveniently illustrated in the main frame of Fig. 5.18, where we plot the power spectrum of the normal mode shown in the inset. In fact, we observe two main contributions in the power spectrum. In the low frequency region, a major peak located at $\nu = 0.00921$ ($\lambda \simeq 108.5$ sites) describes the overall periodic-like pattern. On the other hand, starting at about $\nu = 0.09$, we observe a series of nested, subsidiary features, characterized by the *twin peaks* labelled by the letters a_i, b_i, c_i, and d_i ($i = 1, 2$). Each couple of peaks groups around a frequency value given by

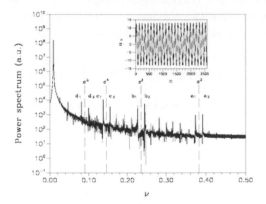

FIGURE 5.18

Power spectrum of the normal mode shown in the inset and atomic displace-ments in a Fibonacci lattice with $N = 2584$ and $\alpha_1 = 1597/987$ at the reso-nance frequency $\lambda^* = 2$ (inset). ([11] Reprinted figure with permission from Maciá E 1999 *Phys. Rev. B* **60** 10 032 © 1999 by the American Physical Society.)

some of the successive powers of the inverse golden mean $\sigma = 1/\tau$. These fea-tures arrange according to a self-similar pattern, which extends through the entire high frequency region of the power spectrum up to $\nu \simeq 0.4$. This self-similar component of the power spectrum reveals the quasiperiodic nature of the corresponding critical normal mode when it is observed at shorter scales. The relative importance of the periodic-like versus the quasiperiodic-like con-tribution can be roughly measured by the height ratio of their related peaks in the power spectrum, i.e., $I_{QP}/I_P \simeq 10^{-4}$. Therefore, we are considering a critical normal mode which behaves as an extended, transparent state, but *still preserves a significant degree of quasiperiodic order* in its inner structure.

Now we shall consider the following question. According to the expression (5.31) the transparency condition $T_N = 1$ is achieved when $\alpha = F_{n-1}/F_{n-2}$, which corresponds to the *best* rational approximant to τ for a given Fibonacci lattice of length N. Let us consider the case where we assign to the parameter α the successive values of the series $\alpha_m \equiv \{F_{n-m}/F_{n-(m+1)}\}$, with $m = 1, 2 \ldots$ giving progressively *worse* rational approximants of τ. What will the spatial structure and related transport properties of the corresponding critical states be? To study this question, we consider the transmission and Lyapunov coefficients, respectively given by (see Sections 9.5.2 and 9.5.4)

$$T_N(\lambda^*) = \frac{1}{1 + (\alpha F_{n-2} - F_{n-1})^2}, \tag{5.32}$$

and

$$\Gamma_N(\lambda^*) = \frac{1}{N} \ln \left[2 + 4(\alpha F_{n-2} - F_{n-1})^2 \right], \qquad (5.33)$$

where, without any loss of generality, we have adopted the reference values $m_A = K_{AA} \equiv 1$. Then assigning different α_m values into (5.32) and (5.33) we can study the mass ratio dependence of T_N and Γ_N coefficients for different system lengths.

TABLE 5.1
Systematic variation of the transmission and Lyapunov coefficients with the mass ratio parameter α_m for the resonant normal mode $\lambda^* = 2$ corresponding to a Fibonacci lattice with $N = 2584$.

m	α_m	$T_N(\lambda^*)$	$L/2584$
1	1597/987	1.00000	1.44269
2	987/610	0.99999	1.44268
3	610/377	0.99999	1.44266
4	377/233	0.99993	1.44239
5	233/144	0.99957	1.44089
6	144/89	0.99685	1.42971
7	89/55	0.97928	1.36131
8	55/34	0.87245	1.05302
9	34/21	0.50000	0.55811
10	21/13	0.12755	0.29588
11	13/8	0.02072	0.19038

In Table 5.1 we summarize the results for a Fibonacci lattice with $N = 2584$, where $L = \Gamma^{-1}$ estimates the localization length of the corresponding states. In the first place, as α_m progressively worsens as a τ approximant, we observe a systematic *degradation* of the transport properties of the resonant state, which evolves from an extended character ($T_N \simeq 1$, $L/N > 1$) to a clearly localized one ($T_N \simeq 0.1$, $L/N < 1$). In the second place, we observe that the extended-localized transition is a relatively sudden episode, taking place in a narrow window of mass ratio values around the critical value $\alpha^* = \alpha_8$. This transition also occurs for other system lengths, although the precise value of α^* depends on N.

In Fig. 5.19 we show the power spectrum and the amplitude distributions of a critical normal mode undergoing this transition. The critical normal mode shown at the left-hand inset (α_7) has a high value of the transmission coefficient ($T_N \simeq 0.97$), and uniformly spreads through the Fibonacci lattice ($L/N \simeq 1.36$). Conversely, the transmission coefficient of the critical normal mode shown at the right-hand inset (α_8) has significantly decreased $T_N \simeq 0.84$

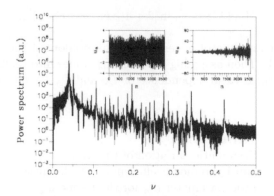

FIGURE 5.19

Power spectrum of the critical normal mode shown in the left-hand inset and atomic displacements in a Fibonacci lattice with $N = 2584$ at the resonance frequency $\lambda^* = 2$ for $\alpha_7 = 89/55$ (left-hand inset) and $\alpha^* = 55/34$ (right-hand inset). ([11] Reprinted figure with permission from Maciá E 1999 *Phys. Rev. B* **60** 10 032 © 1999 by the American Physical Society.)

and $L/N \simeq 1$, indicating a sudden stretching of its spatial extent. The overall structure of the power spectrum is analogous to that shown in Fig. 5.18, but a closer inspection reveals some interesting differences. Thus we observe a shift of the periodic-like peak position towards higher frequencies describing the presence of the long-range modulation amplitude. Conversely, the nested twin peak features broaden, undergoing a substantial shift towards the lower frequency region of the spectrum. Finally, the ratio $I_{QP}/I_P \simeq 10^{-3}$ increases by an order of magnitude, indicating the progressive relevance of the role played by the quasiperiodic contribution.

It is worth to note that the spatial structure of the critical normal mode shown in the left-hand inset exhibits a long-range (about 900 sites) amplitude modulation containing a series of higher frequency quasiperiodic oscillations of minor amplitude. This complex spatial modulation has been previously reported as a characteristic signature of wave propagation on quasilattices in a few experimental studies dealing with Rayleigh surface acoustic waves propagating on the quasiperiodically corrugated surface of a piezoelectric substrate ($LiNbO_3$),[74] and coherent acoustic phonons in GaAs/AlAs Fibonacci superlattices.[75]

Finally, we will briefly comment on the interesting behavior of the critical normal mode when the Fibonacci lattice satisfies the condition $\alpha F_{n-2} - F_{n-1} = 1$. In this case the amplitude distribution exhibits a peculiar signa-

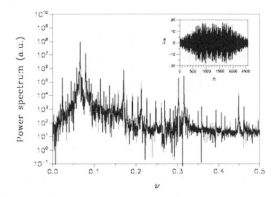

FIGURE 5.20
Power spectrum of the critical normal mode shown in the inset and atomic displacements in a Fibonacci lattice with $N = 2584$ and $\alpha_9 = 34/21$ at the resonance frequency $\lambda^* = 2$ (inset). ([11] Reprinted figure with permission from Maciá E 1999 *Phys. Rev. B* **60** 10 032 © 1999 by the American Physical Society.)

ture, where a complex arrangement of self-similar fluctuations of the normal mode amplitudes seems to be *modulated* by a broad, smooth envelope covering the entire system's length, as shown in the inset of Fig. 5.20. The overall structure of the corresponding power spectrum exhibits an intricate pattern, where a significant overlapping of different nested peaks occurs as a consequence of their progressive broadening. Notwithstanding this, we can clearly appreciate the significant influence of the quasiperiodic contribution over the periodic-like one, as indicated by the relatively high value of the ratio $I_{QP}/I_P \simeq 0.01$.

These results properly illustrate the *rich physical behavior* of critical states and the way the different spatial structures they display can affect their related transport properties. In fact, when studying band structure effects in the thermal conductivity of Fibonacci quasicrystals a great variety of *critical normal modes* are found. [10, 76] These modes exhibit quite different physical behaviors, which range from highly conducting extended states to critical states whose transmission coefficient oscillates periodically between two extreme values, depending on the system's length.[10, 51] In this sense, it is quite reasonable to assume that the transport properties of these critical normal modes are substantially affected by the quasiperiodic order of the underlying lattice.

Similar results concerning the existence of extended states in other kinds of self-similar structures, like Thue-Morse chains and hierarchical lattices, have been reported in the literature,[4, 77] and its role in the transport properties has been analyzed in detail in terms of multifractal formalism on the basis that fractal dimension is directly associated to the localization degree of the eigenstates.[78, 79] More precisely, the phonon diffusivity, D, is related to the spectrum effective bandwidth $S \equiv \bigcup_i \Delta\omega_i$, where $\Delta\omega_i$ denotes the length of each subband in the spectrum, by a power law of the form

$$S \sim L^{-D}, \tag{5.34}$$

where $L = F_n$ for Fibonacci lattices and $L = 2^n$ for Thue-Morse and period-doubling lattices, respectively, with n denoting the generation index of the sequences. Note that S is nothing but the Lebesgue measure of the energy spectrum (see Section 5.2), so that this result provides another example of the significant influence of the aperiodic order of the lattice on the transport properties.

5.6.3   Critical states in Koch fractal lattices

Prior to the discovery of quasicrystals it was suggested by some authors that fractal structures, which instead of the standard translation symmetry exhibit *scale invariance* (see Section 1.5), may be suitable candidates to bridge the gap between crystalline and disordered materials.[80] Such a possibility was further elaborated by subsequent works on inhomogeneous fractal glasses,[81] a class of structures which are characterized by a scaling distribution of pore sizes and a great variety in the site environments. On the other hand, the unexpected finding of quasicrystalline alloys exhibiting forbidden crystallographic symmetries was originally thought as corresponding to a phase intermediate between a crystal and a liquid, but subsequently interpreted as a natural extension of the notion of a crystal to structures with quasiperiodic, instead of periodic, translational order, as it has been explained in Chapter 2. From this perspective it is interesting to compare the physical properties related to these two novel representatives of the orderings of matter, namely, quasicrystals and fractals.

Albeit both kinds of structures possess peculiar electronic spectra supported by Cantor sets of zero Lebesgue measure [25, 80], it was earlier pointed out that electron dynamics on fractal substrates may be richer and more complex than that encountered in simpler quasiperiodic structures, like those described in terms of the on-site or transfer Fibonacci chains.[82] Thus, both numerical and analytical evidences of localized, critical, and extended wave functions alternating in a complicated way have been reported for several fractal models.[51, 83, 84, 85, 86, 87] In addition, it was reported that the interplay between the local symmetry and the self-similar nature of a fractal gives rise to the existence of persistent superlocalized modes in the frequency spectrum.[88] This class of states arises as a consequence of the fact that

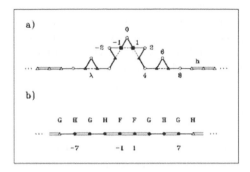

FIGURE 5.21
a) Sketch of the model Koch lattice considered. The different sites are labeled by integers. b) Sketch of the renormalization scheme mapping the Koch lattice into a linear chain. (Reprinted figure with permission from Maciá E 1998 *Phys. Rev. B* **57** 7661 © 1998 by the American Physical Society.)

the minimum path between two points on a fractal does not always follow a straight line.[89] Consequently, the general question regarding whether the nature of the states might be controlled by the fractality of the substrate is an interesting open question, well deserving further scrutiny in both quasi-periodic and fractal lattices.

As a suitable model example we shall consider the tight-binding model on the Koch lattice introduced by Andrade and Schellnhuber.[90] The motivation for this choice stems from the fact that this model Hamiltonian can be exactly mapped onto a linear chain, and the corresponding electron dynamics expressed in terms of just two kinds of transfer matrices. In this way we can use the same algebraic approach discussed in the study of electron and phonons in general Fibonacci systems in the previous sections. The model is sketched in Fig.5.21, and its tight-binding Hamiltonian is given by [90]

$$H = \sum_n \left\{ |n\rangle\langle n+1| + |n\rangle\langle n-1| + \lambda f(n) \left[|n-1\rangle\langle n+1| + |n+1\rangle\langle n-1| \right] \right\},$$

(5.35)

where λ is the cross-hopping integral introduced by Gefen [91] (indicated by dashed lines in Fig.5.21(a)) and

$$f(n) = \delta(0,n) + \sum_s^{k-1} \delta(\frac{4^s}{2}, n(\text{mod}4^s)),$$

(5.36)

with $k \geq 2$ and $-4^k/2 \leq n \leq 4^k/2$, describes the effective next-nearest-neighbor interaction in the kth stage of the fractal growth process. We shall consider a *finite* fractal lattice embedded in an infinite periodic arrangement of identical sites connected by hopping integrals h. The main effect of allowing electron hopping across the folded lattice is the existence of sites with different coordination numbers along the lattice, a characteristic feature of fractals which is not shared by quasiperiodic lattices. Depending on the value of their respective coordination numbers we can distinguish two-fold (circles), three-fold (full triangles), and four-fold (squares) sites. We then notice that even sites are always two-fold, a fact which allows us to renormalize the original lattice mapping it into the linear form sketched in Fig.5.21(b).[90] The hopping integrals represented by single bonds appear always isolated from one another. The hopping integrals represented by double bonds can appear either isolated or forming trimers. Consequently, there are three possible site environments in the renormalized Koch lattice which, in turn, define three possible types of transfer matrices, labelled **F**, **G**, and **H** in Fig.5.21(b). Now, by introducing the matrices $\mathbf{A} \equiv \mathbf{HG}$ and $\mathbf{AB} \equiv \mathbf{FF}$, it can be shown by induction that the global transfer matrix at any given arbitrary stage k of the fractal growth process, $\mathcal{M}_k$, can be iteratively related to that corresponding to the previous stage, $\mathcal{M}_{k-1}$, by the expression [90]

$$\mathbf{A}^{-1}\mathcal{M}_k = \mathcal{M}_{k-1}^2 \mathbf{B} \mathcal{M}_{k-1}^2, \qquad (5.37)$$

with $k \geq 2$ and $\mathcal{M}_1 \equiv \mathbf{A}$. In this way, the topological order of the lattice is translated to the transfer matrices sequence describing the electron dynamics in a natural way. The matrices $\mathbf{A}$ and $\mathbf{B}$ are *unimodular* (i.e., their determinant equals unity) for *any* choice of λ *and* for *any* value of the electron energy, E, and the commutator reads [51]

$$[\mathbf{A}, \mathbf{B}] = \frac{\lambda E(E^2 - 2)(2 + \lambda E)}{r^3} \begin{pmatrix} (2 - E^2)r & r^2 \\ (1 - E^2)(E^2 - 3) & (E^2 - 2)r \end{pmatrix}, \qquad (5.38)$$

where $r \equiv 1 + \lambda E$, and we have defined the energy scale in such a way that the hopping integrals along the chain equal unity. The commutator (5.38) vanishes in four different cases. i) The choice $\lambda = 0$ reduces the original Koch lattice to a trivial periodic chain, so that all the allowed states, $-2 \leq E \leq 2$, are extended. ii) The center of the energy spectrum, $E = 0$, which corresponds to an extended state.[90] iii) $E = \pm\sqrt{2}$. iv) The family of states satisfying $E = -2/\lambda$. For these energies the condition $[\mathbf{A}, \mathbf{B}] = 0$ is fulfilled, so that the global transfer matrix of the system, $\mathcal{M}_k \equiv \mathbf{A}^{n_A}\mathbf{B}^{n_B}$, with $n_A = 4^{k-1}+1$, and $n_B = 4^{k-2} + 1$, can be explicitly evaluated. From the knowledge of $\mathcal{M}_k$ the condition for the considered energy value to be in the spectrum, $|\text{tr}[\mathcal{M}_k]| \leq 2$, can be readily checked and, afterwards, its transmission coefficient can be explicitly determined.

Let us consider, in the first place, the energies $E = \pm\sqrt{2}$. In this case we get

$$\text{tr}[\mathcal{M}_k] = -\frac{1}{(1 \pm \sqrt{2}\lambda)^{n_A - n_B}} - (1 \pm \sqrt{2}\lambda)^{n_A - n_B}. \tag{5.39}$$

A detailed study of the condition $|\text{tr}[\mathcal{M}_k]| \leq 2$ in (5.39) indicates that the only allowed states correspond to $\lambda = \mp\sqrt{2}$, for which $\text{tr}\mathcal{M}_k = -2$. Consequently, these states are just two particular cases of the more general family (iv) which we shall discuss next.

By taking $E = -2/\lambda$ we get $\mathbf{B} = -\mathbf{I}$, where $\mathbf{I}$ is the identity matrix, so that $\mathcal{M}_k = -\mathbf{A}^{n_A}$. Making use of the Cayley-Hamilton theorem, the global transfer matrix corresponding to the energies $E = -2/\lambda$ can be expressed in terms of Chebyshev polynomials of the second kind, $U_m(x)$, with $x \equiv \cos\phi = -1 + 8\lambda^{-2} - 8\lambda^{-4}$, in the closed form

$$\mathcal{M}_k = \begin{pmatrix} U_{n_A-1} - U_{n_A} & -qU_{n_A-1} \\ qU_{n_A-1} & U_{n_A-2} - U_{n_A-1} \end{pmatrix}, \tag{5.40}$$

where $q \equiv 2(\lambda^2 - 2)/\lambda^2$. From expression (5.40) we get $\text{tr}[\mathcal{M}_k] = -2\cos(n_A\phi)$ (where we take into account the relationship $U_n - U_{n-2} = 2T_n$ between Chebyshev polynomials of the first and second kinds (see Section 9.3)) and, consequently, we can ensure that these energies belong to the spectrum in the fractal limit ($k \to \infty$). The transmission coefficient at a given iteration stage reads

$$T_k(\lambda) = \frac{1}{1 + \left[\frac{\lambda(\lambda \pm 2)}{2(\lambda \pm 1)} \sin(n_A\phi)\right]^2}, \tag{5.41}$$

where the plus (minus) sign in the factor of $\sin(n_A\phi)$ corresponds to the choices $h \equiv 1$ and $h \equiv r = -1$, respectively, for the hopping integral of the periodic leads. From expression (5.41) we realize that the transmission coefficients corresponding to the family (iv) are always bounded below for *any* stage of the fractal growth process, which proves their *extended* nature in the fractal limit. In addition, the choices $\lambda = \pm 2$ ($E = \mp 1$) correspond to states which are transparent at every stage of the fractal growth process, a fact which ensures their *transparent* nature in the fractal limit as well. Furthermore, it is possible to find a number of cross-hopping integral values satisfying the transparency condition $T_k = 1$ at certain stages of the fractal growth given by the condition $n_A\phi - p\pi$, which allows us to label the different transparent states at any given stage k, in terms of the integer p.

In Fig.5.22 we plot the transmission coefficient (5.41) at two successive stages $k = 2$ and $k = 3$, as a function of the cross-hopping value. In the first place, we note that the number of λ values supporting transparent states, ν_λ, progressively increases as the Koch curve evolves toward its fractal limit, according to the power law $\nu_\lambda = 2(4^{k-1} + 1)$. It is interesting to compare this figure with the number of sites, $N = 4^k + 1$, present, at the stage k, in the Koch lattice. Thus, we obtain $\nu_\lambda = (N + 3)/2$, indicating that the number of Koch lattices able to support transparent states increases linearly with the

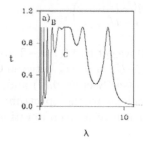 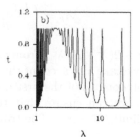

FIGURE 5.22

Transmission coefficient as a function of the cross-hopping integral at two different stages (a) $k = 2$, and (b) $k = 3$. Peaks are labelled from left to right starting with $p = -4$ in (a). Label B corresponds to $p = 0$. Label C indicates the transparent state at $\lambda = 2$. (Reprinted figure with permission from Maciá E 1998 *Phys. Rev. B* **57** 7661 © 1998 by the American Physical Society.)

system size and, consequently, that the fractal growth *favors* the presence
of extended states in Koch lattices. In particular, we can state that, in the
fractal limit, there exist an *infinitely numerable* set of cross-hopping integrals
supporting transparent extended states in the Koch lattice.

Another general feature shown in Fig.5.22 is the presence of a broad plateau
around $\lambda = 2$ where the transmission coefficients take values significantly close
to unity. In addition, as λ separates from the plateau the local minima in the
transmission coefficient, T_{min}, take on progressively decreasing values which
tend to zero in the limits $\lambda \to \infty$ and $\lambda \to 1$. This behavior suggests that
the *best* transport properties in the family (iv) should be expected for those
states located around the plateau.

Up to now we have shown that, as the Koch lattice approaches its fractal limit, an increasing number of cross-hopping integrals are able to support
transparent states in the $E = -2/\lambda$ branch of the phase diagram. But, for
a given value of λ, how many of the related extended states at an arbitrary
stage, say k, will prevail in the fractal limit $k \to \infty$? From a detailed analysis
of expression (5.41) it was found that the considered states can be classified
into two separate classes. In the first class we have those states which are
transparent at any stage k. In the second class we find states whose transmission coefficient *oscillates periodically* between $T = 1$ and a limited range of
$T_{min} \neq 1$ values depending on the value of the label integer p and the fractal
growth stage. Two general trends have been observed in this second class of
extended states. First, the values of T_{min} are significantly lower for states
corresponding to $p < 0$ than for states corresponding to $p > 0$. Second, at
any given fractal stage, the values of T_{min} are substantially higher for states
associated to cross-hopping integral values close to the plateau than for states
corresponding to the remaining allowed λ values. We must note, however, that
not all these almost transparent states are expected to transport in much the
same manner, as suggested by the diversity observed in the values of T_{min}.

In Fig.5.23 we provide a graphical account of the most relevant results obtained so far. In this Figure we show the phase diagram corresponding to
the model Hamiltonian given by Eq.(5.35) at the first stage of the fractal
process (shadowed landscape) along with two branches corresponding to the
states belonging to the family $E = -2/\lambda$ (thick black lines). In the way along
each symmetrical branch we find three particular states whose coordinates are
respectively given by $(\pm 1, \mp 2)$, $(\pm\sqrt{2}, \mp\sqrt{2})$, and $(\pm 2, \mp 1)$. Three of them,
corresponding to the choice $\lambda > 0$, are indicated by full circles labelled A,
B, and C in Fig.5.23. These states correspond to *transparent states* whose
transmission coefficients equal unity *at every* k. The remaining states in the
branches correspond to *almost transparent states* exhibiting an oscillating behavior in their transmission coefficients. By comparing Figs.5.22 and 5.23 we
realize that the positions of the transparent states A-B-C allow us to define
three different categories of almost transparent states according to their related transport properties. The first class (I) includes those states comprised
between the state A, at the border of the spectrum, and the state B, located

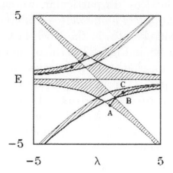

FIGURE 5.23

Phase diagram showing the Koch lattice spectrum at $k = 1$ (shadowed areas) and the branches corresponding to the extended states family $E = -2/\lambda$. (Reprinted figure with permission from Maciá E 1998 *Phys. Rev. B* **57** 7661 © 1998 by the American Physical Society.)

at a vertex point separating two broad regions of the phase diagram. The second class (II) includes those states comprised between the state B and the state C close to the plateau in the transmission coefficient around $\lambda = 2$. Finally, the third class (III) comprises those states beyond the state C. The states exhibiting *better* transport properties belong to the classes II and III, and correspond to those states grouping around the plateau near the state C for which the values of T_{min} are very close to unity. As we move apart from state C, the transport properties of the corresponding almost transparent states become progressively worse, particularly for the states belonging to the class III, for which values of T_{min} as low as 10^{-3} can be found.

In conclusion, we realize that in the study of general, aperiodic systems, eigenstates can be classified attending to two complementary criteria, namely, their spatial structure and the value of their related transmission coefficient. In the case of periodic crystals or amorphous materials both criteria are directly correlated. In periodic crystals we have Bloch states exhibiting a periodic spatial structure and $T = 1$ (transparent state). On the other hand, uncorrelated random systems have exponentially localized wavefunctions with $T = 0$ (localized states). Quasiperiodic crystals display a richer class of wavefunctions, generically referred to as critical states, exhibiting a modulated spatial structure and a broad diversity of transmission coefficient values $(0 < T \leq 1)$. Two points should be highlighted regarding the transmission coefficient values: i) since $T \neq 0$ in general, strictly localized states are not present in quasiperiodic systems; and ii) the very possibility of having $T = 1$ in some particular cases indicates that the notion of transparent state must be widened to include eigenstates which are not Bloch functions at all. Finally, fractal systems generally possess an even richer variety of wavefunctions, stemming from their inherent combination of self-similarity (a property shared with quasiperiodic chains) and fractal dimensionality (manifested by the presence of sites with different coordination numbers along the chain).

5.7 Transport properties of Fibonacci superlattices

5.7.1 dc conductance

The rich fractal structure displayed by the energy and frequency spectra of self-similar aperiodic systems should be reflected, to some extent, in its transport properties. Evidences supporting this to be the case were reported in some experimental works dealing with FSLs (see Section 4.3). Generally speaking there are two factors which must be taken into account in order to evaluate the relative importance of typical quasiperiodic effects on the perpendicular electronic transport of FSLs. On the one hand, since these effects are essentially quantum in nature, we must consider systems with strong coupling between adjacent blocks. For instance, in the FSL model introduced in Section 5.1 the effective coupling threshold between nearest-neighbor blocks is given by the condition $d < 3$ in Eq.(5.6). This condition is fulfilled by GaAs-GaAlAs superlattices (electron effective mass $m^* = 0.067$) with periods ranging from 70 Å to 340 Å and height barriers in the interval 4 meV to 100 meV, respectively. On the other hand, we should consider electron-phonon scattering effects which tend to disrupt coherent quantum transport. These effects crucially depend on the sample temperature and it may be confidently expected that their influence can be neglected at very low temperatures. In this case the relationship between the electrical conductance at zero temperature and the transmission coefficient, $T_N(E)$, is given by the well-known dimensionless single-channel Landauer formula (Eq.(9.87) in Section 9.5.3). The energy dependence of the transmission coefficient can be obtained in a straightforward manner in the transfer-matrix formalism by embedding the finite FSL in an infinite periodic lattice of identical blocks.

A representative example of the obtained results is shown in Fig.5.24, where one clearly appreciates a well differentiated trifurcation structure of conductance peaks, which mimics that corresponding to the energy spectrum approximating a prefractal Cantor-set structure.

In order to obtain realistic outcomes from the model, it is convenient to include finite-temperature effects. To this end we shall consider the expression given by Eq.(9.88) in terms of the transmission coefficient (Section 9.5.3). The

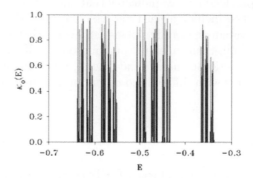

FIGURE 5.24

Landauer conductance at zero temperature for a FSL with $N = 987$, $d =$ 2.5, and $\lambda_B/\lambda_A = 0.9$. Energy measured in eV. ([16] Reprinted figure with permission from Maciá E, Domínguez-Adame F, and Sánchez A 1994 *Phys. Rev. B* **49** 9503 © 1994 by the American Physical Society.)

temperature dependence of the conductance curves will depend on the precise value of the chemical potential of the sample. The Fermi energy, E_F, of the free electron gas is related to the chemical potential μ, by the expression

$$\mu = E_F \left[1 - \frac{\pi^2}{12} \left(\frac{k_B T}{E_F} \right)^2 - \frac{\pi^4}{80} \left(\frac{k_B T}{E_F} \right)^4 + \ldots \right], \qquad (5.42)$$

where k_B is the Boltzmann constant. At low enough temperatures the chemical potential essentially coincides with the Fermi level, and we will use it as a free parameter in order to explore the main transport features of the considered FSL. An illustrative example is shown in Fig.5.25.

At high temperatures the $\kappa(T, \mu)$ curves saturate, reaching stable asymptotic values at about $k_B T \simeq 0.4$. This value is of the order of the contact's bandwidth $\Omega = 4e^{-d} \simeq 0.33$ eV, so that all available electrons contribute to the electronic transport in the superlattice growth direction in the high temperature regime, independently of the adopted μ value. On the contrary, the general form of the $\kappa(T, \mu)$ curve strongly depends on the adopted μ value at low temperatures. If the chemical potential is close to a set of transmission peaks the conductance exhibits several characteristic humps. This is the case of curve I. Conversely, if the chemical potential is located in a main gap region, the conductance monotonically increases with temperature to reach a limiting value and no relevant features are present at all, as seen in curve III. Finally, if the chemical potential lies in an intermediate region a pronounced broad

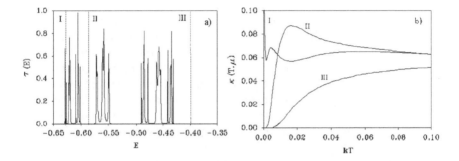

FIGURE 5.25

Influence of the energy spectrum on the finite-temperature conductance of a FSL. In (a) we show the transmission coefficient for a superlattice with $N = 55$, $d = 2.5$, and $\lambda_B/\lambda_A = 0.85$. The vertical dashed lines indicate the location of the chemical potential in three different cases: (I) $\mu = -0.628$, (II) $\mu = -0.585$, and (III) $\mu = -0.4$. The corresponding conductance curves as a function of the temperature are shown in (b). Energy is measured in eV. ([16] Reprinted figure with permission from Maciá E, Domínguez-Adame F, and Sánchez A 1994 *Phys. Rev. B* **49** 9503 © 1994 by the American Physical Society.)

hump occurs due to the contribution of nearest transmission peaks. Such a hump is shown in curve II. Thus, it is clear that the occurrence of conductance humps at low temperatures is intimately related to the fragmented nature of the energy spectrum. These different kinds of conductance behaviors can be related to the position of the chemical potential μ by means of the following expression in the low temperature limit [16]

$$\kappa(T, \mu) \simeq \tilde{\beta} \coth\left(\tilde{\beta}\Delta\right) \sum_{i=1}^{3} \frac{\gamma_i}{\cosh^2\left(\tilde{\beta}\delta_i\right)}, \qquad (5.43)$$

where $\tilde{\beta} \equiv (2k_B T)^{-1}$, $\delta_i \equiv E_i - \mu$, E_i denotes the position of the main peaks of conductance triplets, γ_i is the characteristic strength of each transmission peak, and Δ is the width of the triplet. Thus, once the chemical potential has been fixed, the evolution of the conductance curve is determined by four basic parameters. In consequence, Eq.(5.43) provides a suitable tool to obtain relevant parameters characterizing the fragmentation splitting pattern of the electronic structure from experimentally measured conductance curves fits. In the case of actual superlattices typical bandwidths are in the range $\Omega \simeq 100$ meV, so that the high temperature regime will be attained at $T \simeq \Omega/10k_B \approx 1100$ K (see Fig.5.25). Accordingly, conduction humps due to the first level of fragmentation may be observable at about $T \approx 200$ K, whereas conduction spikes due to the third level of splitting could be appreciated at temperatures below $T \approx 20$ K.

5.7.2   Disorder effects

From an experimental point of view, however, two major limitations appear to fully appreciate the characteristic fingerprints of fractal energy spectra from transport measurements in aperiodic superlattices. In the first place, it is not possible to fabricate *perfect* aperiodic structures. Interface roughness appears during growth in actual aperiodic superlattices: Protrusions of one semiconductor into the other cause in-plane disorder and break translational invariance parallel to the layers. Although x-ray diffraction studies show that the characteristic structural order of aperiodic superlattices is preserved under moderately large growth fluctuations (see Section 4.1), the way this robust aperiodic order can influence the transport properties of actual, defective systems deserves a close inspection. In fact, the observation of inhibition of vertical transport in periodic superlattices with intentional disorder[92] (in agreement with the theory of localization in one-dimensional disordered systems) suggests the possible existence of a competition between the long-range aperiodic order and the unintentional short-range disorder. On the other hand, only *finite* arrangements with a limited number of layers can be usually manufactured, even in the most favorable experimental conditions. In this sense, the observation of fragmentation patterns in the energy spectra of short aperiodic superlattices using different experimental techniques (see Section 4.3) spurs further analyses regarding whether fractal-like spectra are to be expected in actual systems with an increasingly large number of layers.

Two important questions then follow quite naturally: First, what are the effects of unintentional disorder in the splitting scheme of the energy spectrum of aperiodic superlattices? Second, what are the finite size effects on their fractal-like properties? As we will see, the presence of small fluctuations in the sequential deposition of layers considerably smear out the self-similarity property of the energy spectra on increasing the system size, as the coupling between electronic states is progressively reduced due to the loss of quantum coherence. Thus, fractal-like spectra with a richness of finer details such as those obtained from numerical analyses of ideal systems are not to be expected in realistic superlattices (see Section 5.8.2), though quantum coherence is still strong enough in short systems to give rise to a measurable hierarchical set of subminibands in the electronic spectra and to influence their related transport properties to some extent.

As a representative model system we consider a quantum-well based GaAs-$Ga_{0.65}Al_{0.35}As$ superlattice with the same barrier thickness $b = 32\,\text{Å}$ in the whole sample and two different well widths $a = b$ and $a' = 35\,\text{Å}$, respectively, arranged according to the Thue-Morse or Fibonacci sequences. The height of the barrier for electrons is given by the conduction-band offset ($250\,\text{meV}$) at the interfaces and the origin of electron energies is at the GaAs conduction-band edge. We will focus on electronic states close to the bandgap and neglect non-parabolicity effects hereafter, so that a one-band Hamiltonian suffices to describe those states. The effective masses are $m^*_{GaAs} = 0.067m$ and $m^*_{GaAlAs} = 0.096m$, m being the free-electron mass. Physical magnitudes of interest can be easily computed using a transfer-matrix formalism in this simple picture. Because the in-plane average size of structural defects depends on the growth conditions and it is unknown in most cases, one is forced to develop a simple approach, describing local excess or defect of monolayers by allowing the quantum-well thicknesses, Δz_n, to fluctuate uniformly around the nominal values $a + b$ or $a' + b$, according to the expressions $\Delta z_n = a(1 + W\varsigma_n) + b$ or $\Delta z_n = a'(1 + W\varsigma_n) + b$, where W is a positive parameter measuring the maximum fluctuation and ς_n's are distributed according to a uniform probability distribution $P(\varsigma_n) = 1$ if $|\varsigma_n| < 1/2$ and zero otherwise. Note that ς_n is a random uncorrelated variable, even when the lattice is constructed with the constraint that the mean values of Δz_n follow the aperiodic sequences. This approximation should be valid whenever the mean-free-path of electrons is much smaller than the in-plane average size of protrusions as electrons only detect micro-quantum-wells with small area and uniform thickness. Therefore, each micro-quantum-well presents a slightly different value of its thickness and, as a consequence, resonant coupling between electronic states of neighboring GaAs layers is decreased. To get an accurate description of electron dynamics, average over all possible configurations of disorder is indeed required because the number of interface defects as well as their mean

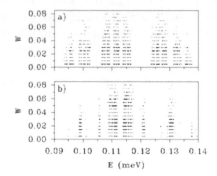

FIGURE 5.26

Phase diagram for a (a) GaAs-Ga$_{0.65}$Al$_{0.35}$As Fibonacci superlattice with $N = F_{11} = 144$ wells and (b) GaAs-Ga$_{0.65}$Al$_{0.35}$As Thue-Morse superlattice with $N = 128$ wells. The energy is measured from the conduction-band edge in GaAs. (From [94]. With permission from IOP Publishing Ltd.)

thickness vary from layer to layer.

Figure 5.26(a) shows the dependence of the energy spectrum structure on the amount of disorder, W. For a perfect ($W = 0$) FSL the overall structure of the energy spectrum is characterized by the presence of four main subbands. Inside each main subband the fragmentation pattern follows a trifurcation scheme in which each subband further splits from one to three subsubbands. Therefore, the energy spectrum of perfect and finite FSLs presents distinct pre-fractal signatures. The situation changes when randomness is introduced. In fact, although the tetrafurcation pattern of the perfect FSL still remains, the finer details corresponding to successive steps in the hierarchical splitting scheme are progressively smeared out on increasing the disorder due to growth fluctuations. Figure 5.26(b) shows the dependence of the spectrum structure with the degree of disorder present in a Thue-Morse superlattice. For a perfect system the fragmentation scheme agrees with that previously discussed in the framework of the Kronig-Penney model,[93] and displays pre-fractal signatures as well. Accordingly, the presence of the short-range disorder reduces the resonant coupling between quantum-wells and, consequently, it weakens the physical mechanism giving rise to the self-similar pattern.

To investigate the effects due to the competition between long-range aperiodic order and short-range disorder on the transport properties, one can evaluate the localization length, ℓ, as a function of the energy (Fig.5.27). For the perfect ($W = 0$) case one obtains a very spiky structure. Each peak cor-

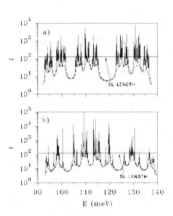

FIGURE 5.27

Localization length for (a) a perfect GaAs-Ga$_{0.65}$Al$_{0.35}$As Fibonacci superlattice with $N = F_{11} = 144$ wells (solid line) as compared to a disordered one ($W = 0.05$) of the same length (dashed line) and (b) the same comparison for a Thue-Morse superlattice with $N = 128$ wells. Averages are taken over 50 realizations of the superlattice. (From [94]. With permission from IOP Publishing Ltd.)

responds to a quasi-level, and most of them completely extend through the superlattice, their localization length being one order of magnitude greater than the system length. Note that the distribution of peaks reflects the overall fragmentation of the energy spectrum. On the contrary, when growth fluctuations are introduced, the localization length distribution becomes smoother and its value remains *always* smaller than the superlattice length, clearly revealing the onset of localization effects. This general trend is observed for both Fibonacci and Thue-Morse superlattices (TMSL), though some remarkable differences are also observed between both kinds of aperiodic superlattices. Thus, one observes that, in the perfect TMSL, there exists a significant number of states whose localization length is several orders of magnitude greater than the system size. This fact is related to the presence of extended states in the TMSL.[93] In addition, the mean value of ℓ is lower for the defective TMSL, hence indicating that the effects of localization are more intense in FSLs than they are for TMSLs with the same amount of disorder. Therefore, in presence of the same degree of growth fluctuations, TMSLs should exhibit better transport properties than FSLs. Finally, notice that the localization onset is more pronounced at the edges of the energy spectrum; meanwhile in the central regions of the spectrum (about 110 meV) the localization length almost equals the system size.

In summary, from Figs. 5.26 and 5.27 we see that moderate fluctuations in the sequential deposition of layers have significant effects on both the energy spectrum and the spatial extension of wave functions. The fractal-like nature of an arbitrary spectrum is determined by two complementary features. In the first place, the energy spectrum becomes more and more fragmented as the aperiodic superlattice length grows. The physical origin for this fragmentation stems from resonant tunneling effects between electronic states of neighboring quantum-wells (short-range effects). In the second place, the splitting scheme of the energy spectrum must display a self-similar pattern. The physical origin for this self-similarity can be traced back to the structural self-similarity of the superlattice itself which, in turn, is imposed by the aperiodic ordering of the system (long-range effects). Taking both facts into account we conclude that the purported robustness of the aperiodic order present in aperiodic superlattices is not sufficient, by its own, to guarantee the fractal-like nature of the energy spectrum *in the presence of disorder*, because the main effect of growth fluctuations is precisely to weaken the resonant coupling between electronic states. Hence, albeit the structural aperiodic order is preserved in the presence of moderate fluctuations, the self-similarity related to it can not be properly expressed, in its finer details, in the energy spectrum due to the loss of quantum coherence as a consequence of short-range effects.

The relative importance that this competition between long-range order and short-range disorder has on the transport properties depends critically on the length of the system. The values of the localization length shown in Fig.5.27 indicate that wave functions no longer spread over the whole aperiodic superlattice, as they do in the perfect case, but their degree of extension amounts to a significant fraction of the system size in contrast with the usual view of localized states extending over just a few wells. Thus, one can safely conclude that distinctive features of a fractal-like spectrum can be experimentally observed in aperiodic superlattices of practical interest, whose length is smaller than certain threshold length. The value of this threshold length will depend on the heterostructure quality attained during the growth process.

5.8 Beyond one-dimensional models

Most theoretical analyses discussed so far have been concentrated on simple single-band models in one dimension, and the effects of the full crystalline structure of each layer in a given aperiodic superlattice have remained unexplored. This topic will be briefly addressed in this concluding section.

5.8.1 Two-band models

An extended version of the single-band on-site model was considered in Ref.[95] according to the Hamiltonian given by

$$H = \sum_i |i\rangle \begin{pmatrix} \varepsilon_1^\alpha & 0 \\ 0 & \varepsilon_2^\alpha \end{pmatrix} \langle i| - \sum_{i,j} |i\rangle \begin{pmatrix} t_{11} & t_{12} \\ t_{21} & t_{22} \end{pmatrix} \langle j|, \qquad (5.44)$$

which describes a two-band nearest-neighbor tight-binding system. The on-site energies describing the two levels associated to each lattice site, ε_1^α and ε_2^α, are arranged according to the Fibonacci sequence, where the label α denotes the respective sites being A or B. The transfer integrals between the orbitals of nearest-neighbors sites are taken to be independent of the atomic species, in analogy with the single-band on-site model given by Eq.(5.3). Making use of the transfer-matrix formalism the corresponding Schrödinger equation can be expressed in the 4×4 matrix form

$$\begin{pmatrix} \psi_{1,n+1} \\ \psi_{2,n+1} \\ \psi_{1,n} \\ \psi_{2,n} \end{pmatrix} = \begin{pmatrix} \frac{t_{22}}{\Delta}(E - \varepsilon_1) & -\frac{t_{12}}{\Delta}(E - \varepsilon_2) & 1 & 0 \\ -\frac{t_{21}}{\Delta}(E - \varepsilon_1) & \frac{t_{11}}{\Delta}(E - \varepsilon_2) & 0 & 1 \\ 1 & 0 & 0 & 0 \\ 0 & 1 & 0 & 0 \end{pmatrix} \begin{pmatrix} \psi_{1,n} \\ \psi_{2,n} \\ \psi_{1,n-1} \\ \psi_{2,n-1} \end{pmatrix}, \qquad (5.45)$$

where E is the electron energy and $\Delta = t_{11}t_{22} - t_{12}^2$ describes an effective transfer integral which plays a role analogous to that played by the actual transfer integral t in the single-band transfer matrix element $(E - \varepsilon)/t$. This effective hopping term describes band hybridization effects involving the bands related to the $\varepsilon_1^{A,B}$ and $\varepsilon_2^{A,B}$ atomic levels, and it vanishes when the resonance condition $t_{12} = \sqrt{t_{11}t_{22}}$ is satisfied. Physically, this situation corresponds to

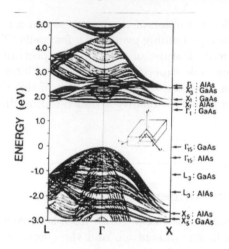

FIGURE 5.28

Band structure of the GaAs/AlAs Fibonacci superlattice composed of 144 GaAs layers and 89 AlAs layers. The position of the energy levels in the bulk GaAs and AlAs is indicated by arrows on the right-hand side. (From ref.[95]. With permission from IOP Publishing Ltd.)

the creation of an almost flat region in the energy dispersion curve and the corresponding increase in the electronic effective mass.

In Chapter 6 we will see that two-band Hamiltonians of the form given by Eq.(5.44), leading to 4×4 transfer matrices like Eq.(5.45), can be fruitfully applied to study charge migration in double-stranded DNA chains. In turn, double-chain quantum models have been recently proposed for studying some basic features of charge transport in icosahedral QCs.[96]

5.8.2   Three-dimensional effects

Hirose and co-workers considered a GaAs/AlAs Fibonacci superlattice composed of single monolayers of GaAs and AlAs by using a semi-empirical tight-binding method. The electronic state at the position **r** in the crystal lattice is expressed as a linear combination of atomic orbitals as[95]

$$\Psi_k^l(\mathbf{r}) = \frac{1}{\sqrt{N}} \sum_j^{2F_n} \sum_\alpha^5 C_{j,k}^{\alpha,l} \, e^{i\mathbf{k}\cdot\mathbf{R}_j} \phi_\alpha(\mathbf{r} - \mathbf{R}_j), \qquad (5.46)$$

where α denotes the orbital states s, p_x, p_y, p_z, and s*, j runs over the site number, N is the number of atoms in a unit cell, and l indicates the band index. The parameters were chosen to reproduce the experimentally determined band structures of bulk GaAs and AlAs, and second-nearest neighbor interactions were included in order to reproduce the L-point energies. The resulting band structure (in a plane perpendicular to the superlattice direction) is shown in Fig.5.28 for a FSL with 233 monolayers. As one appreciates both the valence and conduction bands are greatly fragmented. By comparing the superlattice band structure with the energy levels in the bulk one can distinguish two types of dispersion in the valence band which are respectively related to the bands originating from heavy and light holes, respectively. The heavy hole band (reaching the X point in the diagram) has a smaller dispersion and is divided into three subbands around the Γ point, as expected from the study of simpler FSL models. On the contrary, the light hole band is much wider and no clear fragmentation pattern can be observed. A similar band structure is appreciated for the conduction band, though the trifurcation pattern is not easily resolved in this case. Above 3.0 eV this splitting is clearest, but strong hybridization between GaAs s and AlAs p bands occur in the energy interval 2.5-3.0 eV, significantly smearing out the finer details. Accordingly, one concludes that self-similar structure is destroyed by hybridization effects in more realistic descriptions of the superlattice structure. This general trend was confirmed by subsequent studies considering several monolayers in each FSL slab (thickness effect) as well as including spin-orbit coupling in the treatment (in order to break the degeneracy in the heavy hole band).[97] In general, a characteristic trifurcation scheme is only found at some points of the superlattice Brillouin zone, mainly in the vicinity of the Γ point, and for some energy ranges. Certainly, the main reasons for the absence of such clear fragmentation patterns in the energy spectrum of more realistic systems stems from their intrinsic *complexity*. In fact, these models include a large number of model parameters: different on-site energies and transfer integrals for the different orbitals, the valence band offset and the three-dimensional crystal geometry of the layers, and the effect of quasiperiodicity have a different impact on each one of them.

The absence of clear self-similar features in the energy spectrum of realistic Fibonacci heterostructures has been extensively analyzed by considering the propagation of different kinds of elementary excitations like sagittal elastic waves or polaron optical modes in three-dimensional FSLs.[98] In these cases the waves have several components, whose dynamics couples with the aperiodic structure of the substrate in a complex way. For instance, sagittal waves have two different components in the sagittal plane formed by the propagation direction and the FSL growth direction. As a consequence of their mixing the energy spectrum becomes blurred, which makes difficult the possible identification of clear self-similar patterns, as those shown, for instance, in Fig.5.7 for simpler models.

The nature of the wave functions was analyzed by comparing some representative states of the sp^3s^* model with those corresponding to a single-band on-site model. In this way, it was concluded that the localization is enhanced by the interband hybridization in the sp^3s^* model, although the phase correlation characteristic of Fibonacci systems is not destroyed so that the wave function remains critical.[95] A more detailed study on thickness effects in the wavefunction localization in terms of the local DOS for some selected states showed the presence of selective spatial localization in the thickest slabs of the lowest conduction and highest valence band states, respectively.[97, 99]

References

[1] Lu J, Odagaki T and Birman J L 1986 *Phys. Rev B* **33** 4809

[2] Machida K and Fujita M 1986 *J. Phys. Soc. Jpn.* **55**, 1799

[3] Kohmoto M and Banavar J R 1986 *Phys. Rev. B* **34** 563

[4] Liu Y and Riklund R 1987 *Phys. Rev. B* **35** 6034

[5] Würtz D, Schneider T, and Politi A 1988 *Phys. Lett. A* **129** 88

[6] Wang C and Barrio R A 1988 *Phys. Rev. Lett.* **61** 191

[7] Chakrabarti A, Karmakar S N, and Moitra R K 1992 *Phys. Lett. A* **168** 301; 1994 *Phys. Rev. B* **50** 13276

[8] Oh G Y, Ryu C S, and Lee M H 1992 *J. Condens. Matter* **4**, 8187

[9] Kumar V 1990 *J. Phys. Condens. Matt.* **2** 1349

[10] Maciá E 2000 *Phys. Rev. B* **61** 6645

[11] Maciá E 1999 *Phys. Rev. B* **60** 10032

[12] Ashraff J A and Stinchcombe R B 1988 *Phys. Rev. B* **37** 5723

[13] Chakrabarti A, Karmakar S N, and Moitra R K 1989 *Phys. Rev. B* **39** 973; Ghosh A and Karmakar S N 1998 *Phys. Rev. B* **57** 2834

[14] Kats E I and Muratov A R 2005 *J. Condens. Matter* **17**, 6849

[15] Ghosh A 2001 *Eur. Phys. J. B* **21** 45

[16] Maciá E, Domínguez-Adame F, and Sánchez A 1994 *Phys. Rev. B* **49** 9503

[17] Maciá E, Domínguez-Adame F, and Sánchez A 1994 *Phys. Rev. E* **50** 679

[18] Domínguez-Adame F, Méndez B, Maciá E, and González M A 1991 *Mol. Phys.* **74** 1065

[19] Maciá E and Domínguez-Adame F 2000 *Electrons, Phonons, and Excitons in Low Dimensional Aperiodic Systems* (Editorial Complutense, Madrid)

[20] Albuquerque E L and Cottam M G 2003 *Phys. Rep.* **376** 225

[21] Albuquerque E L and Cottam M G 2004 *Polaritons in Periodic and Quasiperiodic Structures* (Elsevier, Amsterdam)

[22] Bellissard J, Bovier A, and Ghez J M 1993 *Differential Equations with Application to Mathematical Physics*, Eds. Ames W F, Harrell II E M, and Herod J V (Academic Press, Boston)

[23] Damanik D 1998 *Lett. Math. Phys.* **46** 303; Damanik D 1998 *Helv. Phys. Acta* **71** 667; Damanik D 1998 *Comm. Math. Phys.* **196** 477

[24] Baake M, Grimm U, and Joseph D 1993 *Int. J. Mod. Phys. B* **7** 1525

[25] Süto A 1989 *J. Stat. Phys.* **56** 525

[26] Bellissard J, Iochum B, Scoppola E, and Testard D 1989 *Commun. Math. Phys.* **125** 527

[27] Iochum B and Testard D 1991 *J. Stat. Phys.* **65** 715

[28] Bellissard J, Bovier A, and Ghez J M 1991 *Commun. Math. Phys.* **135**, 379; Bovier A and Ghez J M 1993 *Commun. Math. Phys.* **158** 45

[29] Bellissard J, Iochum B, and Testard D 1991 *Comm. Math. Phys.* **141** 353

[30] Bellissard J, Bovier A, and Ghez J M 1992 *Rev. Math. Phys.* **4** 1

[31] Luck J M 1989 *Phys. Rev. B* **39** 5834

[32] Dulea M, Johansson M, and Riklund R 1992 *Phys. Rev. B* **45** 105; 1992 *Phys. Rev. B* **46** 3296; 1993 *Phys. Rev. B* **47** 8547

[33] Maciá E 2006 *Rep. Prog. Phys.* **69** 397

[34] Bursill L A, Lin P J, and Xudong F 1987 *Mod. Phys. Lett. B* **1** 195

[35] Xudong F, Bursill L A, and Lin P J 1988 *Int. J. Mod. Phys. B* **2** 131

[36] Bursill L A 1990 *Int. J. Mod. Phys. B* **4** 2197

[37] Vekilov Y K, Isaev E I, and Arslanov S F 2000 *Phys. Rev. B* **62**, 14040; Vekilov Y K and Isaev E I 2002 *Phys. Lett. A* **300** 500

[38] Bovier A and Ghez J M 1995 *J. Phys. A: Math. Gen.* **28** 2313

[39] Halsey T, Jensen M H, Kadanoff L P, Procaccia I, and Shraiman B I 1986 *Phys. Rev. A* **33** 1141

[40] Kohmoto M 1988 *Phys. Rev. A* **37** 1345

[41] Rüdinger A and Piéchon F 1998 *J. Phys. A: Math. Gen.* **31** 155

[42] Kohmoto M 1983 *Phys. Rev. Lett.* **51** 1198

[43] Kohmoto M, Sutherland B, and Iguchi K 1987 *Phys. Rev. Lett.* **58** 2436

[44] Maciá E and Domínguez-Adame F 1995 *Physica B* **216** 53

[45] Domínguez-Adame F, Maciá E, Méndez B, Roy C L and Khan A 1995 *Semicond. Sci. Technol.* **10** 797

[46] Ostlund S, Pandit R, Rank D, Schellnhuber H J, and Siggia E D 1983 *Phys. Rev. Lett.* **50** 1873

[47] Kohmoto M, Sutherland B, and Tang C 1987 *Phys. Rev. B* **35** 1020

[48] Severin M, Dulea M, and Riklund R 1989 *J. Phys. Condens. Matter* **1** 8851

[49] Ryu C S, Oh G Y, and Lee M H 1992 *Phys. Rev. B* **46** 5162

[50] Gellermann W, Kohmoto M, Sutherland B, and Taylor P C 1994 *Phys. Rev. Lett.* **72** 633

[51] Maciá E 1998 *Phys. Rev. B* **57** 7661

[52] Maciá E and Domínguez-Adame F 1996 *Phys. Rev. Lett.* **76** 2957

[53] Kumar V and Ananthakrishna G 1987 *Phys. Rev. Lett.* **59** 1476

[54] Xie X C and Das Sarma S 1988 *Phys. Rev. Lett.* **60** 1585; Ananthakrishna G and Kumar V 1988 *Phys. Rev. Lett.* **60** 1586

[55] Sil S, Karmakar S N, Moitra R K, and Chakrabarti A 1993 *Phys. Rev. B* **48** 4192

[56] Niu Q and Nori F 1990 *Phys. Rev. B* **42**, 10 329; 1986 *Phys. Rev. Lett.* **57**, 2057

[57] Aubry S and André S 1980 *Ann. Isr. Phys. Soc.* **3** 133

[58] Mott N and Twose W D 1961 *Adv. Phys.* **10** 107

[59] Schreiber M and Grussbach 1992 *Phil. Mag. B* **65** 707

[60] Aoki H 1982 *Physica A* **114** 538; 1983 *J. Phys. C: Solid State Phys.* **16** L205

[61] Soukoulis C and Economou E N 1984 *Phys. Rev. Lett* **52** 565

[62] Castellani C and Peliti L 1986 *J. Phys. A: Math. Gen.* **19** L429

[63] Schreiber M 1985 *Phys. Rev. B* **31** 6145

[64] Aoki H 1986 *Phys. Rev. B* **33** 7310

[65] Oono Y, Ohtsuki T, and Kramer B 1989 *J. Phys. Soc. Jpn.* **58** 1705

[66] Schreiber M and Grussbach 1992 *Mod. Phys. Lett. B* **6** 851

[67] Ostlund S and Pandit R 1984 *Phys. Rev. B* **29** 1394

[68] Fujiwara T, Kohmoto M, and Tokihiro T 1989 *Phys. Rev. B* **40** 7413

[69] Sire C 1994 *Lectures on Quasicrystals*, Eds. Hippert F and Gratias D (Les Editions de Physique, Les Ulis)

[70] Roche S, Trambly de Laissardière G, and Mayou D 1997 *J. Math. Phys.* **38** 1794

[71] Huang X and Gong C 1998 *Phys. Rev. B* **58** 739

[72] Kalman D and Mena R 2003 *Mathematics Magazine* **76** 167

[73] Oviedo-Roa R, Pérez L A, and Wang Ch. 2000 *Phys. Rev. B* **62** 13805; Sánchez V, Pérez L A, Oviedo-Roa R, and Wang Ch. 2001 *ibid.* **64** 174205; Sánchez V and Wang Ch 2004 *ibid.* **70** 144207

[74] Desideri J P, Macon L, and Sornette D 1989 *Phys. Rev. Lett.* **63** 390

[75] Mizoguchi K, Matsutani K, Nakashima S, Dekorsy T, Kurz H, and Nakayama M 1997 *Phys. Rev. B* **55**, 9336

[76] Maciá E and Domínguez-Adame F 1998 *Proceedings of the International Conference on Aperiodic Crystals*, Eds. de Boissieu M, Currat R, and Verger-Gaugry J L (World Scientific, Singapore)

[77] Chakrabarti A, Karmakar S N, and Moitra R K 1995 *Phys. Rev. Lett.* **74** 1403

[78] Anselmo D H A L, Dantas A L, and Albuquerque E L 2005 *Physica A* **349** 259

[79] Naumis G G 1999 *Phys. Rev. B* **59** 11315

[80] Rammal R 1983 *Phys. Rev. B* **28** 4871

[81] Schwalm W A and Schwalm M K 1989 *Phys. Rev. B* **39**, 12 872; 1993 *Phys. Rev. B* **47**, 7847

[82] Andrade R F S and Schellnhuber H J 1991 *Phys. Rev. B* **44** 13 213.

[83] Kappertz P, Andrade R F S, and Schellnhuber H J 1994 *Phys. Rev. B* **49** 14 711

[84] Lin Z and Goda M 1994 *Phys. Rev. B* **50** 10 315

[85] Chakrabarti A 1999 *Phys. Rev. B* **60** 10576; 1996 *J. Phys.: Condens. Matter* **8** 10951

[86] Sengupta S, Chakrabarti A, and Chattopadhyay S 2004 *Physica B* **344** 307

[87] Chakrabarti A and Bhattacharyya B 1996 *Phys. Rev. B* **54** 12625; 1997 *Phys. Rev. B* **56** 13768

[88] Jayanthi C S and Wu S Y 1993 *Phys. Rev. B* **48** 10 188

[89] Lévy Y E and Souillard B 1987 *Europhys. Lett.* **4** 233

[90] Andrade R F S and Schellnhuber H J 1989 *Europhys. Lett.* **10** 73

[91] Gefen Y, Mandelbrot B B, and Aharony A 1980 *Phys. Rev. Lett.* **45** 855

[92] Chomette A, Deveaud B, Regreny A, and Bastard G 1986 *Phys. Rev. Lett.* **57** 1464; Pavesi L, Tuncel E, Zimmermann B, and Reinhart F K 1989 *Phys. Rev. B* **39** 7788

[93] Ryu C S, Oh G Y, and Lee M H 1993 *Phys. Rev. B* **48** 132

[94] Maciá E and Domínguez-Adame F 1996 *Semicond. Sci. Technol.* **11** 1041

[95] Hirose K, Ko D Y K, and Kamimura H 1992 *J. Phys. Condens. Matter* **4** 5947

[96] Landauro C V and Janssen T 2007 *J. Non-Cryst. Solids* **353** 3192

[97] Arriaga J and Velasco V R 1997 *J. Phys. Condens. Matter* **9** 8031; *Physica A* **241** 377

[98] Fernández-Álvarez L and Velasco V R 1998 *Phys. Rev. B* **57** 14141; Zárate J E, Fernández-Álvarez L and Velasco V R 1999 *Superlatt. & Microstruct.* **25** 519; Pérez-Álvarez R, García-Moliner F, Trallero-Giner C and Velasco V R 2000 *J. Raman Spectrosc.* **31** 421

[99] Velasco V R and García-Moliner F 2003 *Prog. Surf. Sci.* **74** 343

6

The aperiodic crystal of life

The detailed mechanism by means of which a gene or a virus molecule produces replicas of itself is not yet known. (...) It might happen, of course, that a molecule could be at the same time identical with and complementary to the template on which it is moulded. (...) If the structure that serves as a template (the gene or virus molecule) consists of, say, two parts, which are themselves complementary in structure, then each of these parts can serve as the mould for the production of a replica of the other part, and the complex of two complementary parts thus can serve as the mould for the production of duplicates of itself. (Linus Pauling, 1948 [1])

There was an even more remarkable suggestion by Linus, which he made some time in the late 1940s — that the gene might consist of two mutually complementary strands. (Francis Crick, 1992 [2])

Despite the clear picture of how the gene may be duplicated, Pauling failed to discover the structure of DNA. Instead he proposed an incorrect three-stranded model of DNA structure with phosphates in the middle and bases on the outside. He had used as a clue for this proposal the manner in which phosphate groups form a helical array in certain minerals. (Alexander Rich, 1995 [3])

It was really partly a matter of bad luck, because although he was hoping to get some good x-ray patterns himself, he used the old x-ray patterns that Astbury had taken of DNA. We know now that they were a mixture of the two forms: the A and the B form. So he was using data which did not correspond to any real single structure. (Francis Crick, 1992 [4])

FIGURE 6.1
An early x-ray photograph of DNA taken by W. T. Astbury in the 1930s. Two characteristic features stemming from the stacked bases are indicated by arrows. (Adapted from [5, 6]).

6.1 The double helix

6.1.1 Diffraction by helices

Shortly after the pioneering introduction of x-ray techniques by van Laue and Bragg, structural analyses were soon devoted to the study of complex organic polymers able to condense in a partially ordered phase. Thus, cellulose (the main component of the cell wall in most plants) or keratin (a protein present in natural hair or wool fibers) were considered in the 1920s and 1930s; and diverse DNA samples in the late 1940s and early 1950s. In this way, it was progressively realized that helical arrangements may play a significant role in a growing number of fibrous polymers. In Fig.6.1 we show one of the first diffraction pictures of a nucleic acid taken by William T. Astbury (1898-1961) in 1936. Although the image is quite blurry, two characteristic features at the north and south regions of the meridian axis suggest a spacing of about 3.34 Å along the fiber axis, which was properly interpreted as

> revealing a close succession of flat nucleotides standing out perpendicularly to the long axis of the molecule to form a relatively rigid structure. [5]

This was not still a helix, however, but in 1949 Sven Furberg, a research student working under John Bernal, proposed a DNA structure based on a single-strand helix (with the sugar units perpendicular to the bases) in his Ph. D. thesis.[7] This orientation of the units in a nucleotide subsequently proved correct and was a considerable help to Francis H. C. Crick (1916-2004) and James D. Watson (1928) in their final model.[8] A specific helical diffraction theory was then required in order to properly interpret the obtained x-ray diffraction patterns. The demand for a theoretical framework was pressing when quite specific models, based on a judicious combination of x-ray diffraction data and stereochemical information about acceptable bond lengths, bond angles, and hydrogen bonding interactions, were proposed for several filamentous proteins by Pauling and his collaborators.

The full theory of diffraction by helices, including a proper formula for the form factor, was first reported and successfully applied to explain the diffraction patterns of certain synthetic polypeptides by W. Cochran, F. Crick, and V. Vand.[9] A version of the formalism was independently developed by Alexander R. Stokes (1919-2003), but it was published in 1955,[10] after the double-helix crucial discovery years.[11] In what follows, we shall briefly introduce some basic results of the helix diffraction theory, which are convenient to properly understand the most significant features of DNA structure and biological functionality. Let us consider a wire of infinitesimal thickness curled around to form a uniform helix of infinite length, constant radius R, and pitch P, given by the equations

$$x = R\cos\varphi(z) \tag{6.1}$$
$$y = R\sin\varphi(z)$$
$$z = z$$

where $\varphi(z) = 2\pi z/P$, and z measures the distance along the helix axis. The value of the Fourier transform at a point $\mathbf{q}$ in the reciprocal space is given by

$$\mathcal{F}(\mathbf{q}) = A\int f(\mathbf{r})e^{i\mathbf{q}\cdot\mathbf{r}}d\mathbf{r}, \tag{6.2}$$

where A is an appropriate constant, $f(\mathbf{r})$ is the electron density through the scattering helix, $\mathbf{r} = x\mathbf{i} + y\mathbf{j} + z\mathbf{k}$ is the position variable given by Eq.(6.1), and the integral is taken over all space. Assuming $f(\mathbf{r}) \equiv f_0$ for a uniform helix, and expressing the volume element as $d\mathbf{r} = \pi R^2 dz$, Eq.(6.2) can be written in the form,

$$\mathcal{F}(\mathbf{q}) = Af_0\pi R^2 \int \exp\left[i(q_x R\cos\varphi + q_y R\sin\varphi + zq_z)\right]dz. \tag{6.3}$$

In order to exploit the cylindrical symmetry of the helix it is convenient to express the components of the reciprocal space vector $\mathbf{q}$ in cylindrical coordinates, so that $q_x = q_r\cos\psi$ and $q_y = q_r\sin\psi$, where $q_r = \sqrt{q_x^2 + q_y^2}$ and $\tan\psi = q_y/q_x$. In this way, Eq.(6.3) can be rewritten as[9]

$$\mathcal{F}(\mathbf{q}) = Af_0\pi R^2 \int \exp\left[i\left\{Rq_r\cos(\varphi - \psi) + zq_z\right\}\right]dz. \tag{6.4}$$

A helix curve can be seen as a composition of a translation along the Z axis plus a uniform rotation about this axis. Since the helix repeats itself at a distance P in the Z direction, the scattering is analogous to that of a one-dimensional diffraction grating with spacing P. Therefore, the diffraction pattern of a uniform helix in reciprocal space occurs along a series of equidistant lines (whose spacing is determined by the helix pitch) rather than the Bragg spots one obtains from a three-dimensional crystal. These lines, referred to as layer lines, are at right angles to the Z axis in reciprocal space, and they are given by the series (Fig.6.2),

$$q_z(n) = 2\pi \frac{n}{P}, \tag{6.5}$$

with $n \in \mathbb{Z}$. Accordingly, Eq.(6.4) can be rearranged in the form [12]

$$\mathcal{F}(\mathbf{q}) = F_0 \sum_n \mathcal{I}_n \delta \left(q_z - 2\pi \frac{n}{P} \right), \tag{6.6}$$

where $F_0 \equiv A f_0 P R^2 / 2$, and we have introduced the auxiliary integral

$$\mathcal{I}_n = e^{in\psi} \int_0^{2\pi} e^{iu \cos(\varphi - \psi)} e^{in(\varphi - \psi)} d\varphi, \tag{6.7}$$

where $u \equiv R q_r$. This integral can be evaluated by using the identity

$$\int_0^{2\pi} e^{ix \cos \theta} e^{in\theta} d\theta = 2\pi i^n J_n(x), \tag{6.8}$$

where $J_n(x)$ denotes the nth-order Bessel function of the first kind.[13] Thus, adopting $\theta \equiv \varphi - \psi$ in Eq.(6.7), and taking into account the identity $i^n = e^{in\pi/2}$, one finally obtains[9]

$$\mathcal{F}(\mathbf{q}) = 2\pi F_0 \sum_n J_n(u) e^{in(\psi + \pi/2)} \delta \left(q_z - 2\pi \frac{n}{P} \right). \tag{6.9}$$

For a given value of n, Eq.(6.9) gives the amplitude and phase of the x-ray scattering on the nth layer line. The intensity of the diffraction peaks is given by $I_n(\mathbf{q}) = |\mathcal{F}_n(\mathbf{q})|^2 \propto |J_n(u)|$, which is independent of the angle ψ and the resulting pattern has cylindrical symmetry. In addition, from the mathematical properties $J_{-n}(u) = (-1)^n J_n(u)$, and $J_n^2(-u) = J_n^2(u)$, $n = 0, 1, 2 \ldots$, one realizes that the diffraction pattern $I_n(\mathbf{q}) \propto |J_n(u)|$ will be symmetric with respect to the q_r and q_z axes, respectively (Fig.6.2).

Thus, the presence of Bessel functions is a natural consequence of the cylindrical symmetry in helical diffraction, where they replace the trigonometric functions one finds in usual diffraction by three-dimensional lattices. Bessel functions characteristically begin with a strong peak and then oscillate like a damped sine wave as the argument increases. The position of the first strong

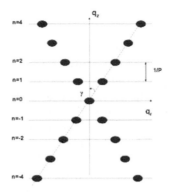

FIGURE 6.2

Diffraction pattern of a continuous helix of radius R and pitch P. Main diffraction spots are arranged along a series of lines (layer lines) labeled by integer values n. The lines are perpendicular to the meridian axis q_z and are separated by a distance $1/P$. The characteristic cross-shaped pattern stems from the symmetry properties of Bessel functions.

peak depends on the order of the Bessel function. For $n = 0$ the Bessel function attains its maximum in the middle of the diffraction pattern (i.e., for $u = 0$). For $n \neq 0$ the position of the first strong peak occurs when the argument is approximately equal to n, so that

$$q_r(n) \simeq \frac{n}{R}. \tag{6.10}$$

On the other hand, one finds that the order of the Bessel function occurring on a certain layer line, say k (counted from the middle of diffraction pattern), is just the same as the layer line, that is, J_k. Then, since the order of the Bessel function progressively increases with the layer line, the first strong peak progressively shifts from the meridian axis in the diffraction pattern giving rise to a cross-like pattern, which is a characteristic feature of diffraction by a helix (Fig.6.2). The semi-angle γ of the cross can be obtained from the ratio

$$\tan \gamma = \frac{q_r(n)}{q_z(n)} = \frac{P}{2\pi R}, \tag{6.11}$$

which takes the same value for any value of n. Eq.(6.11) relates the helix radius with its pitch value (which can be directly determined from the layer lines separation), so that one can obtain R from the γ value measured from the diffraction picture. In addition, Eq.(6.11) has the following geometrical meaning: as a point on the helix moves through one pitch, its z value changes by

P, and its projection on the XY plane travels round a circle of circumference $2\pi R$, so that the rise angle φ of the helix in physical space is directly related to the diffraction pattern cross geometry through the relationship $\varphi \simeq \gamma$.[8]

Now, actual DNA helices are not continuous at the atomic level, but they consist of repeating groups of atoms clustered in a number of basic structural motives (i.e., phosphate groups, sugars, nucleobases, counterions, and water molecules), so that they must be properly regarded as helical polymers rather than helical coils. A helical polymer is one in which the monomer units are aligned along a regular circular helix and are all equivalent under the symmetry operations of the helical screw axis. The symmetry properties of a discontinuous helix can be characterized by measuring the distance between two successive repeating motives (e.g., sugar-phosphate groups) in the macromolecule (the so-called rise per residue p) along with the angle you must turn between one motif and the next. As a consequence, the structure is characterized by two superimposed periods, namely P and p. For instance, for B-DNA with eleven nucleotides in the basic unit cell, one has $P = 34$ Å and $p = 3.4$ Å. The diffraction properties of the resulting structure can be described in terms of two one-dimensional diffraction gratings aligned along the same direction (namely, the Z axis), but with a fixed relative displacement between them. Accordingly, the corresponding layer-lines are now given by

$$q_z(n, m) = 2\pi \left(\frac{n}{P} + \frac{m}{p} \right), \tag{6.12}$$

instead of Eq.(6.5), and Eq.(6.9) generalizes to

$$\mathcal{F}(\mathbf{q}) = 2\pi F_0 \sum_{n,m} J_n(u) e^{in(\psi + \pi/2)} \delta \left(q_z - 2\pi \frac{n}{P} - 2\pi \frac{m}{p} \right). \tag{6.13}$$

When P/p can be expressed as a ratio of whole numbers the main effect of shifting from a continuous to a discontinuous helix is to introduce new helix crosses in the diffraction pattern, with their origins displaced up and down the meridian axis of the diffraction pattern by a distance $1/p$ (Fig.6.3). Thus, we obtain a denser set of layer-lines which are mutually separated a distance l/P' in reciprocal space. For a given vale of l, the intensity along the corresponding layer-line is governed by Eq.(6.13) where the various integer couples (n, m) must be chosen to satisfy the selection rule

$$\frac{l}{P'} = \frac{n}{P} + \frac{m}{p}. \tag{6.14}$$

This creates a diamond-like shape in the diffraction pattern, where the meridian diamonds are nearly void of intensity (Fig.6.3). For instance, in the case corresponding to B-DNA, $P' = P = 10p$, and Eq.(6.14) reads $l = n + 10m$. The layer-lines labeled by the integer n describe the contribution due to the sugar-phosphate backbones, which can be regarded as a nearly continuous

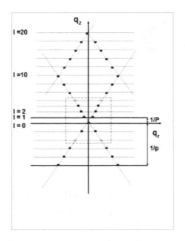

FIGURE 6.3

Ideal diffraction pattern of a DNA double-helix with $p = 0.34$ nm, $P = 3.4$ nm, $R = 1$ nm, and $\varphi = \pi/5$. The layer-lines satisfy the condition $l = n + 10m$, where allowed n and m values are determined by the sugar-phosphate and nucleobase electronic density distributions, respectively. The dashed area covers the reciprocal space region shown in Fig.6.2. Note the empty features arising from destructive interference effects due to the presence of two simultaneously diffracting helices in double-stranded DNA (see Fig.6.4).

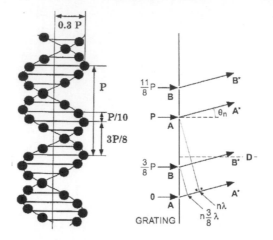

FIGURE 6.4

The basics of diffraction by two linear gratings mutually displaced by a fixed amount explains the extinction of the $n = 4$ layer-lines corresponding to the B-DNA molecule sketched on the left. (Adapted from [14].)

helical distribution of electron density. Although the bases are irregular in their chemical composition and geometrical structure, they scatter coherently because they behave like thin slabs of almost constant electronic density seen edge-on by the x-ray beam propagating perpendicularly to the fiber axis. The discrete contribution due to the bases is given by the layer-lines labeled by m integers, so that the spectra due to the bases are much more widely spaced than the layer-lines of the backbone, and in practice only the $m = 1$ spectrum is observed (Fig.6.3). Therefore, x-ray patterns do not give the sequence of bases, but only the average separation between successive bases along the helix axis.

On the other hand, when P/p cannot be expressed as a ratio of whole numbers, layer-lines for all values of n and m are present, filling the whole reciprocal space.[9] It should be noted that this situation corresponds to an incommensurate structure, such as those described in Section 1.3, and it is remarkable that the very possibility of a reciprocal-space filling diffraction pattern was explicitly mentioned in the original paper by Crick and co-workers two decades before the systematic study of incommensurate structures by de Wolff (see Section 1.3).

Previous calculations can be readily extended to the case of two identical helices of pitch P sharing a common symmetry axis, mutually separated by a distance αP along the helical direction. The diffraction by this double-helix structure can be described in terms of two identical diffraction gratings on the same straight line, but with a fixed relative displacement (Fig.6.4). In order to constructively interfere, successive pairs of rays scattered from A centers in Fig.6.4 must have a path length difference equal to an integer multiple of their common wavelength, i.e., $P \sin \theta_n = n\lambda$, where n indicates the order of the considered layer-line. Similarly, the grating composed of B centers will produce exactly the same diffraction pattern. Now, although both gratings give the same amplitude function, they are shifted with respect to each other. Therefore, there is a varying phase difference between the two functions, so that interference phenomena occur.[14] As a consequence, the intensities of the resulting x-ray diffraction pattern vary from one layer-line to another. In the case of B-DNA $\alpha = 3/8$ (note the presence of two Fibonacci numbers in the quotient), so that the path length difference between the rays scattered by A and B centers is $\Delta = 5n\lambda/8$. For $n = 4$, this amounts to $5\lambda/2$, so that the resulting amplitude of the waves scattered by one grating is π out of phase with that of the other, and the resulting total amplitude is zero. Thus in the x-ray pattern of a double-stranded helix there would be a characteristic zero of intensity for the $n = 4$ line. In actual DNA samples, however, the intensity of this layer is not exactly zero, because the two helices run in opposite sense and the cancellation is not complete (Fig.6.6).

So far we have considered the diffraction by a single double-helix molecule. In actual fiber diffraction experiments one usually has an ensemble of helical molecules rather than an isolated one in the volume inside the x-ray beam. In that case one must consider the location of each molecule with respect to the laboratory frame XYZ, say $\mathbf{R}_v$. The intensity of the diffraction pattern due to the phosphate groups from a double-stranded helical molecules ensemble is then given by [15, 16]

$$I(\mathbf{q}) \propto N \sum_{n,m=-\infty}^{\infty} \delta_{q_z,l} \cos^2(n\phi) J_n^2(u) + \delta_{q_z,0} J_n^2(u) \sum_{v \neq \mu} \left\langle e^{i\mathbf{Q}.(\mathbf{R}_v - \mathbf{R}_\mu)} \right\rangle \quad (6.15)$$

where N is the number of molecules in the x-ray beam, l is given by Eq.(6.14), and 2ϕ is the azimuthal angle between strands in the double helix. The first term in Eq.(6.15) describes intramolecular scattering and its physical information has been previously described. The second term in Eq.(6.15) describes the intermolecular scattering, and it only contributes in the equatorial plane ($q_z = 0$) in practice.

6.1.2 Unveiling double-stranded DNA

When one takes x-ray data one gets essentially the three-dimensional Fourier components of the electron density. But, in so doing, one only gets half of the information, since experiments provide the amplitudes of the Fourier components but not their phases. In order to surmount such a difficulty Pauling realized that one may fruitfully combine data provided by diffraction techniques with information gained from the study of suitable chemical constraints such as bond distances and angles. In fact, Pauling nicely illustrated

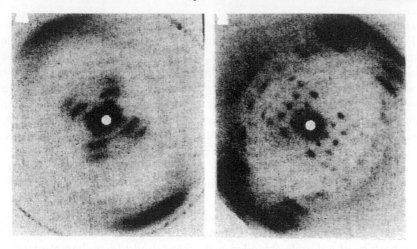

FIGURE 6.5

X-ray diffraction patterns from fibers at 98% (left) and 79% (right) relative humidity, illustrating typical B- and A- patterns, respectively. (From Ref.[17]. With permission.)

the capabilities of this approach by solving the three-dimensional structure of the so-called α-helix, present in a number of proteins. Watson and Crick, who developed the duplex model for DNA, used the methods that Pauling had developed, paying careful attention to bond angles, distances, and hydrogen bonding interactions. The chemical clues obtained in this way were associated with the diffraction pattern produced by DNA molecules, ultimately leading them to the correct structure. The method of solution adopted by Crick and Watson was essentially of trial and error, since the number of diffraction data was much fewer than the number of parameters needed to fully describe the structure, and an explicit solution was an impossibility. It was necessary instead to postulate plausible models and calculate their diffraction patterns, comparing these with observation.

> Rosalind Franklin was reluctant to use (our) approach because she wanted to use the method by Patterson, which is a difficult one. For that reason she concentrated on the A form of DNA which gave much better spots and much more information, and put aside the B form, even though she had produced a picture which essentially gave the game away and showed that the structure was helical. (Francis Crick [18])

A major discovery reported by Rosalind Franklin (1920-1958) and Raymond Gosling (1926) was that there existed two main structural forms of fibrous DNA in the biological samples considered by them, which were referred to as A and B, respectively (Fig.6.5).[19] The so-called A-DNA type patterns correspond to low hydrated samples and were first obtained by Maurice H. F. Wilkins (1916-2004) and Raymond Gosling in 1950, though they were not initially published in print. Franklin and Gosling also discovered that it was possible to reversibly pass from A to B form and vice versa by changing the relative water content of the fiber, the A form being drier (i.e., 75% relative humidity) and the B form wetter (i.e., 90-95% relative humidity).[19] The spectacular reversible change in the overall pattern of spots corresponding to the A and B forms of DNA was taken as reflecting a reversible change of the internal structure of DNA molecules themselves, mediated by the interaction of the DNA molecules with surrounding water molecules. Indeed, a more perfect crystallinity could be frequently attained by strong drying followed by re-wetting, a process which mimics usual thermal treatments (annealing) to remove defects in crystals. In this way, it was realized that Astbury pictures (Fig.6.1) were much more blurred than those obtained by Franklin and Gosling because it turned out that his fibers were in a mixed A and B state.

The A form consists of microcrystalline regions, in each of which the DNA molecules are arranged in a periodic fashion. Thus, the A form DNA fiber exhibits long-range order leading to interference effects between the x-rays scattered by different molecules in the same microcrystal, as prescribed by Eq.(6.15). The resulting diffraction patterns display a significant number of relatively sharp spots near the center of the pattern (Fig.6.5). On the contrary, in the B form the helix axes are all aligned, but the relative orientations of the molecules around the helix axis are randomly distributed through the sample, so that there are not constructive interference effects between the x-ray scattered by different molecules, and the observed pattern is that from a single molecule. As a consequence, the diffraction pattern of the B form is simpler and can be readily interpreted in terms of the theory introduced in Section 6.1.1. In fact, a clear succession of layer-lines can be clearly appreciated, and the intensity along the nth line is proportional to the square of the Bessel function J_n, as prescribed by Eq.(6.13). Quite ironically, the now celebrated B-DNA type pattern, taken in May 1952 by Franklin and Gosling (Fig.6.6), was set aside for nearly a year, because it was considered to be too simple to provide enough crystallographical information!

Notwithstanding this, four basic structural parameters of the B-DNA helix can be obtained by closely inspecting Photo 51 in the light of the results presented in Section 6.1.1:

1. from the relative distance between two successive layer-lines one gets the helix pitch ($P = 3.4$ nm)

2. making use of Eq.(6.11) one gets the helix radius ($R = 1$ nm) from the meridian angle of the cross, γ

3. from the vertical diagonal of the meridian diamonds (see Fig.6.3) one gets the nucleotide repeat ($p = 0.34$ nm)

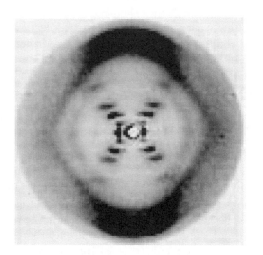

FIGURE 6.6

Diffraction pattern of the sodium salt of DNA fibers extracted from a thymus gland and purified, known as Photo 51. The material in the fibre comprises a macroscopic number of long, roughly parallel, negatively charged DNA strands, Na^+ counter ions, and a relatively high quantity of solvating water (up to ten water molecules per nucleotide). The solvated cations and water molecules are in an amorphous state and cause diffuse, background scattering. ([11] Reprinted by permission from Macmillan Publishers Ltd.: *Nature* **171** 740 ©1953.)

4. by combining the results obtained in 1 and 3 one gets the number of nucleotides by helix turn $n_b \equiv P/p = 10$

In addition, the two strong spots on the equator ($q_z = 0$) in the diffraction pattern shown in Fig.6.6 correspond to the first order of intermolecular scattering described by the second term in Eq.(6.15). They give the smallest Q at which $\mathbf{Q}.(\mathbf{R}_\nu - \mathbf{R}_\mu) = 2\pi s$, $s = 0, \pm1, \pm2, \ldots$. Their location provides the interaxial spacing d between nearest neighbors helices in the x-ray beam according to the expression

$$Q = \frac{4\pi}{\sqrt{3}d}, \tag{6.16}$$

which corresponds to an hexagonal packing of the helices. Higher order diffraction peaks cannot be seen because this packing does not extend to cover a very long-range and due to the relatively small number of molecules in the fiber as well.[16]

On the other hand, the study of the A-DNA patterns, exhibiting a significant number of sharper diffraction spots, indicated that the A-DNA crystal lattice pertained to the monoclinic space group ($a = 2.2$ nm, $b = 4.0$ nm, $c = 2.8$ nm, $\beta = 97°$, centered on ab faces) with n_c molecules per unit cell oriented along the c axis. The structure was nearly a hexagonal close packing of long rod-like molecules, the deviations from exact hexagonal symmetry arising from interactions between the grooves of the backbones of neighboring molecules pressed to each other by the partial removal of water (in comparison with the hydrated B-DNA form).[19] In fact, the water content of a sample at 75% relative humidity shows that there are 8 molecules of water per nucleotide, whereas one finds up to 20 molecules of water per nucleotide in the B-DNA form. From the knowledge of the unit cell volume $V = 1.22 \times 10^{-26}$ m^3, determined from x-ray measurements, Franklin and Gosling determined the value of n_c by using the measured density of the A-DNA fiber $\rho = 1471$ kg m^{-3}, through the relationship

$$n_c = \frac{\rho V}{nM}, \tag{6.17}$$

where $n = 11$ is the number of nucleotides by helix turn, and $M \simeq 475 \times 1.66 \times 10^{-27}$ kg is the mean molecular weight of a nucleotide (including a nucleobase, a sugar, a NaPO$_4$ phosphate, and eight water molecules). By plugging the numerical values in Eq.(6.17) one obtains $n_c = 2.07$, hence indicating the presence of two DNA molecules per unit cell. This fact, along with the presence of a binary symmetry axis perpendicular to the ac plane of a monoclinic crystal, led Crick to conclude that the DNA molecule itself should also have a two-fold symmetry axis perpendicular to the backbone. But since the sugar-phosphate backbone of nucleic acids was known to be an oriented polymer (due to the asymmetric attachments of each pentose to its two phosphate neighbors), the only possible way to have a two-fold symmetric molecule with such polar strands was to assume that it possessed an *even* number of strands, arranged along the same direction in complementary sense. In this way, Pauling's conjecture opening this Chapter, originally based on purely stereochemical motivations, was addressed on an entirely crystallographical basis.

As it has been previously indicated, in the study of biomolecules the solution of the structure cannot be completely determined from x-ray measurements alone, but one usually also needs some help from chemistry. By 1950 Erwin Chargaff (1905-2002) discovered an important regularity: although the sequence of bases along the DNA chains was complex, and the base composition of different DNA's varied considerably among different species, the numbers of adenine and thymine bases were always equal in all studied cases, and so were the numbers of guanine and cytosine bases.[20] Inspired by this remarkable finding, one of the keys to DNA molecular structure was the discovery that, if the bases were joined in pairs by hydrogen-bonding, the overall dimensions of the pairs of adenine and thymine and of guanine and cytosine were identical. This meant that a DNA molecule containing these pairs would be structurally regular (i.e., there would exist a common and fixed radius value) in spite of the sequence of bases being aperiodically distributed. In addition, the chemical constraint imposed by this base pairing implies that the sequence of bases in one chain runs in opposite direction to that in the other. As a result, one chain is complementarily identical to each other if turned upside down. This symmetry operation nicely fits with the crystallographical data obtained from a close analysis of A-DNA microcrystals leading to the presence of a two-fold symmetry axis, as previously mentioned.

Finally, the base-pairing restriction requires that A in one chain must be linked to T in the other, and similarly G to C. Thus, the sequence along one chain can vary without restriction (providing a physical realization of the one-dimensional aperiodic crystal, endowed with full coding capabilities, originally proposed by Schrödinger), but the sequence in the other chain is completely determined by Chargaff's rule, and is said to be complementary to it. The deep biological implications of this complementarity in order to allow for efficient self-similar assemblies was readily noted by the scientific community. Thus, biological functionality relies on a nice combination of chemical diversity (the different chemical flavours of the bases allowing for huge informative capabilities when properly ordered in specific aperiodic sequences) and chemical specificity due to the complementary interaction between purines and pyrimidines mediated by hydrogen-bonding.

In summary, two different kinds of order coexist at the same structural level in DNA: the helicoidal arrangement of sugar-phosphate groups (periodic order component) and the stacking of nucleobase pairs along the helix axis (aperiodic order component). X-ray diffraction techniques are able to reveal most structural features related to the periodic component, but the possibility of determining the base sequence by x-ray diffraction analysis is prevented by several factors. First, the blurry appearance of the broad spots at the north and south regions of the meridian axis does not allow for an accurate enough analysis. Second, diffraction techniques lack the chemical resolution necessary to properly discriminate the chemical nature of the different nucleobases. In fact, the commonly used Sanger sequencing method relies on chemistry to read the bases G, A, C, and T in DNA. Nevertheless, this method is still much too slow and costly for reading the personal genetic codes.[21] Thus, the proliferation of the large-scale DNA-sequencing projects for applications in clinical medicine, health care, and criminal research has driven the quest for alternative methods to reduce time and cost. This quest has spurred new perspectives in nanotechnology looking for methods entirely based on physical principles allowing for non-invasive analysis of a huge number of nucleotides along the DNA strands. Accordingly, new technologies for low-cost genome sequencing must be evaluated not only by their accuracy but also by the length of the genome fragments which can be sequenced at once. Among the different proposals which have been discussed in the literature during the last few years, in the following sections we will explore those approaches based on the study of the electrical and thermal response of nucleotides for potential DNA sequencing.[22, 23]

6.2 Electronic structure of nucleic acids

Physicists are used to considering a number of properties of matter such as electrical and thermal conductivity, optical or magnetic responses following the application of electromagnetic fields, generation of electrical currents in response to external thermal gradients, and so on. In most considered cases the samples studied in these experiments have no significant biological role, but in some instances (e.g., when considering proteins or DNA), they certainly have. In that case, when macromolecules of biological interest are considered from the viewpoint of condensed matter physics, a fundamental question naturally arises regarding the potential role of certain physical properties on the biological functions of these macromolecules. Thus, the role of charge migration in DNA mutation repair has been extensively discussed during the last decade, and the possible existence of correlation effects in electrical conductivity due to the presence of long-range spatial correlations in DNA has been also explored in detail. In this way, the field of transport properties in aperiodic systems brings useful concepts and approaches to the fundamental study of the possible relationship between information storage (determined by the order of appearance of nucleotides) and physical properties directly related to the electronic structure of nucleic acids.

To this end, a number of *ab initio* calculations based on the density functional theory have been performed during the last decade.[24, 25, 26, 27] The case of the homopolymers poly(dG)-poly(dC) and poly(dA)-poly(dT) has

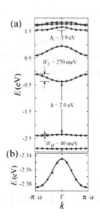

FIGURE 6.7

Energy bands close to the Fermi level as a function of the wave vector k of a polyG-polyC molecule in dry conditions. In the plot results obtained from *ab initio* calculations (dots) are compared to those derived from a one-dimensional tight-binding model with one orbital per unit cell (curve). Δ indicates the HOMO-LUMO gap, Δ_i the gap between closest orbitals in the guanine system (relevant to optical transitions), $W_{H(L)}$ are the HOMO (LUMO) bandwidths, respectively.[24] (Courtesy of Emilio Artacho).

FIGURE 6.8
Surfaces of constant charge density for the states corresponding to the lowest unoccupied band and highest occupied band of a polyG-polyC molecule in the A form in dry conditions.[24] (Courtesy of Emilio Artacho).

been extensively considered, along with some related structures like poly(GC)-poly(CG).[28] In order to reduce the computational effort earlier calculations did not take explicitly into account either the water shell or the cations around the sugar-phosphate backbone. Accordingly, these preliminary works focused on the dry A-DNA electronic structure. Close to the Fermi level it shows well defined, narrow bands separated by a broad gap (2-3 eV). The valence bands in A-poly(dG)-poly(dC) and A-poly(dA)-poly(dT) consist of 11 states, that is, one per base pair in the unit cell. In the case of poly(dG)-poly(dC) the topmost valence band has a very small bandwidth (Fig.6.7). This band is associated with the π-like highest occupied molecular orbital (HOMO) of the guanines. The charge density of the states associated with this band appears almost exclusively on the guanines, with negligible weight either in the backbones or in the cytosines (Fig.6.8). The lowest conduction band is significantly broader and it is made of the lowest unoccupied molecular orbital (LUMO) of the cytosines. Similar results are obtained for A-poly(dA)-poly(dT) chains, where the charge density appears concentrated on the HOMO orbitals of the adenines, and exhibit a broader valence band width (~ 0.25 eV).[26]

The spatial separation of the HOMO and LUMO orbitals in the purines and pyrimidines, respectively, can be understood as a direct consequence of the Watson-Crick hydrogen-bonding interaction, which is by far not a weak interaction. In fact, let us suppose we have two DNA bases infinitely separated from each other. Their HOMO and LUMO will both reside on the respective residues, and one can consider these orbitals are degenerate. Now, let us

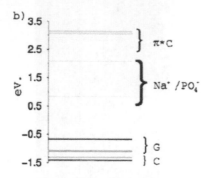

FIGURE 6.9

Schematic energy level diagram around the Fermi level of a fully hydrated double-stranded polyG-polyC molecule in the Z conformation. The Fermi level positioned in the middle of the gap has been chosen as the zero of energy. The HOMO (located at about -0.6 eV) is associated to guanines. The states immediately below the top of the valence band are also related to G. The first C localized state is located at 0.78 eV below the top. The bottom of the conduction band is a charge transfer state related to Na^+ counterions and PO_4^- groups. The first excited state with a strong C base character (π^*) is located at 2.85 eV above the Fermi level and the first π^* G state is at 3.18 eV. The $\pi \rightarrow \pi^*$ gap is 3.94 eV for cytosine and 3.82 eV for guanine bases. ([25] Reprinted figure with permission from Gervasio F, Carloni P and Parrinello M 2002 *Phys. Rev. Lett.* **89** 108102 © 2002 by the American Physical Society.)

bring the bases nearer to each other, so as to switch on the complementary H-bonding. This interaction will lift the level degeneracy, so that the frontier orbitals will be shifted towards each other. As a consequence, the HOMO of the whole complex will reside on one of the bases, whereas the LUMO on its Watson-Crick partner.[29]

The phosphate groups of the DNA molecule are negatively charged. Hence positive protons or metal cations (usually referred to as counterions) are necessary to neutralize and stabilize DNA in physiological conditions. Water also plays a crucial role to this end. Hydrophobic forces compel DNA to adopt the B-form, and the polarity of the water molecules helps screen DNA's charges. The comparison of the electronic structures corresponding to dry DNA structures with those obtained for wet conditions shows that the LUMO location is quite sensitive to the environment conditions. Thus the inclusion of Na^+ cations evenly distributed through the backbone gives rise to the presence of a

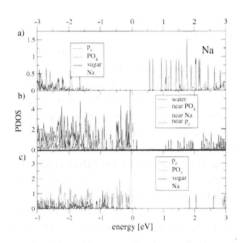

FIGURE 6.10

Effects of sodium counterions and water on the molecular orbitals of a short (5'-GAAT-3') B-DNA molecule. Panels (a) and (b) show the projected DOS of wet DNA on the bases' molecular orbitals (p_z), on the phosphates (PO_4), and on the sugars. Panel (c) contains the same for dry DNA (i.e., all water molecules removed). A rather large $\pi-\pi^*$ gap is observed between the guanine HOMO and the thymine LUMO. Some energies associated with water and counterions orbitals appear in the $\pi - \pi^*$ gap, which are rather close to the LUMO. (From ref.[27]. Courtesy of Robert G. Endres.)

band related to the Na-phosphate groups between the π-electron bands of the base molecules, so that the LUMO moves from cytosines to the phosphate-cations system when in presence of Na^+ for both A-poly(dA)-poly(dT) and A- poly(dG)-poly(dC).[26] A similar effect takes place when water is included along with cations, and the gap value is increased due to the significant role of water in shielding the DNA from the electrostatic field of counterions (Figs.6.9 and 6.10).

Accordingly, the water shell and the counterions can lead to the presence of a number of states in the main $\pi - \pi^*$ energy gap (which can be regarded as impurity states), hence effectively doping the DNA molecule. This effect is substantially reduced in longer B-type DNA chains, since the Na-Na nearest-neighbor distance is about three times as long as that of the A-type DNA. In addition, in the A-type, the Na ion is located inside the backbone structure, whereas in the B-type it lies in the outermost part.

The main features of the electronic structure obtained from numerical results have been experimentally confirmed by means of some spectroscopic techniques.[30, 31, 32] In particular, it has been confirmed that the HOMO originates in the DNA bases, in agreement with numerical calculations, for both polyG-polyC and polyA-polyT duplexes forming a mixture of A- and B-DNA forms.[30] It has been also demonstrated that when holes are doped in polyG-polyC by chemical oxidation the doped hole charge is localized on G, but not on cytosine, deoxyribose, or phosphates.[32]

Before concluding this section some words are in order regarding the possible role of aperiodicity in the electronic structure of DNA. In fact, as we commented at the end of Section 6.1.2, the spatial arrangements of the different nucleobases can not be determined from x-ray measurements, so that it would be appealing to explore the possibility of gaining useful information about the nucleobases sequence from electronic structure related properties. Numerical calculations comparing the DOS of double-stranded DNA (containing up to 200 bps) of a human oncogene and a random sequence constructed in the proportion A:C:G:T = 1:1:1:1 indicated that the overall electronic structure of aperiodic DNA chains is rather sequence independent, and the valence and conduction bands originate from the HOMO and LUMO orbitals of the single bases, in complete analogy with the results obtained from the study of periodic chains.[33] The main difference between both cases is the large value of the $\pi - \pi^*$ energy gap obtained for aperiodic chains, probably related to the presence of four different nucleobases which mimics alloying effects in solids. Thus, attending to their electronic structure one should expect a semiconductor-like behavior for both periodic and aperiodic DNA chains, though the particular outcome of a given experiment will significantly depend on environmental conditions (determining the presence and distribution of doping levels due to solvated counterions in the sugar-phosphate backbone) as well as the relative alignment between the DNA molecular orbitals and the Fermi level at the contacts. As we will discuss in Section 6.4.5 these states can play a significant role for understanding room temperature thermally activated charge transfer in the DNA molecule.

6.3 Charge transfer in DNA: What experiments say

A. Szent-Györgyi (...) has come to the idea in 1941 that proteins have to be conductors. This hypothesis was first not accepted because of the too large gap and non-periodic nature of a protein chain (János Ladik [34])

Following a suggestion by Szent-Györgyi (1893-1986),[35] a number of experiments measuring the electrical conductivity of dry proteins and nucleic acids were performed in the 1960s yielding positive results that were interpreted in terms of the classical expression for semiconductor conduction

$$\sigma = \sigma_0 \exp\left(-\frac{E_g}{k_B T}\right) \quad (\Omega \text{ cm})^{-1}, \tag{6.18}$$

with the fitting parameters listed in Table 6.1.

TABLE 6.1
Energy gap width, E_g, asymptotic electrical conductivity, σ_0, and room temperature conductivity values for some biopolymers.[36, 37]

SAMPLE	E_g (eV)	σ_0 $(\Omega \text{ cm})^{-1}$	$\sigma(300 \text{ K})$ $(\Omega \text{ cm})^{-1}$
DNA	0.90	$\sim 10^8$	8×10^{-8}
myosin	0.88	5×10^2	8×10^{-13}
collagen	0.45	4×10^{-3}	1×10^{-10}

For the sake of comparison we recall that the conductivity values of different conducting polymers range from almost metallic values for heavily doped polyacetylene ($\sim 10^4$ $(\Omega \text{ cm})^{-1}$) to semimetallic values for polythiophene (40-500 $(\Omega \text{ cm})^{-1}$), and semiconducting values for polysquarines ($1\text{-}10^{-5}$ $(\Omega \text{ cm})^{-1}$) at room temperature.[38, 39] In those early measurements some specimens were in crystalline state and others in the powdered and fibrous state. In the later case the samples were prepared in the form of pellets compressed between brass electrodes. In this case, the results are certainly influenced by charge transport between polymer strands in close proximity to each other, which cast some doubts about the reliability of these experiments in order to extract information about the intrinsic conduction properties of the polymers themselves.[40]

First attempts to infer charge transfer from fluorescence measurements on double-stranded DNA spurred a great interest. In those studies a donor and an acceptor moiety are attached to the DNA at a given distance, and upon photo-excitation a single charge carrier (usually a hole) is injected into the chain, travels the distance, and finally recombines at the acceptor site. Thus, the DNA conductivity was assessed from charge transfer rates as a function of the distance between the donor and acceptor sites, suggesting that duplex DNA is somewhat more effective than proteins as a medium for charge transfer but it does not function as a molecular wire.[41]

In recent years, a number of experimental measurements aimed to directly probe the electric current as a function of the potential applied across the DNA molecules have been reported.[42, 43, 44, 45, 46, 47, 48, 49, 50, 51, 52, 53] These experiments are performed in a variety of conditions, where important factors including DNA-substrate interaction, contact effects with the electrodes, relative humidity, the spatial distribution of counterions, and the nucleotide sequence nature (i.e., periodic or aperiodic one) are not kept constant. This state of affairs considerably makes difficult a proper comparison among different experimental reports, which range from completely insulating to semiconducting and even superconducting behaviors.[27] For the sake of illustration in Table 6.2 we list some results reported for the same kind of DNA sample by different research groups.

TABLE 6.2
Different electrical
conductivity values
reported in the
literature for $\lambda-$
phage DNA samples
measured under
different conditions.

$\sigma \ (\Omega \ \mathrm{cm})^{-1}$	REF.
800	[44]
$1 - 3$	[52]
$10^{-4} - 10^{-5}$	[43]
$< 10^{-6}$	[53]
$< 10^{-6}$	[50]

From the collected data three main conclusions can be drawn. First, long DNA samples of biological origin are typically more insulating than short synthetic oligomers, generally exhibiting a semiconducting behavior. Second, by all indications the structure of the DNA helix when deposited on dry surfaces may be very different from that found by crystallization of DNA in solution,

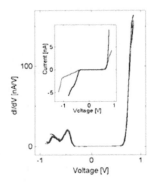

FIGURE 6.11
Differential conductance versus applied voltage at 100 K for a 10.4 nm long double-stranded poly(dG)-poly(dC) DNA chain trapped between two platinum electrodes that are 8 nm apart. The differential conductance shows a clear peak structure. Good reproducibility can be seen from the six nearly overlapping curves. The inset shows two typical I-V curves. ([48] Reprinted by permission from Macmillan Publishers Ltd.: *Nature* **403**, 635 © 2000.)

so that the DNA-substrate interactions are critical in determining the conductivity of an immobilized molecule, generally leading to poor conductivity. Third, the role of contacts deserves a particular attention. In many measurements, contact with metal electrodes was achieved by laying down the molecules directly on the electrodes. In this case, it is rather difficult to prove that the DNA molecule is in direct contact with the electrodes. Even so, the weak physical adhesion between DNA and metal may produce an insulating contact.

Figure 6.11 shows the characteristic I-V curve for a 10.4 nm-long (30 bp), double stranded poly(G)-poly(C) DNA molecule connected (by physical adhesion) to two metal (Pt) nanoelectrodes. The measurements were performed at temperatures ranging from room temperature down to 4 K. The general shape of the obtained current-voltage curves was preserved for tens of samples (though their details varied from curve to curve) and they indicated a relatively large-bandgap (∼2 eV) semiconducting behavior. The possibility of ionic conduction was ruled out by measurements that were performed in vacuum and at low enough temperature, where no ionic conduction is possible. The voltage dependence of the differential conductance exhibits a peaked structure, which is suggestive of the charge carrier transport being mediated by the molecular energy bands of DNA.[48]

A more elaborated experimental approach allows for measurement of the electrical conductivity in a systematic way. The experimental set-up is formed by well-characterized monolayers of 5' end thiol-modified single stranded DNA over a gold surface. The complementary DNA chain, which is also modified with a thiol at its 5' end, is attached onto a gold nano particle of 10 nm in diameter (Fig.6.12a). Hybridization of the two strands yields an insulating ssDNA monolayer in which some dsDNA chains can be easily identified by the gold nanoparticle connected to them. A conductive tip is then used to form a contact to the gold nano-particle, and through this contact the I-V curves are measured. It is estimated that up to 10 dsDNA molecules can connect simultaneously between the gold-nanoparticle and the underlying gold surface although it is likely that the number of connected DNA molecules is smaller. The measured I-V curves show a characteristic semiconductor behavior (Fig.6.12b), and show current values of the order of 220 nA at 2.0 V. Subsequent experiments with smaller gold nanoparticles (about 5 nm in diameter), able to simultaneously attach two or three molecules only, have confirmed that single short dsDNA chains can support up to $\sim$ 70 nA.[54]

In Fig.6.13 measurements of the electrical transport through short ds-DNA chains (8-14 bps) in aqueous solution (where the B-DNA native form is favored) are shown. The contact is formed through a thiolated chemical bond between the electrode (Au) and the DNA molecule, whose 3' end has been modified with a C_3H_6SH linker. In the same buffer solution a gold STM tip, which is covered with an insulating layer over most of the tip surface except for its end, is brought into contact (Fig.6.13a). Once contact is formed the tip is pulled backwards and the resulting current is monitored with a piezo-electric transducer. Distinct steps at integer multiples of 1.3×10^{-3} G_0 (where $G_0 = 1/12906$ Ω^{-1} is the conductance quantum) were observed in the current when the tip is pulled away, which were interpreted as consecutive breaking of DNA strands that connect both electrodes. Then, to measure transport through a single DNA molecule, the tip retraction is halted at the position of the last step and one measures the characteristic I-V curve. The curves obtained through three different single molecules show a rather smooth ohmic profile, coinciding with the average values of conductivity obtained from the pulling experiments. The measurement approach allows to accumulate larger statistics (i.e., over 500 individual measurements) than most previous experiments. Similar measurements were performed on DNA duplexes of the form 5'-CGCG(AT)$_m$CGCG-3', where some GC bps are replaced by AT ones, in order to analyze sequence effects on the transport properties. The conductance histograms also reveal well-defined peaks, which are located near integer multiples of 7.5×10^{-5} G_0 and 3.6×10^{-6} G_0, for $m = 1$ and $m = 2$, respectively. The conductance data can be described by an expression of the form $G = Ae^{-\beta L}$, where L is the length of the AT bridge, with $A = (1.3 \pm 0.1) \times 10^{-3}$ G_0 and $\beta = 0.43 \pm 0.01$ Å^{-1}.[56] These findings are consistent with a tunneling process across AT regions between the GC domains and properly demonstrate the existence of sequence dependence effects.

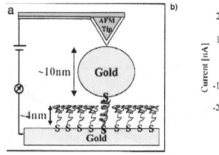
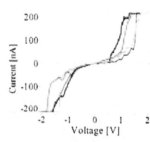

FIGURE 6.12

(a) Schematic diagram of the system layout. A smoothed gold substrate is covered with a packed monolayer (∼ 1 nm thick) of 26-bp long single-stranded DNA molecules of a non-periodic sequence (5'-CAT TAA TGC TAT GCA GAA AAT CTT AG-3'), chemically connected to the substrate via a propyl-thiol end group. Complementary strands, similarly connected to a 10 nm gold particle, are hybridized with the monolayer single strands. The gold particles are contacted by means of a metal covered AFM tip to close the electrical circuit. (b) A set of current-voltage curves that were measured on different gold nano particles on different samples, tips, and dates. One can appreciate as a common feature the presence of a small semiconducting gap.(From ref.[55]. With permission © 2005 National Academy of Sciences, USA.)

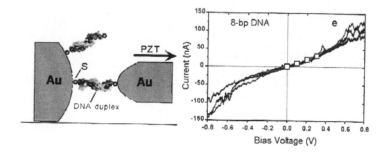

FIGURE 6.13

(a) Schematic illustration of a single ds-DNA conductance measurement. (b) Current-voltage characteristic curves of three different ds-DNA samples containing 8 bps [5'-(GC)$_4$-3'-thiol linker]. Lines are obtained by recording current versus bias voltage. The open squares are obtained from the peak positions of the conductance histograms at different bias voltages. (Reprinted with permission from ref.[56]. Copyright (2004) American Chemical Society.)

Making use of the same experimental condition (i.e., all measurements carried out in aqueous solution with the B-DNA chemically bounded to the electrodes) the change of conductance that occurs if a single base, a single base-pair, or two separate bases in the stack are modified was investigated. These intentional mismatches give rise to the so-called single-nucleotide polymorphisms and their study can shed some light onto the causes of mutation-related cancers. The measurements of the transport properties of 11 and 12 bp long duplexes showed that the alteration of a single base in the stack can either increase or decrease the conductivity of the dsDNA helix, depending on the type of the mismatched base. Therefore, the presence of a single base pair mismatch can be identified from conductance measurements and can cause a change in the conductance of short dsDNA by as much as an order of magnitude depending on the specific sequence of the DNA chain. For instance, the sequence 5'-CGCGAATTGCGCG-3' was hybridized first with its complement and showed a conductance value of 3.6×10^{-6} G_0, in accordance with that previously reported for the closely related 5'-CGCG(AT)$_2$CGCG-3' chain. Now, upon replacement of two of the thymines by guanines, two mismatched A-G pairs were introduced in the duplex and the conductance of the mutated DNA dropped to a value of 1.7×10^{-6} G_0.[57] Conversely, the conductance of the sequence 5'-GGAGCCCGAGG-3', containing a triplet G-C bp in the central position, is significantly enhanced ($G = 1 \times 10^{-5}$ G_0). This result is in good agreement with the idea that HOMO guanine orbitals favour charge migration, whereas short A-T sequences create a tunneling barrier for charge hopping through guanines along the DNA stack. However, although duplexes containing C-G and T-G pairs would have similar pathways in a guanine-hopping scheme, they actually have quantifiable conductivity differences, hence indicating that the coupling of an individual base to its neighbors and the structural stability of the duplex itself (T-G is not a Watson-Crick bp) are extremely important in the charge transfer dynamics.[57] In this regard, a clear enhancement of conductivity observed in moist conditions has been interpreted as properly illustrating the interplay between the structural conformation of DNA molecules and their conductivity.[58]

These promising results spur the interest in exploring the possible identification of specific genes, or at least to distinguish coding regions from non-coding ones, on the basis of differential change in electrical signals. To this end, a nanoelectronic platform based on single-walled carbon nanotubes was fabricated for measuring electrical transport in single-molecule ssDNA and dsDNA samples of a 80 bp long DNA fragment encoding a portion of the H5N1 gene of the avian Influenza A virus.[59] To enhance the contact efficiency a covalent bonding between an amine-terminated DNA molecule and a carboxyl-functionalized carbon nanotube was established and the DNA molecule was suspended over a nanotrench in order to mitigate the problem of compression-induced perturbation on the charge transport. A non-linear I-V characteristic curve was observed indicating a semiconducting behavior (gap width ∼ 1 eV, p-type conduction) in both aqueous (sodium acetate buffer)

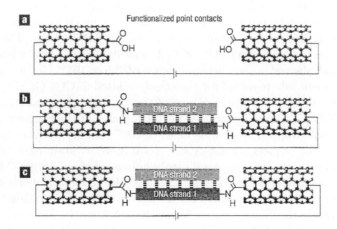

FIGURE 6.14

Schematics illustrating a method to chemically attach single-walled nanotubes (SWNT) with duplex DNA strands. (a) Functionalized point contacts made through the oxidative cutting of a SWNT wired into a device (b) bridging by functionalization of both strands with amine functionality (c) bridging by functionalization of one strand with amines on either end. (From ref.[60]. Reprinted by permission from Macmillan Publishers Ltd.: *Nature Nanotechnology* **3** 163 © 2008.)

and vacuum (10^{-5} torr) conditions. About a 25-40 pA (0.5-1.5 pA) current at 1V was measured for dsDNA (ssDNA) duplexes, respectively, at ambient conditions. The resistance increased in both cases in vacuum, presumably due to the depletion of water molecules in the hydration shell surrounding DNA (as well as possible conformational changes in the double helix at high vacuum). The obtained current values are about three orders of magnitude lower than those reported in refs.[48, 54] for three times shorter dsDNA chains, which indicates a nearly exponential current dependence with the chain length. In fact, following a similar approach currents within the range $\sim 15 - 100$ nA (comparable to those previously reported from STM measurements) were measured for 15 bp long DNA duplexes with aperiodic sequences depending on the semiconducting (lower currents) or metallic (higher currents) nature of the single-walled carbon nanotube electrodes.[60] In these latter experiments a special attention was paid to (i) the method of DNA attachment (Fig.6.14), and (ii) the influence of mismatched base-pairs in the charge transfer efficiency. It was reported that no significant differences were appreciated in the conductance measurements when using the two connection strategies shown in Fig.6.14. This result clearly indicates that efficient charge transfer takes place from one strand to the other during the charge migration process. On the other hand, the resistance of duplexes with a single GT or CA mismatch increases by about 300-fold relative to the well-matched ones, lending support to the strategy of detecting point-mutation regions through electrical signaling techniques.

From basic principles it is expected that a ssDNA molecule will carry only a feeble current due to lack of structural integrity. This has been experimentally confirmed in a systematic way by comparing the single-molecule conductance of short thiolated ssDNA and dsDNA homopolymers in aqueous solution (sodium phosphate) at room temperature. In this way, it has been reported that the conductance measured for 5'-C_6S-$(dG)_{15}$-$(dC)_{15}$-C_6S-3' duplexes ($G = 1.4 \times 10^{-6} \ G_0$) between gold metallic contacts compares well with that measured for 5'-C_6S-$(dG)_7$-C_6S-3' single-stranded chains ($G = 1.6 \times 10^{-6} \ G_0$) at 0.2 V bias potential. Accordingly, the conductance of the double-stranded structures is about an order of magnitude higher than that for single-stranded ones with similar number of bases (such a conductivity difference is significantly greater for oligo-dC, oligo-dT, and oligo dA chains). This observation clearly demonstrates that the interactions between the base pairs and stacking effects play a vital role in electron transport through DNA.[61]

In summary, the reported experiments demonstrate the high sensitivity of DNA electrical conductivity to several factors. Firstly, we have the structural complexity of nucleic acids, which is significantly influenced by its close surrounding chemical environment (humidity degree, counterions distribution) affecting the integrity of the base-pair stack, as well as by the unavoidable presence of thermal fluctuations. Secondly, the kind of order present in the DNA macromolecule plays an important role in determining their transport characteristics: periodically ordered polyG-polyC chains exhibit semiconducting behavior, whereas biological λ-phage chains are more insulating. From general considerations one expects that the presence of four different types of nucleotides favors localization of charge carriers, reducing transfer rate due to backscattering effects stemming from a larger chemical diversity. Nevertheless, the presence of long-range spatial correlations among different bases in certain fragments of the genome in biological samples will favor transport via resonant effects. Accordingly, a proper description of charge migration through these samples would require a balanced treatment of both physical effects. Finally, measuring charge transport in a DNA chain is strongly biased by the invasive role of contacts, the charge injection mechanism, the quality of the DNA-electrode interface, and the possible interaction with some inorganic substrate, or other components of the experimental layout. In consequence, undivided caution has to be paid when interpreting results of a particular approach in order to properly discriminate the relative role of intrinsic properties (v.g. base pair sequence effects) from both contact conditions and environmental effects.

6.4 Modeling charge migration in DNA

> One of the dreams of the theoreticians is to solve the Schrödinger equation with a potential that is given by a one-dimensional array of the real DNA and protein sequences. (...) This problem has been a main theme in quantum molecular biology, which I would like to call *the Schrödinger's dream.* (Kazumoto Iguchi, [62])

As we have seen, the fundamental question regarding the intrinsic conducting nature of different kinds of DNA samples is far from being definitively settled down. In the first place, nucleic acids can be classified in two broad classes, namely, duplex and single-stranded molecules of either DNA or RNA. In turn, each class can be further split into biological (i.e., samples extracted from living organisms, like viruses, bacteria or eucaryotic cells) and artificially engineered molecules (e.g., polyG-polyC, polyA-polyT, or polyGC-polyCG chains). In addition, fragments of biological DNAs can be further split into coding (the so-called introns) and non-coding ones (exons). In general, synthetic nucleic acids considered so far comprise short oligonucleotides where relatively few base pairs (bps) are *periodically* arranged. These structures are quite different from the biological ones, in which several thousands to millions of bps are *aperiodically* distributed, exhibiting characteristic scale invariant properties due to the presence of long-range correlations in certain regions.[63, 64, 65, 66, 67, 68, 69] Accordingly, biological DNAs exhibit a greater chemical complexity, determined by their bp sequencing. As a general trend, experimental results reported in Section 6.3 indicate that i) periodic synthetic DNAs transport charge better than aperiodic biological samples, and ii) double-stranded DNA exhibits electrical conductance values orders of magnitude higher than single-stranded chains of comparable length.

The question whether DNA is an insulator, a semiconductor, or a metal is often raised. This terminology originates from the field of solid-state physics and it is intimately related to the electronic structure of the considered system. As it has been discussed in Section 6.2, the electronic structure of periodic DNA chains resembles that of doped semiconductors, in agreement with most current-voltage curves obtained to date. From a theoretical point of view the question is not so clear in the case of aperiodic DNA duplexes of biological interest (see Section 6.5), though experimental measurements for relatively short chains have shown similar results to those obtained for periodic chains of comparable size.

Another important fundamental question refers to the physical mechanisms responsible for charge migration through DNA. In organic crystals based on stacked planar molecules the characteristic time scales for charge motion from molecule to molecule by means of a coherent tunneling mechanism are determined by the single particle transfer integral t according to the expression $\hbar/t$, with typical values of the order of fs.[70] Depending on the DNA sequence composition, its length and effective temperature, the value of the transfer integral between stacked bases can vary over a relatively broad interval, ranging from $t_0 = 0.01$ to $t_0 = 0.4$ eV.[24, 53, 71, 72, 73, 74, 75] These figures yield charge migration time scales within the range $1.6 - 65.8$ fs for coherent tunneling. On the other hand, long-time molecular dynamics simulations of DNA lead to typical fluctuation angles for roll and twist modes of the bps of order 5-9 deg in the ps to ns time window,[76] in agreement with dynamic Stokes shifts experimentally observed in the fluorescent spectrum of DNA.[77] The strength of these local fluctuations compares with those due to thermally excited phonon modes in linear-chain materials. Accordingly, strong interaction between the electronic degrees of freedom and molecular vibrations may reduce the coherent tunneling time scale from fs to ps.

Nevertheless, femtosecond spectroscopy experiments aimed at determining the rates of DNA charge-transport processes unveiled an unusual two-step decay process with characteristic time scales of 5 and 75 ps, respectively.[78] Although the 5 ps time scale roughly accommodates to the above mentioned electron-phonon coupling scenario, the second-stage step is still about one order of magnitude larger than the typical ps time scale figure for lattice or intramolecular dynamical degrees of freedom, and can not be accounted for in terms of a coherent tunneling mechanism. Rather, a reversed situation, characterized by localized electronic states undergoing thermally induced hopping migration, has been suggested. In this way, large-scale structural fluctuations would interfere with the π orbital overlap mediated charge transfer, leading to longer relaxation times.[79] To this end, one should either consider twist oscillations of successive base pairs or radial oscillations stemming from hydrogen bonds stretching vibrations. However, since the dynamics of angular twist and radial vibrational modes evolve independently on distinct time scales they can be regarded as decoupled degrees of freedom in the harmonic approximation.[80] Over the longest (ns) timescales currently accessible by full-atomistic modular dynamics simulations the helix undergoes large-scale global oscillatory motions dominated by rise and twist oscillations of the bp planes as a whole. These motions dominate the range of molecular conformations generated by thermal agitation and are of high relevance to any biological process which relies on shape recognition.

Generally speaking, any current measured through a DNA molecule results from the carrier injection onto the stack of bases, combined with the intrinsic conduction along the DNA sequence. Charge transfer in DNA has been proven to be mainly conveyed by intrastrand $\pi - \pi$ coupling, through either sequential incoherent hopping or coherent tunneling.[81, 82] The latter mechanism might be expected to dominate the conduction in the very low temperature regime, specially in the case of periodic oligonucleotides. At low voltage, the main contribution to the resistance comes from the metal-DNA junction potential mismatch (barrier), whereas for high enough voltage, new conduction channels are provided by the molecular states. The current intensity versus applied voltage I-V characteristics are thus somehow inferred from the energy difference between the metallic work function and the lowest ionization energy levels of the DNA (in the case of hole transport).[83]

The study of the DNA conductance over a temperature range also provides information about the transport mechanisms (activated charge hopping, intramolecular thermal fluctuations) and relevant energy scales. The role of vibration modes on the temperature dependence of the conductance is determined by the Debye temperature of the system, which has been estimated as $\Theta_D \simeq 166$ K.[84] Thus, DNA acoustic modes will significantly affect the conductance at low temperatures, but at room and higher temperature (the melting point of DNA is located within the range $T_M = 320-340$ K, depending on the sequence composition and environmental conditions) one expects dynamical effects will have no noticeable temperature dependence, in agreement with experimental measurements performed in polyG-polyC and polyA-polyT oligomers composed of 15 bps.[84]

In order to properly describe the charge transfer mechanisms we must consider a model Hamiltonian accounting for different scales of time and space by means of an adequate choice of generalized coordinates describing both electronic and dynamic DNA degrees of freedom. In the following we will first present a basic formalism to compute transmission coefficients of DNA chains connected to two external metallic leads.[85] Some basic properties of current-voltage characteristics in the coherent regime will be also addressed. Then, in Section 6.5, we will investigate the effect of long range correlations on charge transfer in several types of DNA, from artificially aperiodic to long range correlated genomic sequences.

6.4.1 Effective single-stranded chains

Figure 6.7 illustrates the excellent agreement between the DNA band structures calculated from detailed, fully atomistic, *ab initio* calculation (points) and a one-dimensional, tight-binding chain model (solid line). This result demonstrates that each base pair contributes with one single orbital which interacts negligibly with other orbitals in the pair.[24] In fact, quantum mechanical studies show that hydrogen bonding interaction gives rise to a spatial separation of the HOMO and LUMO in the nucleobase system, so that hole (electron) transfer proceeds through the purine (pyrimidine) bases, where the HOMO (LUMO) carriers are located in polyG-polyC (polyA-polyT), respectively.[24, 29] For instance, in the case of a polyG-polyC chain the central sites will model the G nucleotide only, with the effect of the C bases being neglected as not so relevant for transport due to their different on-site HOMO/LUMO energies (see Section 6.2). Thus, as a first approximation, the basic physics of charge transport in DNA molecules can be addressed in terms of a simple model with one single orbital per base pair plus a transfer parameter describing the coupling between successive neighbors. Accordingly, neglecting charge transfer within the Watson-Crick base pairs, one can consider an effective one-dimensional tight-binding model given by the Hamiltonian,

$$\mathcal{H}_{1D} = \sum_{n=1}^{N} \varepsilon_n c_n^{\dagger} c_n - \sum_{n=1}^{N-1} t_{n,n+1}(c_n^{\dagger} c_{n+1} + c_{n+1}^{\dagger} c_n), \tag{6.19}$$

where N is the number of nucleotides, $c_n^{\dagger}$ (c_n) is the creation (annihilation) operator of a charge at site n, the on-site energies ε_n describe the energetics of a charge located at bp site n, and $t_{n,n+1}$ is the hopping integral simulating the $\pi - \pi$-stacking between adjacent nucleotides. Since base-pairs are modelled by a single site, the DNA is effectively described as a binary sequence of G:C (identical to C:G) and A:T (or T:A) pairs with links between like (G:C-G:C or A:T-A:T) or unlike (G:C-A:T, A:T-G:C) pairs.

The so-called fishbone model (Fig.6.15) retains the central conduction channel in which individual sites represent a base-pair.[86] However, these central sites are now further linked to upper and lower sites, representing the backbone. The backbone sites themselves are *not* interconnected along the backbone. In some previous models a transport channel associated to the possible hopping of charge carriers between successive phosphate groups along the backbone was considered.[87, 88, 89] However, first principle calculations, showing that the phosphate molecular orbitals are systematically below the base related ones, do not favour the presence of such a transport channel.[27, 90] The Hamiltonian of the fishbone model is given by

$$\mathcal{H}_F = \sum_{n=1}^{N} \sum_{q=\uparrow,\downarrow} (\varepsilon_n + \varepsilon_n^q) c_n^{\dagger} c_n - \sum_{n=1}^{N-1} \sum_{q=\uparrow,\downarrow} t_{n,n+1}(c_n^{\dagger} c_{n+1} + H.c.) + t^q(c_n^{\dagger} c_{n,q} + H.c.), \tag{6.20}$$

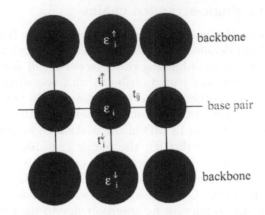

FIGURE 6.15

The fishbone model for electronic transport along DNA corresponding to the Hamiltonian given by Eq.(6.19). Every link between sites implies the presence of a hopping amplitude.[91] (Courtesy of Rudolf A. Römer).

where *H.c.* stands for the Hermitian conjugate and the transfer integral t^q with $q = \uparrow, \downarrow$ gives the hopping from each site on the central branch to the upper and lower backbone respectively. The on-site energy at the sites of the upper and lower backbone is given by ε_n^q, with $q = \uparrow \downarrow$.

The backbone sites can be decimated in order to reduce Eq.(6.20) to an effectively renormalized Hamiltonian of the form given by Eq.(6.19) where the on-site energies of the renormalized bases, $\tilde{\varepsilon}_n$, are now energy-dependent

$$\tilde{\varepsilon}_n(E) = \varepsilon_n + \frac{t^{\uparrow 2}}{E - \varepsilon_n^{\uparrow}} + \frac{t^{\downarrow 2}}{E - \varepsilon_n^{\downarrow}}. \tag{6.21}$$

Alternatively, one may consider Eq.(6.19) as describing a single-stranded DNA chain. In that case ε_n are usually chosen according to the ionization potentials of the respective bases. By considering nearest-neighbors interactions only, $\mathcal{H}_{1D}$ can be cast in terms of the unimodular matrices

$$Q_n \equiv \begin{pmatrix} 2x_n & -1 \\ 1 & 0 \end{pmatrix}, \tag{6.22}$$

where $x_n = (E - \varepsilon_n)/2t_{n,n+1}$, with $n = \{G,A,C,T\}$. There exist different kinds of transfer matrices, depending on the kind of nucleotides being considered, and the global transfer matrix is obtained from the product $M_N(E) = \prod_{n=N}^{1} Q_n$. As an illustrative example which can be analytically solved let us consider a periodic chain with the unit cell GACT. If one assumes all the transfer integrals to be the same, the dispersion relation is given by (see Section 9.2),[92]

$$4t^4 \sin^2(2k) - t^2(2E - \varepsilon_T - \varepsilon_A)(2E - \varepsilon_C - \varepsilon_G) + \prod_{i=G,A,C,T} (E - \varepsilon_i) = 0, \quad (6.23)$$

where k is the wave vector. Thus, the energy spectrum of a polyGACT single-stranded DNA (see Fig.6.21) is composed of two relatively wide bands ($W_A = 0.15$ eV centered at 8.198 eV, and $W_C = 0.16$ eV centered at 8.844 eV), related to the adenine and cytosine bases respectively, plus two narrower bands ($W_G = 0.04$ eV centered at 7.422 eV, and $W_T = 0.05$ eV centered at 9.535 eV), which are related to the guanine and thymine bases respectively. These allowed bands are separated by the relatively broad gaps $\Delta_{GA} = 0.830$ eV, $\Delta_{AC} = 0.488$ eV, and $\Delta_{CT} = 0.583$ eV.

6.4.2 Double-stranded chains

A central simplification of the one-dimensional, wire model is the description of a DNA base-pair as a single site. By doing so, one ignores the possible role of H-bonds in the charge transfer process, and one also loses the distinction between a pair with, say, a G on the 5' strand of the DNA and a C on the 3' side, and one where C sits on the 5' and G on the 3', i.e., G:C is equal to C:G. This deficiency of the wire model may be overcome by modelling each DNA base as an independent site. The hydrogen-bonding between base-pairs is then described as an additional transfer integral perpendicular to the DNA helix. This two-channel model (usually referred to as the ladder model [87, 91]) is a planar projection of the structure of the DNA with its double-helix unwound.

Earlier ladder models considered a quantum system in which there are two main chains (describing each DNA strand) and bridges between face-to-face sites along the chains (describing the H-bonds between complementary bases, Fig.6.16a). Each site is decorated with an on-site energy value describing the HOMO or LUMO orbital associated to that site and charge carriers hop to the nearest-neighbor sites along the same strand (intrachain hopping) as well as from one strand to the other (interstrand hopping, Fig.6.16b). Another version of this model consists of an alternate repetition of sugar-phosphate sites in the two main chains so that there are bridges only between the base sites in the main chains (Fig.6.16c). Subsequent, more elaborated works have included the sugar-phosphate contribution by analogy with the treatment given in the fishbone model, as it is illustrated in Fig.6.17. The ladder model Hamiltonian corresponding to the system shown in Fig.6.16b can be expressed in

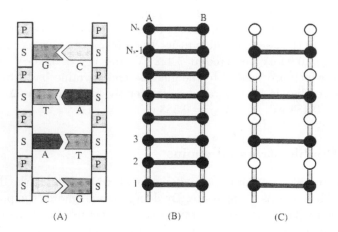

(A)   (B)   (C)

FIGURE 6.16

Ladder models of a double-stranded DNA. (a) A schematic diagram of DNA basic building blocks: P (phosphate groups), S (sugar), A (adenine), T (thymine), G (guanine), and C (cytosine). (b) A simple ladder model where each site represents an entire nucleotide unit (phosphate-sugar-base). (c) A decorated ladder model where sugar-phosphate units are separated from the bases which, in turn, are connected through H-bonds forming Watson-Crick base-pairs.[87] (Courtesy of Kazumoto Iguchi).

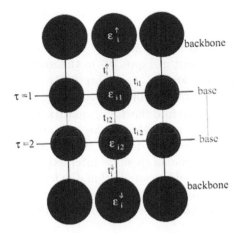

FIGURE 6.17

Fishbone ladder model for the electronic transport through DNA. There are two central branches, linked with one another, with interconnected sites where each one represents a complete base. Additional links describe the coupling of each base to the backbone.[91] (Courtesy of Rudolf A. Römer).

the form[93]

$$\mathcal{H}_{2D} = \sum_{n=1}^{N} \hat{\boldsymbol{\varepsilon}}_n \mathbf{c}_n^\dagger \mathbf{c}_n - \sum_{n=1}^{N-1} \hat{\mathbf{t}}_{n,n+1}(\mathbf{c}_n^\dagger \mathbf{c}_{n+1} + \mathbf{c}_{n+1}^\dagger \mathbf{c}_n),$$ (6.24)

where

$$\mathbf{c}_n \equiv \begin{pmatrix} c_{n,A} \\ c_{n,B} \end{pmatrix}$$ (6.25)

are the charge destruction operators at site n in the chain $m \in \{A, B\}$, and

$$\hat{\boldsymbol{\varepsilon}}_n \equiv \begin{pmatrix} \varepsilon_{n,A} & t_{AB} \\ t_{AB} & \varepsilon_{n,B} \end{pmatrix}, \qquad \hat{\mathbf{t}}_{n,n+1} \equiv \begin{pmatrix} t_{n,n+1}^A & 0 \\ 0 & t_{n,n+1}^B \end{pmatrix},$$ (6.26)

where $\varepsilon_{n,m}$ is the on-site energy of site n in the strand m, $t_{n,n+1}^m$ is the nearest-neighbor transfer integral between nth and $(n+1)$th sites in strand m, and $t_{AB} = t_{BA}$ are the interstrand hopping integrals. A similar expression is obtained for the system shown in Fig.6.17 by properly decimating the backbone sites,[85, 91] to get

$$\hat{\boldsymbol{\varepsilon}}_n' \equiv \begin{pmatrix} \tilde{\varepsilon}_{n,A} & t_{AB} \\ t_{AB} & \tilde{\varepsilon}_{n,B} \end{pmatrix},$$ (6.27)

where

$$\tilde{\varepsilon}_{n,m} = \varepsilon_{n,m} + \frac{t_n^{q\,2}}{E - \varepsilon_{n,m}^q}.$$ (6.28)

We note that the 2D Hamiltonian given by Eq.(6.24) is formally analogous to the 1D Hamiltonian given by Eq.(6.19), so that the higher dimensionality of the ladder model is properly described by generalizing the scalar nature of the model parameters present in Eq.(6.19) to the matrix form corresponding to the magnitudes $\mathbf{c}_n$, $\hat{\boldsymbol{\epsilon}}_n$, and $\hat{\mathbf{t}}_{n,n+1}$ in Eq.(6.24). It is also interesting to note that while the intrastrand hopping terms $t^m_{n,n+1}$ (describing the $\pi - \pi$ orbital coupling) are included into the $\hat{\mathbf{t}}_{n,n+1}$ matrix, the H-bonding mediated interstrand coupling is incorporated into the $\hat{\boldsymbol{\epsilon}}_n$ matrix instead. In this way, the algebraic formalism adopted to describe the dsDNA Hamiltonian properly distinguishes the different types of chemical interactions in the macromolecule. Finally, we see that the possible contribution due to $\pi - \pi$ couplings between neighboring bases belonging to complementary strands can be easily included into the description by assigning non-zero values to the non-diagonal terms in the matrix $\hat{\mathbf{t}}_{n,n+1}$.

Making use of the transfer-matrix formalism the two coupled Schrödinger equations related to the Hamiltonian given by Eq.(6.24) can be expressed in the compact form

$$(E\mathbf{I} - \hat{\boldsymbol{\epsilon}}_n)\boldsymbol{\psi}_n = \hat{\mathbf{t}}_{n,n-1}\boldsymbol{\psi}_{n-1} + \hat{\mathbf{t}}_{n,n+1}\boldsymbol{\psi}_{n+1}, \tag{6.29}$$

where

$$\boldsymbol{\psi}_n \equiv \begin{pmatrix} \psi_{n,A} \\ \psi_{n,B} \end{pmatrix}, \tag{6.30}$$

$\psi_{n,m}$ is the wave function amplitude at site n in the strand m, and $\mathbf{I}$ is the 2×2 identity matrix. Eq.(6.29) properly generalizes the canonical motion equation for elementary excitations moving in one-dimensional lattices (Eqs.(5.3) and (5.17)) to the two-dimensional case. Explicitly, Eq.(6.29) reads[87, 94, 95]

$$(E - \varepsilon_{n,A})\psi_{n,A} = t^A_{n,n-1}\psi_{n-1,A} + t^A_{n,n+1}\psi_{n+1,A} + t_{AB}\psi_{n,B} \tag{6.31}$$
$$(E - \varepsilon_{n,B})\psi_{n,B} = t^B_{n,n-1}\psi_{n-1,B} + t^B_{n,n+1}\psi_{n+1,B} + t_{AB}\psi_{n,A},$$

which, in turn, can be expressed in terms of 4×4 matrices of the form

$$\mathbf{T}_n(E) = \begin{pmatrix} \frac{E-\varepsilon_{n,A}}{t^A_{n,n+1}} & -\frac{t^A_{n,n-1}}{t^A_{n,n+1}} & -\frac{t_{AB}}{t^A_{n,n+1}} & 0 \\ 1 & 0 & 0 & 0 \\ -\frac{t_{AB}}{t^B_{n,n+1}} & 0 & \frac{E-\varepsilon_{n,B}}{t^B_{n,n+1}} & -\frac{t^B_{n,n-1}}{t^B_{n,n+1}} \\ 0 & 0 & 1 & 0 \end{pmatrix}, \tag{6.32}$$

from which one can obtain the electronic structure and transport properties in terms of the transmission and Lyapunov coefficients (see Sections 9.2, 9.5.2, and 9.5.4).[87, 88] Making use of this approach the charge transfer in DNA has been studied in a number of works. In this way, the very weak distance dependence of charge rates experimentally measured for the DNA sequence $(G{:}C)(T{:}A)_m(G{:}C)_3$, for different m values, was explained in terms of the ability of charge carriers to bridge from one strand to the other depending on the ratio of intra- and interstrand neighboring base-base couplings.[94] These results properly illustrate the importance of increasing the coordination degree among nearest neighbors when going from one-dimensional to two-dimensional quantum models. The role of such a dimensionality effect on the very nature of the wave functions in DNA ladder models was discussed in detail and it was concluded that the intrinsic DNA correlations, arising from the Watson-Crick base pairing, do not suffice to give rise to the presence of extended states when four values of the on-site energies corresponding to the G, C, T, and A bases are randomly assigned in one of the strands (say $\varepsilon_{n,A}$), with the same probability, while the sites of the second strand are set to follow Chargaff's complementary rule.[95, 96]

As we have seen in Section 6.2, quantum chemical and *ab initio* band structure calculations reveal the existence of different subsystems in DNA, each one of them characterized by its own energy scale. Thus, the description of the electronic energetics of more realistic double-stranded DNA chains, as that shown in Fig.6.18a, must take into account three different contributions stemming from (i) the nucleobase system, (ii) the backbone system, and (iii) the environment, as it is sketched in Fig.6.18b. In the nucleobase system one must include the HOMOs associated to one of the strands along with the LUMOs associated to the complementary one, as well as the Watson-Crick H-bonding interactions. Attending to the energies involved in the different interactions, the resulting energy network can be hierarchically arranged, starting from high energy values related to the on-site energies of the bases and sugar-phosphate groups $(8 - 12$ eV),[97, 98, 99] passing through intermediate energy values related to the hydrogen bonding between Watson-Crick pairs $(\sim 0.5$ eV),[97, 100] and the coupling between the bases and the sugar moiety $(\sim 1$ eV),[98] and ending up with the aromatic base stacking low energies $(0.01 - 0.4$ eV).[97, 101, 102, 103] The energy scale of environmental effects $(1 - 5$ eV) is related to the presence of counterions and water molecules, interacting with the nucleobases and the backbone by means of hydration, solvation, and charge transfer processes. It is about one order of magnitude larger than the coupling between the complementary bases, and about two orders of magnitude larger than the base stacking energies.

One can recover the mathematical simplicity of the wire model, keeping at the same time a realistic description of the rich DNA chemistry, making use of a two-step renormalization approach. For the sake of illustration, in Fig.6.19a we introduce a tight-binding model for a double stranded polyGACT-polyCTGA unit cell including four different nucleotides. This unit cell provides a basis for both periodic and aperiodic longer DNA chains, where ε_j, with $j = \{$G,C,A,T$\}$, are the on-site energies of the bases, t_j is the hopping integral between the sugar's oxygen atom and the base's nitrogen atom, and t_{GC} (t_{AT}) respectively describe the hydrogen bonding between complementary bases. The backbone's contribution is described by means of the on-site energies γ_j, introduced in Fig.6.19b. In general, γ_j will depend on

a)   b)

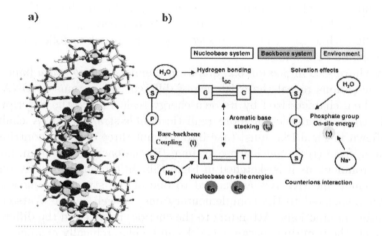

FIGURE 6.18
(a) Quantum chemical description of a realistic double-stranded DNA chain including the sugar-phosphate backbone and the presence of solvated counterions. The isosurface plot of the HOMO of B-poly(dG)-poly(dC) derived from *ab initio* calculations is shown in side-view. The dark and gray surfaces show positive and negative isovalues, respectively. (Adapted from ref.[26].) (b) Diagram illustrating the overall energetics of a double-stranded DNA model shown in (a) including the different parameters considered in the DNA tight-binding model described in the text. ([105]. Reprinted figure with permission from Maciá E 2006 *Phys. Rev. B* **74** 245105 © 2006 by the American Physical Society.)

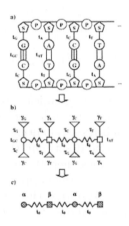

FIGURE 6.19

Sketch illustrating the two step renormalization process mapping a ds-DNA chain into a linear diatomic lattice. a) Starting effective tight-binding model for the polyGACT-polyCTGA unit cell. b) renormalized model after the first decimation step. c) renormalized model after the second decimation step. ([105]. Reprinted figure with permission from Maciá E 2006 *Phys. Rev. B* **74** 245105 ⓒ 2006 by the American Physical Society.)

the nature of the neighboring base as well as the presence of water molecules and/or counterions attached to the backbone, according to the overall scheme illustrated in Fig.6.18b.

In order to obtain a simple mathematical description, containing most of the relevant physical information, we will map the tight-binding model shown in Fig.6.19a into the equivalent binary lattice model shown in Fig.6.19c. To this end, the Watson-Crick bps are first renormalized to obtain the branched tight-binding model shown in Fig.6.19b. The topological structure of the renormalized chain shown in Fig.6.19b coincides with that of the fishbone model (Fig.6.15), but in the renormalized model the parameters $\tilde{\varepsilon}$ and τ_j entail substantial physicochemical information concerning nucleotide interactions and backbone gating effects. In fact, the renormalized on-site energies and transfer integrals are respectively given by $\tilde{\varepsilon}_n = t_{ij}$, and [104]

$$\tau_j = t_j + \frac{\varepsilon_j}{t_j}(E - \gamma_j), \quad j = \{G, C, A, T\}. \tag{6.33}$$

Note that the renormalized on-site energies (given by the hydrogen bonding energy scale) are now about one order of magnitude smaller than the original ones (given by the ionization potentials of the nucleobases), so that the effective $\pi - \pi$ overlap integral describing the aromatic base stacking between adjacent nucleotides becomes energetically relevant and it is explicitly included into the model by means of the hopping integral t_0. Next, the backbone contribution is decimated to obtain the one-dimensional lattice shown in Fig.6.19c, where the renormalized on-site energies are now given by

$$\alpha(E) = t_{CG} + \frac{\tau_G^2(E - \gamma_C) + \tau_C^2(E - \gamma_G)}{(E - \gamma_G)(E - \gamma_C)},$$

$$\beta(E) = t_{AT} + \frac{\tau_A^2(E - \gamma_T) + \tau_T^2(E - \gamma_A)}{(E - \gamma_A)(E - \gamma_T)}. \tag{6.34}$$

In this way, the original polyGACT-polyCTGA chain is mapped into the equivalent diatomic lattice shown in Fig.6.19c, where the renormalized "atoms" correspond to the Watson-Crick complementary pairs in the DNA molecule. The renormalized on-site energies enclose substantial physicochemical information about the Watson-Crick bp energetics, including the nucleobases onsite energies, the transfer integral between backbone and base states, the hydrogen bonding between the complementary bases, and the sugar-phosphate backbone on-site energies. In this way, one obtains a realistic description, including 15 physical parameters, $\{\varepsilon_j, t_j, \gamma_j, t_{GC}, t_{AT}, t_0\}$, fully describing the energetics of the DNA molecule in terms of just three variables (i.e., α, β, t_0) in a unified way in terms of the effective one-dimensional Hamiltonian

$$\tilde{H}_{1D} = \sum_{n=1}^{N} \tilde{\varepsilon}_n(E)c_n^\dagger c_n - t_0 \sum_{n=1}^{N-1}(c_{n+1}^\dagger c_n + c_n^\dagger c_{n+1}), \tag{6.35}$$

where $\tilde{\varepsilon}_n(E) \in \{\alpha(E), \beta(E)\}$. Within the framework of the transfer matrix formalism and considering nearest-neighbor interactions only, the Schrödinger equation of the renormalized binary chain shown in Fig.6.19c can be expressed in terms of the following transfer matrices

$$Q_\alpha \equiv \begin{pmatrix} 2x & -1 \\ 1 & 0 \end{pmatrix}, \qquad Q_\beta \equiv \begin{pmatrix} 2y & -1 \\ 1 & 0 \end{pmatrix}, \tag{6.36}$$

where $x = (E - \alpha)/2t_0$ and $y = (E - \beta)/2t_0$. Assuming periodic boundary conditions the dispersion relation is given by the relationship

$$2\cos(qNa_*) = \mathrm{tr}[(Q_\beta Q_\alpha)^m] \equiv \mathrm{tr}[\mathcal{M}_m(E)], \tag{6.37}$$

where q is the wave vector, N is the bp number, a_* measures the separation between neighboring bps along the helix axis, and $m \equiv N/2$. Since both Q_α and Q_β are unimodular matrices, we can make use of the Cayley-Hamilton theorem (see Section 9.2) to express the global transfer matrix of the DNA chain as

$$\mathcal{M}_m(E) = \begin{pmatrix} U_m + U_{m-1} & -2yU_{m-1} \\ 2xU_{m-1} & -U_{m-1} - U_{m-2} \end{pmatrix}, \tag{6.38}$$

where $U_m(z)$, with $z \equiv \frac{1}{2}\mathrm{tr}[Q_\beta Q_\alpha] = 2xy - 1$, are Chebyshev polynomials of the second kind. In this way one gets,

$$4t_0^2 \cos^2(qa_0) = E^2 - (\alpha + \beta)E + \alpha\beta, \tag{6.39}$$

which has the typical form for a binary chain, though in this case the renormalized on-site energies, $\alpha(E)$ and $\beta(E)$, explicitly depend on the charge carriers energy E after Eq.(6.34), leading to rather involved analytical expressions. In order to grasp the basic energy spectrum structure one can introduce two simplifications. First, according to x-ray experiments the counterions condense around the nucleic acid chain in a tightly bound layer.[106, 107] Accordingly, a homogeneous charge distribution through the backbone can be assumed as a first approximation, so that $\gamma_j \equiv \gamma$. Second, the transfer integral describing the coupling between the sugar and the neighbor base takes on essentially the same values for the different nucleotides,[98] and one can confidently assume $t_j \equiv t$ as well. Thus, Eq.(6.34) simplifies to

$$\alpha(E) = \alpha_0 + \alpha_1 E + \frac{2t^2}{E - \gamma}, \quad \beta(E) = \beta_0 + \beta_1 E + \frac{2t^2}{E - \gamma} \tag{6.40}$$

where $\alpha_0 \equiv a_0 - \gamma\alpha_1$, $\beta_0 \equiv b_0 - \gamma\beta_1$, $a_0 \equiv t_{GC} + 2(\varepsilon_G + \varepsilon_C)$, $b_0 \equiv t_{AT} + 2(\varepsilon_A + \varepsilon_T)$, $\alpha_1 \equiv (\varepsilon_G^2 + \varepsilon_C^2)/t^2$, and $\beta_1 \equiv (\varepsilon_A^2 + \varepsilon_T^2)/t^2$. Plugging Eq.(6.40) into Eq.(6.39) one obtains [105]

$$E^4 + AE^3 + BE^2 + CE + D = 0, \quad (E \neq \gamma) \tag{6.41}$$

where the polynomial coefficients depend on the different model parameters. Therefore, though the renormalized chain includes only two "atomic" species, the energy spectrum is composed of four bands, as one expects for the tetranu-cleotide unit cell shown in Fig.6.19a. This result properly illustrates that the renormalized chain encompasses a full quantum description of the DNA en-ergetics and the detailed structure of the energy spectrum will depend on the adopted model parameters.

TABLE 6.3
Model parameters adopted for the
double-stranded DNA shown in
Fig.6.19.

Model Hamiltonian parameters (eV)	
$\varepsilon_G = 7.77$	$\varepsilon_C = 8.87$
$\varepsilon_A = 8.25$	$\varepsilon_T = 9.13$
$\gamma = 12.27$	
$t = 1.5$	
$t_{GC} = 0.90$	$t_{AT} = 0.34$
$t_0 = 0.15$	

By considering the realistic values listed in Table 6.3 we obtain the energy spectrum shown in the left panel of Fig.6.20. The location of the different allowed bands and their respective bandwidths are listed in Table 6.4.

TABLE 6.4
Locations of the allowed bands centers (E_i), bandwidths
(W_i), and gap widths (Δ_{ij}) in the energy spectrum of the
polyGACT-polyCTGA chain.

Band center (eV)	Band width (meV)	Gap width (eV)
$E_1 = -14.209$	$W_1 = 269$	-
$E_2 = -0.423$	$W_2 = 120$	$\Delta_{12} = 13.591$
$E_3 = +6.440$	$W_3 = 29$	$\Delta = 6.788$
$E_4 = +11.595$	$W_4 = 177$	$\Delta_{34} = 5.052$

As we see, the energy spectrum consists of four narrow bands separated by wide gaps. The wide separation among the different allowed bands stems from hybridization effects between the nucleobase system and the sugar-phosphate backbone.[86, 104] We note that the obtained bandwidths compare well with the values reported for short (5-12 bp) polyG-polyC and polyA-polyT chains from first principles band structure calculations (HOMO bandwidths $\simeq 50 -$ 400 meV; LUMO bandwidths $\simeq 100- 300$ meV, see Section 6.2). Assuming, as it is usual, that each bp contributes one free charge carrier, the HOMO band is centered at $E = -0.423$ eV, yielding an HOMO-LUMO gap width $\Delta =$

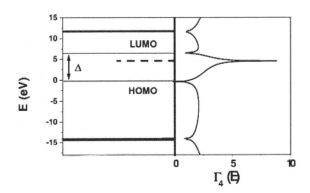

FIGURE 6.20

The band structure (left) and the Lyapunov exponent as a function of the energy (right) for the periodic polyGACT-polyCTGA chain derived from Eqs.(6.41) and (6.43), respectively, making use of the model parameters listed in Table 6.3. The origin of energy is set at ε_G. More details in the text. ([105]. Reprinted figure with permission from Maciá E 2006 *Phys. Rev. B* **74** 245105 © 2006 by the American Physical Society.)

6.79 eV. This figure occupies an intermediate position between numerically obtained values for polyG-polyC chains $(7.4-7.8$ eV$)$,[108] and photoemission spectroscopy measurements $(4.5-5.0$ eV$)$ performed in polyG-polyC and polyA-polyT chains.[109, 110]

The information about the overall structure of the energy spectrum obtained from the dispersion relation is complemented with the density of states (DOS, see Section 9.5.1). From the definition of the variables z, x, and y, and Eq.(6.40), one obtains

$$D(E) = \frac{y(\alpha_1 - 1) + x(\beta_1 - 1) - 2t^2(x + y)(E - \gamma)^{-2}}{4t_0\sqrt{xy(1 - xy)}}. \qquad (6.42)$$

Due to the one-dimensional nature of the considered model the obtained DOS is characterized by a number of sharp features (van Hove singularities) given by the conditions $E = \alpha$, $E = \beta$, and $4t_0^2 = (E - \alpha)(E - \beta)$, which determine the allowed band edges positions. In addition, we also have a resonant feature at $E = \gamma$. This resonance is a characteristic signature of the sugar-phosphate subsystem, which is shown as a dashed line in the left panel of Fig.6.20.

The system transport properties are related to the localization degree of the different states belonging to the spectrum. The Lyapunov coefficient measures the localization length of eigenstates. Therefore, it is quite useful in order to establish a relationship between the energy spectrum and the transport properties of the system (see Section 9.5.4). In the case of the polyGACT duplex one obtains

$$\Gamma(E) = \lim_{m \to \infty} \frac{1}{4m} \ln \left(2 + 4U_{m-1}^2 \left[4x^2y^2 + (x - y)^2\right]\right). \qquad (6.43)$$

The length dependence of the logarithm appearing in Eq.(6.43) is determined by the Chebyshev polynomials $U_{m-1}(z)$, which remain always bounded for $E \neq \gamma$. Accordingly, one gets $\Gamma(E) \to 0$ in the thermodynamic limit, hence indicating the extended nature of these eigenstates. This result is properly illustrated in the right panel of Fig.6.20, where we clearly appreciate the correlation between small Lyapunov coefficient values and the presence of allowed bands. We also note the presence of a localized state corresponding to the resonant state $E = \gamma$. In this case, the product xy diverges in Eq.(6.43), so that we obtain $\Gamma(E) \to \infty$ (i.e., the localization length $\xi(E) \to 0$). The resonant state $E = \gamma$ is determined by the backbone on-site energies, which in turn depend on environmental effects due to solvation and hydration processes involving the cations and the water shell. Therefore, this Hamiltonian model is able of including the existence of localized states in the HOMO-LUMO gap stemming from environmental effects, in agreement with previous results obtained from detailed *ab initio* calculations.[25, 27] In this regard, we note that the number of resonant localized states within the gap region will be increased by properly relaxing the condition $\gamma_j \equiv \gamma$ in the treatment.

6.4.3 The role of contacts

In Section 6.3 we learnt that contacts between the DNA sample and the experimental set-up play a very important role in order to properly determine the intrinsic DNA electrical transport properties. In fact, an increasing number of transport experiments have shown that deliberate chemical bonding between DNA and metal electrodes is a prerequisite for achieving reproducible conductivity results.[49, 50, 51] Accordingly, the study of contact effects on the charge migration efficiency is an important issue to be considered in realistic models of DNA transport.

In general, modeling the geometry and bonding character of the contact at the interface is a very delicate issue, since detailed information on both the metal geometry and DNA chemical bonding at the contacts is poorly known. Consequently, when modeling the DNA-contact interface within the tight-binding approach, one introduces an effective parameter τ dealing with the tunneling probability between the frontier orbitals, thus roughly encompassing bonding effects at the interface. Broadly speaking, one expects the binding to metallic leads would affect the electronic structure of the molecule. If so, we should consider the states belonging to the coupled molecular-metallic system rather than those of the molecular subsystem alone.[111] Thus we shall consider henceforth that the coupling between the contacts and the molecule is weak enough, so that the lead-molecule-lead junction can be properly described in terms of three non-interacting subsystems. Besides, as a first approximation, we neglect the finite cross section of the electrodes (only one channel for charge transfer at the Fermi level). Thus, the lead-DNA global system will be described by means of the tight-binding Hamiltonian

$$\mathcal{H} = \mathcal{H}_{1D} + \mathcal{H}_C + \mathcal{H}_L, \tag{6.44}$$

where $\mathcal{H}_{1D}$ is either given by Eq.(6.19) or Eq.(6.35),

$$\mathcal{H}_C = -\tau \left(c_0^\dagger c_1 + c_1^\dagger c_0 + c_{N+1}^\dagger c_N + c_N^\dagger c_{N+1} \right), \tag{6.45}$$

describes the DNA-metal coupling, where τ measures the coupling strength, and

$$\mathcal{H}_L = \sum_{l=0}^{-\infty} \varepsilon_M c_l^\dagger c_l - t_M (c_l^\dagger c_{l+1} + c_{l+1}^\dagger c_l) + \sum_{l=N+1}^{+\infty} \varepsilon_M c_l^\dagger c_l - t_M (c_l^\dagger c_{l+1} + c_{l+1}^\dagger c_l) \tag{6.46}$$

gives the energetics of the metallic leads at both sides of the DNA chain, where ε_M is related to the metallic work function, while t_M is the hopping term, so that the leads dispersion relation is given by $E = \varepsilon_M + 2t_M \cos k$. Then, sites comprised between $[-\infty, 0] \cup [N+1, +\infty]$ belong to the leads, whereas sites $n = 1, \ldots N$ are associated to a single-stranded DNA chain of size N. The transmission coefficient is computed using the transfer matrix formalism in

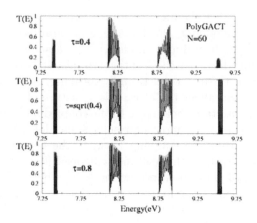

FIGURE 6.21

Transmission coefficient curve for a polyGACT chain with $N = 60$, $t_M = 1.0$ eV, $t = 0.4$ eV, and $\tau = 0.4$ eV (top panel); $\tau = \sqrt{0.4}$ eV (central panel); and $\tau = 0.8$ eV (bottom panel). Energy ε_M is adjusted to simulate a resonance with the G-HOMO energy level, $\varepsilon_M = \varepsilon_G$. ([92] Reprinted figure with permission from Maciá E, Triozon F, and Roche S 2005 *Phys. Rev. B* **71** 113106 © 2005 by the American Physical Society.)

which the time independent Schrödinger equation is projected into a localized basis (see Section 9.5.2).

As a suitable representative example, the charge transport through a periodic polyGACT tetranucleotide chain, connected to metallic leads at both ends, will be considered. In Fig.6.21 we show the transmission patterns for a system with $t_M = 1.0$ eV and $t = 0.4$ eV for several values of the DNA-metal coupling. We can readily see the narrow G and T bands at the edges as well as the relatively broader A and C bands at the central regions of the energy spectrum. By decreasing t these bands progressively stretch, eventually collapsing into the series $\{\varepsilon_j\}$, describing the energy levels of a set of isolated nucleotides. From the top and bottom panels we realize that, depending on the value adopted for τ, the obtained transmission coefficient does not reach in general the full transmission condition $T_N(E) = 1$ due to the symmetry breaking related to the coupling of the G (T) end nucleotides at the left (right) leads, respectively. This transmission degradation is a direct consequence of interference effects between the DNA energy levels and the electronic structure of the leads at the metal-DNA interface. It was obtained that the *optimal* system configuration for efficient charge transfer is determined by the resonance condition $\tau = \sqrt{t * t_M}$.[92] A complete set of such resonant states is shown in the central panel of Fig.6.21. Quite interestingly, one realizes that, due to resonance effects, a stronger coupling to the leads does not always result in a larger conductance through the system.[112] Subsequent works have exploited the existence of this optimal charge injection condition to study the charge migration efficiency through more realistic duplex chains.[94]

Similar results are obtained in the case of a double-stranded DNA chain by replacing the base on-site energies ε_j by the renormalized ones $\alpha(E)$ and $\beta(E)$ in the Hamiltonian $\mathcal{H}_{1D}$ in Eq.(6.44). The Landauer conductance $G_N(E)$ (see Section 9.5.3) can be obtained from the knowledge of the matrix elements of the metal-DNA-metal transfer matrix $\mathcal{M}_N(E) = L_N(Q_\alpha Q_\beta)^{m-1} L_1$, with $m = N/2$, where the contact matrices

$$L_1 = \begin{pmatrix} 2x & -\lambda \\ 1 & 0 \end{pmatrix}, \quad L_N = \lambda^{-1} \begin{pmatrix} 2y & -1 \\ \lambda & 0 \end{pmatrix} \tag{6.47}$$

describe the coupling between the DNA and the metallic leads in terms of the coupling strength $\lambda \equiv \tau/t_0$. After some algebra one gets,

$$G_m(E) = G_0 \left\{ 1 + (x-y)^2 U_{m-1}^2 + t_M^2 \frac{[f_\lambda(E, U_m) - 2(x+y)U_{m-1}\cos k]^2}{(E - E_-)(E_+ - E)} \right\}^{-1} \tag{6.48}$$

where the auxiliary function $f_\lambda(E, U_m) = \lambda^{-1}(U_{m-1} + U_m) + \lambda(U_{m-2} + U_{m-1})$ describes contact effects,[92] and $E_\pm = \varepsilon_M \pm 2t_M$ define the allowed spectral window as determined by the metallic leads bandwidth. The term $(x-y)^2 U_{m-1}^2$ in Eq.(6.48) accounts for the greater chemical diversity of a polyGACT-polyCTGA chain as compared to either polyG-polyC or polyA-polyT chains (for which $x = y$), and its main physical effect is to reduce the

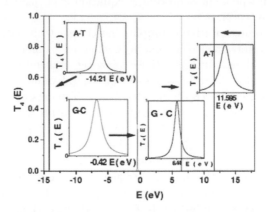

FIGURE 6.22

Transmission coefficient as a function of the energy for a periodic polyGACT-polyCTGA duplex with $N = 4$ bps. ([105] Reprinted figure with permission from Maciá E 2006 *Phys. Rev. B* **74** 245105 © 2006 by the American Physical Society.)

overall conductance of the former with respect to that obtained for the simpler ones. In Fig.6.22, the energy dependence of the transmission coefficient is shown as a function of the injected charges energy at zero bias. In the insets the transmission band profile is magnified. As we see, the full transmission condition is fulfilled for all four bands, indicating the extended nature of their eigenstates. By taking $y = x$ in Eq.(6.48) we obtain the transmission spectra for the polyG-polyC chain (the expression for polyA-polyT is then obtained by simply replacing $x \rightarrow y$). In this way, we can assign the central bands in the energy spectrum to GC bps, while the edge bands in the spectrum are related to the AT bps. Since the central bands are closer to the adopted Fermi energy, we conclude that the charge transfer will be dominated by the HOMO band in the considered system, so that it will exhibit a p-type behavior, in agreement with the experiments reported in Section 6.3.

6.4.4 Helicoidal structure and dynamical models

In physiological conditions DNA double helix exhibits a full-fledged three-dimensional (3D) geometry, so that every two consecutive bases are twisted by a certain angle ($\theta_0 \simeq \pi/5$ in equilibrium conditions). As a result, the orbital overlapping is substantially reduced, yielding smaller values for the $\pi - \pi$ transfer integral values.[27, 113] Therefore, one should expect a significant reduction of the charge transfer efficiency stemming from purely geometrical

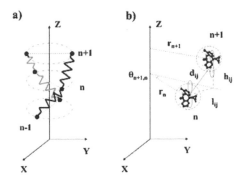

FIGURE 6.23
Sketch illustrating the adopted DNA model at two different scale lengths. a)
At the molecular level the corresponding dynamical degrees of freedom are
described in terms of cylindrical coordinates. b) the p_z atomic orbital overlapping between neighboring nucleobases is described in terms of Cartesian
coordinates referred to the nucleobase center of mass. Its position, in turn, is
determined by the relative twist and radial variables, $\theta_{n,n+1}$, and r_n, at the
molecular scale. ([116] Reprinted figure with permission from Maciá E 2007
Phys. Rev. B **76** 245123 © 2007 by the American Physical Society.)

considerations (dimensionality effect).

At the molecular level the basic dynamical building-blocks are the sugar-
phosphate groups and the nucleobases. As a first approximation the nucleobases can be treated as identical point masses or rigid platelets, helically
arranged and mutually connected by means of elastic rods, which describe
the sugar-phosphate backbone (Fig.6.23).[80, 114, 115] Adopting the reference frame indicated in Fig.6.23a, the position of the nth nucleobase can be
expressed as $x_n = r_n \cos \varphi_n$, $y_n = r_n \sin \varphi_n$, and $z_n = c\varphi_n$, where n labels the
considered bp along the DNA double strand, r_n and φ_n are usual cylindrical
coordinates, and $c = h_0/\theta_0$, $h_0 \simeq 0.34$ nm being the equilibrium separation
between two successive bp planes (B-DNA form). Thus, the Euclidean distance between two neighboring bases can be expressed as

$$d_{n,n\pm1} = \sqrt{c^2\theta_{n,n\pm1}^2 + r_n^2 + r_{n\pm1}^2 - 2r_n r_{n\pm1} \cos \theta_{n,n\pm1}}, \qquad (6.49)$$

where $\theta_{n,n+1} \equiv \varphi_{n+1} - \varphi_n$ ($\theta_{n,n-1} \equiv \varphi_n - \varphi_{n-1}$) measures the relative angular
displacement between two adjacent bps. In equilibrium conditions (i.e., $r_n = R_0 \simeq 1$ nm, $\theta_{n,n\pm1} = \theta_0\ \forall n$) Eq.(6.49) reduces to $l_0 = \sqrt{h_0^2 + 4R_0^2 \sin^2(\theta_0/2)} \simeq$

0.7 nm. The shortest path between two points (not directly above the other) on a cylinder of radius R_0 is given by the arc length between two consecutive points along the helix, according to the formula

$$s_{n,n\pm1} = \int_{\varphi_n}^{\varphi_{n+1}} \sqrt{R_0^2 + c^2} d\varphi. \qquad (6.50)$$

In the limit of small twist oscillations ($r_n = R_0$, $\theta_{n,n+1} \lll 1$) Eq.(6.49) reads

$$d_{n,n\pm1} = \sqrt{R_0^2 + c^2}\, \theta_{n,n\pm1} \equiv \xi\theta_{n,n\pm1} = s_{n,n\pm1}, \qquad (6.51)$$

so that the Euclidean distance given by Eq.(6.49) coincides with the helix arc length in this case.

The effective model Hamiltonian can be expressed as the sum of two main contributions $H = H_e + H_l$, where H_e describes the charge carrier dynamics over the π-stacked electronic system and H_l describes the duplex DNA dynamics. The electronic degrees of freedom of a double-stranded DNA (including sugar-phosphate and environmental effects) are described by properly generalizing the effective Hamiltonian given by Eq.(6.35) in the form

$$H_e = \sum_{n=1}^{N} \tilde{\varepsilon}_n(E)c_n^\dagger c_n - \sum_{n=1}^{N-1} t_{n,n+1}(\theta_{n,n+1})(c_{n+1}^\dagger c_n + c_n^\dagger c_{n+1}) \,. \qquad (6.52)$$

Eq.(6.52) describes the charge carrier propagation through a DNA duplex in terms of an equivalent monatomic lattice, where the renormalized "atoms" correspond to complementary pairs in the original DNA molecule whose on-site energies $\tilde{\varepsilon}_n(E)$ are given by Eq.(6.34), and the transfer integral $t_{n,n+1}$ now explicitly describes the angular dependence of the aromatic base stacking between adjacent nucleotides.

When describing the phonon dynamics in DNA one can disregard the inner degrees of freedom of the bases, since we can separate the fast vibrational motions of atoms about their equilibrium positions from the slower motions of molecular groups. In this way, three characteristic vibrational states have been usually considered in DNA normal mode calculations, namely, the stretch oscillations of each base back and forth with respect to the center of mass of the system located at the helical axis (radial oscillations), longitudinal oscillations of the bps planes along the helix axis, and twist oscillations of each bp as a whole around the helical axis. A treatment of the lattice dynamics, describing the motion of two complementary bases of reduced mass μ_n from their equilibrium position can be introduced in terms of the following lattice Hamiltonian (in cylindrical coordinates)

$$H_d = \sum_n \left(\frac{p_\varphi^2}{2J_n} + \frac{p_r^2 + p_z^2}{2\mu_n} \right) + U_H + U_S + U_B, \qquad (6.53)$$

where J_n is the reduced moment of inertia for the relative motion of the bp, n runs over the number of bps, and U_k denote different elastic potentials describing:

- the radial stretching of the hydrogen bonds connecting complementary bases in the opposite strands of the double helix,[80, 117, 118]

$$U_H = \sum_{n=1}^{N} D_n \left[e^{-\alpha_n(r_n - R_0)} - 1 \right]^2, \tag{6.54}$$

where the adopted Morse potential accurately describes both the attraction due to the H-bonds forming the bps and the repulsion of the negatively charged phosphates in the backbone of the two strands screened by the surrounding solvent. Note that sequence dependence is explicitly considered by adopting a site dependence in the model parameters D_n and α_n which will take on two different values depending on the considered base pairs (i.e., G:C or A:T).[112]

- the stacking interaction between adjacent base pairs:

$$U_S = \frac{k_s}{2} \sum_{n=1}^{N-1} [1 + Ee^{-b(r_n + r_{n+1} - 2R_0)}](r_n - r_{n+1})^2, \tag{6.55}$$

where k_s is an effective elastic constant. This interaction is characterized by the exponential term that effectively modifies a harmonic-like radial oscillation and describes local constraints in nucleotide motions, which result in long-range cooperative elastic effects.[117] Physically, this constraint describes the change of the next-neighboring stacking interaction due to distortion of the H-bonds connecting a given bp.[112] In this regard, this stacking interaction differs from that due to hybridization of the π electronic systems of neighboring bp planes along the helical axis, which will be described in terms of Eqs(6.58)-(6.60) below. The description of the radial degree of freedom in terms of the non-linear potential given by Eqs.(6.54) and (6.55) is more realistic than a purely harmonic approach and has been successful in capturing denaturation effects as well as transcription initiation processes in various DNA chains.[118, 119, 120]

- the harmonic coupling between neighboring bases along the helical strand:

$$U_B = k \sum_{n=1}^{N-1} (d_{n,n+1} - l_0)^2, \tag{6.56}$$

where k is an effective force constant, $d_{n,n+1}$ is given by Eq.(6.49), and l_0 is the equilibrium distance between neighboring phosphate groups, describing harmonic oscillations along the backbone.

In order to get a basic picture of the physical effects related to the coupling between electronic and dynamical degrees of freedom, we shall focus on the low frequency twist mode, hence keeping $r_n = R_0 \; \forall n$. In that case, one can express the lattice Hamiltonian in the form

$$H_l = \frac{1}{4m\xi^2} \sum_{n=1}^{N} p_{\varphi,n}^2 + k \sum_{n=1}^{N-1} \left(\sqrt{c^2\theta_{n,n+1}^2 + 4R_0^2 \sin^2\left(\frac{\theta_{n,n+1}}{2}\right)} - l_0 \right)^2,$$

(6.57)

where m is the base mass, ξ is the effective helix arc length introduced in Eq(6.51), $p_{\varphi,n}$ is the angular momentum, k is an effective force constant, and l_0 is the equilibrium distance.

Hamiltonians (6.52) and (6.57) describe the most relevant physics of the DNA molecule and its environment in terms of the model parameters $\tilde{\varepsilon}_n(E)$ and $t_{n,n\pm1}(\theta_{n,n\pm1})$. As it is illustrated in Fig.6.23b, the overlapping between π-orbitals of stacked bps depends on the euclidean distance between atoms belonging to neighboring nucleobases, given by $d_{ij} = \sqrt{l_{ij}^2 + h_{ij}^2}$, and the transfer integral between successive bps can be expressed in the form,[27]

$$t_{n,n\pm1}(r_n,\theta_{n,n\pm1}) = t_0 \left[1 - \bar{\eta}l_0^{-2}(r_n^2 + r_{n\pm1}^2 - 2r_n r_{n\pm1}\cos\theta_{n,n\pm1}))\right], \quad (6.58)$$

where t_0 is the transfer integral corresponding to the planar geometry and $\bar{\eta} \equiv 1 + |\eta_{pp\pi}|/\eta_{pp\sigma}$, where $\eta_{pp\pi}$ and $\eta_{pp\sigma}$ describe the hybridization matrix elements between neighboring bases p_z orbitals. In this way, one recovers the usual expression $t_{n,n\pm1} = t_0$ for a planar model (i.e., $r_n = R_0$, $\varphi_n \equiv 0 \; \forall n$). If one adopts a rigid helix geometry (i.e., $r_n = R_0$, $\theta_{n,n\pm1} = \theta_0$), Eq.(6.58) takes the form

$$t_{n,n\pm1}(R_0,\theta_0) = t_0 \left[1 - \bar{\eta}\left(\frac{2R_0}{l_0}\sin\frac{\theta_0}{2}\right)^2\right]. \quad (6.59)$$

Since $\bar{\eta} > 0$, we get $t_{n,n\pm1}(R_0,\theta_0) < t_0$. Therefore, the main effect of explicitly considering the helical geometry is to reduce the strength of the $\pi - \pi$ base coupling in the equilibrium configuration, as expected. Let us now relax the equilibrium structure, allowing for the propagation of low frequency twist oscillations (acoustic modes), but keeping the radial variable describing H-bonding stretch oscillations fixed (no optical modes). In that case, Eq.(6.58) can be approximated as

$$t_{n,n\pm1}(R_0,\theta_{n,n\pm1}) \simeq t_0 \left(1 - \chi\theta_{n,n\pm1}^2\right), \quad (6.60)$$

for small enough twists, where the dimensionless parameter $\chi \equiv \bar{\eta}(R_0/l_0)^2 > 0$ measures the coupling strength between the charge and the lattice system.[124, 125, 126, 127, 128] Albeit its approximate nature, Eq.(6.60) reasonably reproduces the main features of the transfer integral versus twist angle dependence derived from detailed quantum-chemistry calculations,[129] as it is illustrated in Fig.6.24.

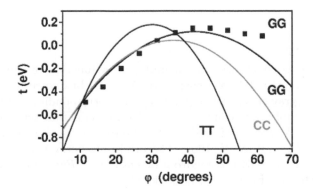

FIGURE 6.24

Transfer integral as a function of the twist angle between neighboring GG (squares) bases in the same strand (after Ref.[129]). Solid lines are obtained from expression $t = t_{jj}[1 - \chi_{jj}(\theta - \theta_{jj})^2]$ with $t_{GG} = 0.119$ eV, $t_{CC} = 0.042$ eV,[129] $\chi_{GG} = 17$, $\chi_{CC} = 63$, $\theta_{GG} = 42$, and $\theta_{CC} = 36$, respectively. ([116] Reprinted figure with permission from Maciá E 2007 *Phys. Rev. B* **76** 245123 © 2007 by the American Physical Society.)

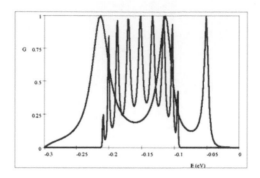

FIGURE 6.25

The Landauer conductance (in G_0 units) as a function of the Fermi level energy of a polyG-polyC oligomer with $N = 10$ bps corresponding to the planar model (gray curve) is compared to that corresponding to the helicoidal model at the equilibrium configuration. The model parameters are $\varepsilon_G = 7.77$ eV, $\varepsilon_C = 8.87$ eV, $t_{GC} = 0.90$ eV,[97] $t = 1.5$ eV [87], $t_0 = 0.15$ eV, $\gamma = 12.27$ eV, and $\tau = t_M = 0.15$ eV [105]. The origin of energy is set at ε_G. ([116] Reprinted figure with permission from Maciá E 2007 *Phys. Rev. B* **76** 245123 © 2007 by the American Physical Society.)

Plugging Eq.(6.60) into Eq.(6.52) one obtains the following nearest-neighbor tight-binding equation of motion

$$(E - \tilde{\varepsilon}_n(E))\psi_n - t_0 \left(1 - \chi\theta_{n,n+1}^2\right)\psi_{n+1} - t_0 \left(1 - \chi\theta_{n,n-1}^2\right)\psi_{n-1} = 0, \quad (6.61)$$

where ψ_n is the electronic wave function at site n. This expression generalizes the equation of motion usually considered within the framework of the Su-Schrieffer-Heeger Hamiltonian,[130] including the quadratic dependence $\theta_{n,n\pm1}^2$ in the off-diagonal terms (instead of a linear one). Thus, non-linearity emerges in Eq.(6.61) as a natural consequence of the three-dimensional DNA geometry.

At very low temperatures the bases remain very close to the equilibrium positions, hence providing a suitable physical scenario to estimate the main contribution of purely geometrical (helical) effects on the charge transport. In that case, one can reasonably assume $\theta_{n,n+1} = \theta_0 \ \forall n$, so that Eq.(6.61) reduces to

$$(E - \tilde{\varepsilon}_n(E))\psi_n - \tau_0(\psi_{n+1} + \psi_{n-1}) = 0, \quad (6.62)$$

where $\tau_0 \equiv t_0 \left(1 - \chi\theta_0^2\right)$. For the sake of simplicity let us consider homopolymer chains, like polyG-polyC or polyA-polyT, so that all the on-site energies are equal. Assuming periodic boundary conditions the dispersion relation can be expressed as $E = \alpha(E) + 2\tau_0 \cos(kl_0)$ or $E = \beta(E) + 2\tau_0 \cos(kl_0)$, respectively, where k is the wave vector. This expression has the typical form for a monatomic chain, though in this case the renormalized on-site energy $\alpha(E)$ or $\beta(E)$ explicitly depends on the charge carrier's energy according to Eq.(6.34). In the case of polyG-polyC chains, the energy spectrum consists of two asymmetric bands of width

$$W_\pm = \frac{\sqrt{\Delta_\pm} - \sqrt{\Delta_\mp} - 4\tau_0}{2(\alpha_1 - 1)}, \tag{6.63}$$

separated by a gap of width

$$\Delta_g = \frac{4\tau_0 - \sqrt{\Delta_+} - \sqrt{\Delta_-}}{2(\alpha_1 - 1)}, \tag{6.64}$$

where $\Delta_\pm = (a_0 \pm 2\tau_0)^2 - 8(\alpha_1 - 1)t^2 + \gamma(\gamma - 2a_0 \mp 4\tau_0)$.[131] Completely analogous expressions are obtained for polyA-polyT chains. By inspecting Eqs.(6.63) and (6.64) one realizes that the overall electronic structure depends on several physical mechanisms, including the aromatic base stacking between neighboring bps, whose value depends on the helicoidal structure in terms of the parameter τ_0. Thus, helicoidal geometry produces two main effects in the electronic structure of DNA: (i) the HOMO and LUMO bandwidths shrink, and (ii) the width of the gap increases with respect to the values obtained for planar models. Both changes degrade charge transport efficiency. In fact, in Fig.6.25 we compare the Landauer conductance corresponding to the helicoidal and planar models. As we see, the net effect of introducing helical structure is to reduce the conductance spectral width as compared to the value obtained for $\theta_0 = 0$.

6.4.5   Thermal and vibrational effects

The relative motion of bases can either occur in a synchronized manner (normal modes propagation at low temperatures) or incoherently (due to thermal motion at higher temperatures). At physiological temperatures the relative orientation of neighboring bases becomes a function of time, thereby modifying their mutual overlapping in a complex way. Accordingly, realistic treatments of charge migration in DNA should take into account (i) the intrinsic three-dimensional, helicoidal geometry of DNA, and (ii) the coupling between charge motion in DNA and its molecular dynamics, as we have described in the previous Section. In so doing, the question arises as to which DNA vibration modes should be taken to be coupled to charge motion and which ones should not. As a first approximation the transfer integral values were usually computed for idealized molecular DNA geometries assuming that t does not significantly change with the molecular geometry during charge migration (the so-called Condon approximation). However, it was subsequently demonstrated that the electronic coupling between nucleobases is very sensitive to structural fluctuations and the Condon approximation is rather limited.[113] Making use of molecular dynamics calculations it was shown that the standard deviation of the coupling of nucleobases is much larger than its average value. This result indicates that charge transport in DNA preferentially occurs in specific conformations, which may considerably deviate from the canonical B-DNA structure.[121] Therefore, the electronic couplings found for idealized B-DNAs can substantially differ from the corresponding values averaged over thermally accessible DNA configurations. Accordingly, a combined quantum mechanical- molecular dynamics approach should be used in order to obtain more accurate estimates of electronic couplings. In fact, recent studies following this approach have revealed that transfer integral values are extremely sensitive to conformational fluctuations, so that the transport efficiency calculated ignoring thermal fluctuations can be underestimated by several orders of magnitude.[122, 123]

6.4.5.1 Incoherent thermal effects

The role of thermal fluctuations on the charge transfer efficiency has been discussed in a number of works, where the structural fluctuations of the DNA double helix are described by sampling the initial angular velocities and twist angles from a Boltzmann distribution at a given temperature.[112, 124, 125, 126, 127, 129, 132, 133] Most of these works considered a Hamiltonian of the form

$$\mathcal{H} = \sum_{n=1}^{N} \varepsilon_n c_n^\dagger c_n - t_0 \sum_{n=1}^{N-1} \cos(\theta_{n,n+1})(c_{n+1}^\dagger c_n + c_n^\dagger c_{n+1}), \qquad (6.65)$$

which describes charge migration along a single-stranded chain. By comparing Eq.(6.65) with the double-stranded DNA Hamiltonian given by Eq.(6.52) we observe that the transfer integral has been particularized to the form $t_{n,n\pm1} \simeq t_0 \cos\theta_{n,n\pm1}$, where $\theta_{n,n+1}$ is the relative twist angle. In addition the Watson-Crick energetics is simplified as $\tilde{\varepsilon}_n(E) \to \varepsilon_n$ in the kinetic energy term. Each $\theta_{n,n+1}$ is an independent random variable that follows a Gaussian distribution with average $\langle \theta_{n,n+1} \rangle = 0$, whereas the variance is taken according to the equipartition law, i.e., $\langle \theta_{n,n+1}^2 \rangle = k_B T / I\Omega^2$, where I is the reduced moment of inertia for the relative rotation of the two adjacent bases and Ω is the oscillator frequency of the mode ($I\Omega^2/k_B = 250K$). In the small fluctuations approximation one further considers $t_0 \cos(\theta_{n,n+1}) \sim t_0(1 - \theta_{n,n+1}^2/2)$, which coincides with Eq.(6.60) for $\chi = 1/2$. Some illustrative examples of the result derived from Eq.(6.65) are shown in the figures below.

In Fig.6.26 we show the temperature dependent transmission coefficient for a periodic polyG-polyC DNA chain, whereas in Figs.6.27 and 6.28 this magnitude is shown for two representative samples of biological interest: a sequence extracted from the sequenced part of the Human chromosome 22 (Ch22), and

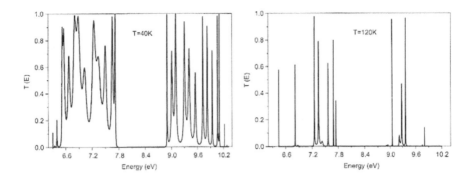

FIGURE 6.26

Transmission coefficient at $T = 40$ K (left) and $T = 120$ K (right) for a polyG-polyC chain with 30 bp.[127] (Courtesy of Ai-Min Guo).

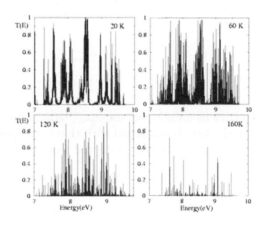

FIGURE 6.27

Temperature dependent transmission coefficient for a 20nm chain sequence extracted from the human chromosome Ch22 at different temperatures. (Adapted from ref.[134]. Courtesy of Stephan Roche).

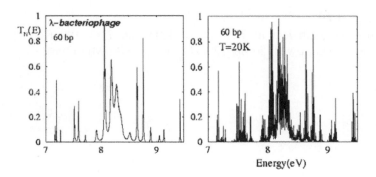

FIGURE 6.28

The transmission coefficient for a λ-bacteriophage based sequence with 60 bps at $T = 0$ K (panel on the left) is compared with that corresponding to a finite temperature one (panel on the right). (Adapted from ref.[134]. Courtesy of Stephan Roche).

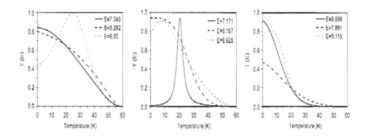

FIGURE 6.29

Temperature dependent transmission coefficient for periodic polyG-polyC (left frame), Fibonacci-GC (middle frame), and random DNA (right frame) for some selected values of the energy (expressed in eV).[127] (Courtesy of Ai-Min Guo).

a short sequence extracted from the complete λ-phage DNA genome (see Section 6.5.3). In Fig.6.27, the temperature dependent transmission coefficient for the chain (CTTCCGGAGG CTGAGGCGGA TGAATCACGA GGTCAGGAGT TCAA-GACCAG CCTGGCCAAC) extracted from the Ch22 (starting site position = 150.000) is shown. As expected, temperature yields misorientations of adjacent bases that result in temperature dependent base-base hoppings. At low temperatures ($T = 20 - 60$ K), the transmission spectrum presents a relatively large number of transmitting states, due to a breaking of level degeneracy. At higher temperatures, the number of transmitting states progressively decreases but interestingly there persist many states with high transmission coefficient at temperatures as high as ~ 160 K. A similar behavior is obtained for the λ-phage chain sequence λ_1-GGGCGGCGAC CTCGCGGGTT TTCGCTATTT ATGAAAATTT TCCGGTTTAA GGCGTTTCCG, as well (Fig.6.28).

Quite interestingly, the robustness of these high transmission states seems to be closely related to the topological order of the underlying lattice. This property is illustrated in Fig.6.29. In the random lattice the transmission coefficient progressively decreases as the temperature is increased as a consequence of reduced quantum coherence. This trend is also observed for most energy states in the periodic and quasiperiodic chains, but in both systems there also exist some energy values exhibiting just the reversed behavior, i.e.,

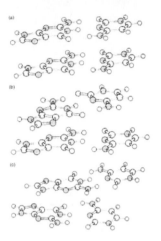

FIGURE 6.30

Geometries of a (G:C)$_2$ stack: (a) symmetric structure and (b) B-form structure. (c) A snapshot from the molecular dynamics trajectory with the largest electronic coupling value between both bps.[123] (Courtesy of Alexander Voityuk).

the value of the transmission coefficient increases with temperature for a certain temperature range. For instance, for Fibonacci-GC (middle frame) the transmission coefficient peaks at $T = 7$ K and $T = 20$ K for the energy values $E = 6.975$ eV and $E = 7.171$ eV, respectively, indicating a thermal enhancement of electrical conductance. Such an enhancement can also be observed in the polyG-polyC chain at a somewhat higher temperature ($T = 25$ K) for the state with energy $E = 6.90$ eV.

Therefore, thermal fluctuations in base stack conformation play an important role in determining charge mobility within DNA. It is then expected that the motion of surrounding water molecules and counterions will play a significant role as well. In order to address this issue one needs to identify which are the DNA helix structural parameters related to a high charge transfer efficiency. In fact, since electronic couplings between nucleobases are very sensitive to conformational changes (Fig.6.24) [113] different conformations must exhibit different conduction properties. Indeed, studies based on molecular dynamics and quantum chemical calculations have pinpointed a number of snapshots showing base stack conformations exhibiting very strong electronic couplings (Fig.6.30). According to the calculations, undertwisted structures ($\theta_{n,n\pm1} < 20^o$) are of special interest because strong couplings are found over a wide range of the remaining structural parameters considered.[123] The finding of these particularly favorable structural configurations may be indicating a relatively general trend, likely related to the presence of some current enhancement due to coupling to certain normal modes. This appealing issue will be analyzed in more detail in the next section.

6.4.5.2 Normal modes dynamical effects

As mentioned before, normal mode calculations have usually considered a substantially reduced number of freedom degrees for DNA, namely, the stretch oscillations of each base back and forth with respect to the center of mass of the system located at the helical axis, and the twist oscillations of each bp as a whole around the helical axis. In this section we will focus on coherent transport due to the coupling between low frequency vibration modes and charge motion through duplex DNA, explicitly taking into account its characteristic helical geometry. To this end, we shall consider the equations of motion, derived from Eq.(6.57) via the canonical equations, which read

$$\ddot{\varphi}_n = \frac{k}{m\xi^2} \left\{ \left(1 - \frac{l_0}{d_{n,n+1}} \right) f_{n+1}(\theta) - \left(1 - \frac{l_0}{d_{n,n-1}} \right) f_{n-1}(\theta) \right\}, \quad (6.66)$$

where $f_{n\pm1}(\theta) \equiv c^2\theta_{n,n\pm1} + R_0^2 \sin\theta_{n,n\pm1}$. The low frequency response is obtained linearizing Eq.(6.66) by considering only linear terms of the Taylor expansion to obtain $\ddot{\varphi}_n \simeq \omega_0^2(\varphi_{n+1} + \varphi_{n-1} - 2\varphi_n)$, where $\omega_0 \equiv \sqrt{k/m}$ is the natural twist frequency of each base. The corresponding dynamical equation for the variables $\theta_{n,n\pm1}$ is then straightforwardly derived to get

$$\ddot{\theta}_{n,n+1} - \ddot{\theta}_{n,n-1} = \omega_0^2(\theta_{n+1,n+2} - 3\theta_{n,n+1} + 3\theta_{n,n-1} - \theta_{n-1,n-2}). \quad (6.67)$$

This expression describes a *correlated* motion involving three consecutive bps (codon unit cell). Searching for solutions in the form of linear waves we plug the ansatz $\theta_{n,m} = \sqrt{2}\theta_0 e^{i\omega t} \cos((n+m)q/2)$, where q is the wave number, into Eq.(6.67) to obtain the dispersion relation $\omega^2 = 4\omega_0^2 \sin^2(q/2)$.[80] Finally, inserting $\theta_{n,m}$ into Eq.(6.61) one can express it in the form

$$(E - \tilde{\varepsilon}_n)\psi_n - (\tau_0 - B\Omega T_n(\tilde{\Omega}))\Psi_{n,+} - 2B\Omega(1 - \Omega^2)U_{n-1}(\tilde{\Omega})\Psi_{n,-} = 0, \quad (6.68)$$

where $B \equiv t_0\chi\theta_0^2$, $\Psi_{n,\pm} \equiv \psi_{n+1} \pm \psi_{n-1}$, $T_k(\tilde{\Omega})$ and $U_{k-1}(\tilde{\Omega})$ are Chebyshev polynomials of the first and second kinds, respectively (see Section 9.5), $\tilde{\Omega} \equiv 2\Omega^2 - 1$, and $\Omega \equiv 1 - \omega^2/2\omega_0^2$ ($\Omega \in [-1,1]$). This expression properly extends Eq.(6.61) by including charge-lattice interaction (the so-called polaron) effects. The effect of twisting on charge transport through DNA has been investigated by considering two types of polarons depending on the coupling between the transfer integral and nucleotides geometry: radial polarons (where charge induced deformations mainly affect the radial variables) and twist polarons. By all indications twist polarons can transport charge in a very efficient way (even in the presence of a base-pair inhomogeneity), whereas radial polarons experience either reflection or trapping.[135]

By inspecting Eq.(6.68) we realize that the motion equation considerably simplifies in the cases $\Omega = 0$ and $\Omega = \pm1$. In the case $\Omega = 0$ (i.e., $\tilde{\Omega} = -1$,

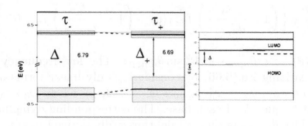

FIGURE 6.31

HOMO-LUMO bands for a polyGACT-polyCTGA chain corresponding to the vibrational states ω_- and ω_+ (left panel). The overall electronic spectrum structure is shown in the right panel on a broader energy scale. The spectra have been derived from Eq.(6.69) making use of the same model parameters used to plot Fig.6.25. ([116] Reprinted figure with permission from Maciá E 2007 *Phys. Rev. B* **76** 245123 © 2007 by the American Physical Society.)

$\omega = \sqrt{2}\omega_0$), Eq.(6.68) reduces to Eq.(6.62). In this way, when a charge couples to the $q = \pi/2$ vibrational state the resulting charge dynamics mimics that corresponding to the equilibrium configuration one. The frequencies corresponding to the cases $\Omega = \pm 1$ (i.e., $\tilde{\Omega} = 1$) are located at the edges of the frequency spectrum ($\omega_+ = 0$ and $\omega_- = 2\omega_0$). Then, using the relationship $T_n(1) = 1$, $\forall n$, Eq.(6.68) takes the form

$$(E - \tilde{\varepsilon}_n(E))\psi_n - \tau_\pm(\psi_{n+1} + \psi_{n-1}) = 0, \qquad (6.69)$$

where $\tau_+ = t_0(1 - 2\chi\theta_0^2)$ and $\tau_- = t_0$ are respectively labelled after frequencies $\omega_\pm$. The mathematical structure of Eq.(6.69) describes a charge propagating through a linear chain with an effective transfer integral whose value depends on the considered frequency. We note that, broadly speaking, an increase in the transfer integral value usually contributes to a lowering of the system energy (see Eq.(6.52)) so that the charge gets localized by its interaction with the lattice (polaronic effect).[130, 136, 137] In the ω_+ case, however, the transfer integral becomes negative ($\tau_+ = -1.3t_0 \simeq -0.2$ eV for $\chi = 2.9$) and charge becomes delocalized instead. The normal modes corresponding to the DNA codon shown in Fig.6.23a are $\omega_1 = 0 = \omega_+$, $\omega_2 = \eta\omega_0$, and $\omega_3 = \sqrt{3}\eta\omega_0$, where $\eta \equiv f(\theta_0)l_0^{-1}\xi^{-1} \simeq 0.961$. Therefore, the lowest frequency state ω_+ is a normal mode describing the simultaneous rotation of all bps around the helical axis by an arbitrary amount. On the other hand, the coupling of the charge to the ω_- vibration state results in a competition between dynamical and helicoidal effects, so that the equation of motion reduces to that of an effective 2D model in this case.

The energy spectra corresponding to the effective hopping $\tau_\pm$ are shown in Fig.6.31 for the polyGACT-polyCTGA DNA chain studied in Section 6.4.2. By comparing both spectra two main features can be observed close to the Fermi level: (i) a significant broadening of the band widths; and (ii) a narrowing of the HOMO-LUMO gap width for the ω_+ spectrum as compared to the ω_- one. Therefore, the coupling with the low band edge vibrational state results in a *significant improvement* of the charge transport efficiency through the DNA chain, as it is illustrated by the broadening of the Landauer conductance spectral window shown in Fig.6.32. A similar broadening effect has been reported for polyA-polyT chains with 15 bps as arising from compressional acoustic modes propagating through the helical axis (although in the case of a polyG-polyC chain of the same length a narrowing effect was observed instead).[84]

Although an ensemble of bps twisting back and forth around the helix axis generally results in a degraded charge transfer efficiency, a significant improvement of charge migration can occur via charge coupling to the lattice modes at low temperatures. In fact, from basic physical principles one expects the acoustic modes will significantly affect the conductance at temperatures below the Debye temperature, which measures the temperature above which all modes begin to be excited, and below which modes begin to be progressively

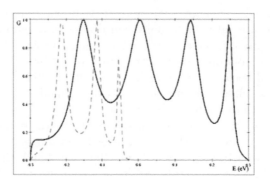

FIGURE 6.32

The Landauer conductance (in G_0 units) as a function of the Fermi level energy of a polyG-polyC oligomer with $N = 10$ bps corresponding to the planar model (dashed curve) is compared to that corresponding to a 3D model coupled to the ω_+ normal mode (solid line). The model parameters are those used in Fig.6.25. The origin of energy is set at ε_G. ([116] Reprinted figure with permission from Maciá E 2007 *Phys. Rev. B* **76** 245123 © 2007 by the American Physical Society.)

inactive. By assuming a speed of sound in B-DNA form of 1900 ms^{-1},[138] and a lattice constant of approximately 0.34 nm, the Debye temperature was estimated as $\Theta_D \simeq 166$ K for longitudinal compression modes.[84] On the other hand, specific heat measurements of biological DNA samples over the temperature range $0.5 - 5$ K suggest significantly smaller values ($\Theta_D \simeq 20 - 40$ K) for torsional modes.[139] Accordingly, it is reasonably expected that charge-vibration coupling effects would require low temperature measurements of the DNA electrical conductance to be clearly appreciated.

6.5 Long-range correlations: The biophysicist viewpoint

6.5.1 General motivations

Amongst the various physical, chemical, or biological phenomena that might be inferred from sequence correlations, charge transfer properties deserve particular attention. Indeed, the nature of DNA-mediated charge migration has been related to the understanding of damage recognition process, protein binding, or with the task of designing nanoscale sensing of genomic mutations, opening new challenges for emerging *nanobiotechnologies*.[63, 64, 65, 82] Short and long range correlations between base pairs further provide valuable information to distinguish between almost random distributions, and more complex sequences, whose long range correlations might also be associated with some biological properties (folding, introns vs exons featuring).[66, 67, 68, 69, 140]

As mentioned in Section 1.8, biological and artificial DNA molecules significantly differ in size, chemical complexity, and their kind of structural order. Consequently, one can hardly expect that results obtained from the study of the oversimplified synthetic molecular systems may be directly extrapolated to understand the physical properties of complex DNA molecules of biological interest. In fact, both the sugar-phosphate backbone and the nucleotide bases sequence are periodically ordered in, say, polyG-polyC chains, whereas in biological DNA the nucleotide bases are aperiodically ordered instead. From general principles one expects the aperiodic nature of the nucleotide sequence distribution would favour localization of charge carriers in biological nucleic acids, reducing charge transfer rate due to backscattering effects. Nevertheless, this scenario must be refined in order to take into account correlation effects among nucleotides reported in biological DNA samples, since these correlations can enhance charge transport via resonant effects.[134, 141, 142]

Scale invariant properties in complex genomic sequences with thousands of nucleotide base pairs have been investigated from a mathematical viewpoint, and in particular wavelet analysis has revealed complex fingerprints. Such studies clearly show that correlations are present at different scales, and that they are strongly sequence dependent. The statistical significance of the regular features in nucleotide sequences should be estimated with respect to the corresponding characteristics for random DNA sequences with the same nucleotide composition. To gather deeper insight of the effect of correlations in electronic transfer, in the following sections we discuss the electronic transmission properties of various correlated and uncorrelated DNA sequences connected to metallic leads.

6.5.2 Fibonacci DNA

In order to get some insight into the nature of charge transport in DNA molecules of biological interest, for which exact or even approximate analytical results are not available, the study of quasiperiodic systems exhibiting characteristic self-similarity properties may be of considerable help. One typical aspect of any quasiperiodic system is the presence of some repeating features which show up as some sort of building blocks of these structures. As an archetypal example, let us consider the Fibonacci-GC sequence, which is constructed starting from a G base as a seed and following the inflation rule G$\rightarrow$ GC and C$\rightarrow$ G. This gives successively, G,GC,GCG,GCGGC,GCGGCGCG, $\cdots$, and so forth. In the thermodynamic limit the ratio of (majority) G bases over (minority) C bases will approach the golden mean value $\tau = (1+\sqrt{5})/2 \sim 1.618$. A symmetrical inflation rule (C$\rightarrow$CG and G$\rightarrow$C) can be also used to generate a Fibonacci chain where the roles of majority and minority bases are exchanged.

Since the basic building block in the Fibonacci inflation rule is the dimer GC, we briefly review some basic properties of the periodic polyGC chain for the sake of illustration. Its electronic structure is given by the dispersion relation (see Section 9.2)

$$4t^2 \cos^2 q = E^2 - (\varepsilon_C + \varepsilon_G)E + \varepsilon_C\varepsilon_G, \qquad (6.70)$$

so that the energy spectrum of a GC chain is composed of two wide bands separated by a gap of width $\Delta_{GC} = \varepsilon_C - \varepsilon_G = 1.12$ eV, as it is illustrated in the left-top panel of Fig.6.33. The transmission coefficient for a GC chain embedded between guanine leads is given by (see Section 9.5.2)

$$T_N(E) = \left[1 + \frac{\Delta_{GC}^2}{4t^2 - (E - \varepsilon_G)^2} U_{\frac{N}{2}-1}^2(v)\right]^{-1}, \qquad (6.71)$$

where $U_{m-1}^2(v)$ is a Chebyshev polynomial of the second kind, and $v \equiv (E - \varepsilon_C)(E - \varepsilon_G)/2t^2 - 1$. The next approximant in the series is the periodic polyGCG chain. In that case the energy spectrum is composed of three bands, as can be readily checked from the dispersion relation

$$2t^3 \cos 3q = (E - \varepsilon_G)^2(E - \varepsilon_C) - t^2(3E - 2\varepsilon_G - \varepsilon_C), \qquad (6.72)$$

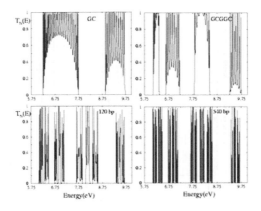

FIGURE 6.33
Energy dependent transmission coefficient for a finite length Fibonacci-GC series approximants of increasing length. The spectrum is defined within $[\varepsilon_M - 2t_M = 5.75, \varepsilon_M + 2t_M = 9.75]$ with $\varepsilon_M = 7.75$ eV and $t_M = 1$ eV. ([134] Courtesy of Stephan Roche).

and its electronic spectrum consists of three bands separated by two gaps. As we proceed by considering higher order approximants to the Fibonacci DNA, new bands and gaps progressively appear in the energy spectrum, showing a hierarchical nested structure (Fig.6.33). In the bottom panels of Fig.6.33 we compare the energy dependence of the transmission coefficient numerically obtained for Fibonacci approximants with an increasing number of bases. It can be clearly appreciated that, as the system grows larger, several peaks with high transmission values remain at certain energy values. In addition, some degree of clustering around these resonant energies can be appreciated. It has been shown that the global structure of the asymptotic electronic spectrum of quantum quasiperiodic lattices can be obtained in practice by considering very short approximants to infinite quasiperiodic chains (see Section 5.3). To check this result in the present context, let us consider the third order approximant of a Fibonacci chain, which corresponds to the periodic poly(GCGGC) chain containing five nucleotides in its unit cell. Its dispersion relation is given by

$$2t^5 \cos(5q) = \frac{\Phi^3}{E - \varepsilon_C} - \Phi t^2 \left(5E - 4\varepsilon_G - \varepsilon_C\right) + t^4 \left(5E - 3\varepsilon_G - 2\varepsilon_C\right), \quad (6.73)$$

where $\Phi(E) \equiv (E - \varepsilon_G)(E - \varepsilon_C)$. The energy spectrum of a GCGGC chain is thus composed of three broad bands, whose centers are located at the energies $E_2 = 6.915$ eV, $E_3 = 8.143$ eV, and $E_4 = 9.527$ eV, plus two

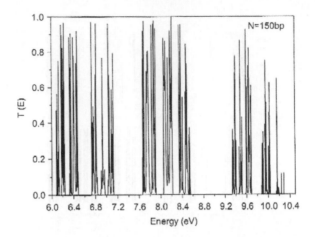

FIGURE 6.34

Transmission coefficient spectrum for a Fibonacci-GC approximant containing 150 base-pairs.[127] (Courtesy of Ai-Min Guo).

narrower bands located at the edges of the spectrum at $E_1 = 6.190$ eV and $E_5 = 10.213$ eV. This electronic structure accurately described the location of the main bands and gaps in the highly fragmented energy spectrum shown in Fig.6.34. Accordingly, there exist a set of resonant energies which are robust enough to persist against backscattering effects due to the presence of C bases interspersed in the Fibonacci-GC chain. Thus, one may be tempted to think that these states should exhibit good transport properties even in the thermodynamic limit.

To further substantiate such a possibility let us consider the transmission coefficient corresponding to the poly(GCGGC) chain embedded between guanine loads, which is given by [134]

$$T_N(E) = \left[1 + q(x, y)U^2_{\frac{N}{5}-1}(w)\right]^{-1}, \tag{6.74}$$

where $w = 16x^2y^3 - 16xy^2 - 4yx^2 + 3y + 2x$, and $q(x, y)$ is a rational function. According to Eq.(6.74) the roots of the Chebyshev polynomial label a full transmission peak series given by $\cos(5k\pi/N) = w$, with $k = 0, ..., N$. Then, as the Fibonacci chain length is increased, less and less states will present good transmissivity, due to the progressive fragmentation of the spectrum. Nonetheless, a significant number of resonant states, satisfying the full transmission condition given by Eq.(6.74), will persist in the thermodynamic limit. For instance, states belonging to the broader central bands around $E_2 \simeq 6.9$ eV and $E_3 \simeq 8.1$ eV are very robust to the progressive fragmentation of the energy spectrum, as it can be seen in the bottom frames of Fig.6.33.

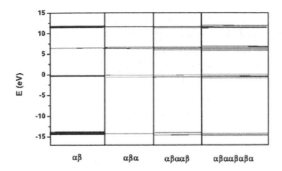

FIGURE 6.35
The energy spectra of successive approximants of the Fibonacci DNA chain
are compared to that corresponding to the polyGACT-polyCTGA chain. The
model parameters are listed in table 6.3. ([105] Reprinted figure with permission from Maciá E 2006 *Phys. Rev. B* **74** 245105 © 2006 by the American
Physical Society.)

These basic results can be generalized to the more realistic case of the
double-helix DNA models considered in Section 6.4. By inspecting the renormalized binary lattice shown in Fig.6.19c we realize that, instead of considering a periodic chain with unit cell $\alpha\beta$, we could arrange the G:C and A:T
complementary pairs according to the Fibonacci sequence, by means of the
substitution rule $\alpha \to \alpha\beta$ and $\beta \to \alpha$. In this way, we obtain the series of unit
cells $\alpha\beta$, $\alpha\beta\alpha$, $\alpha\beta\alpha\alpha\beta$, $\alpha\beta\alpha\alpha\beta\alpha\beta\alpha$, ... The first representative in this series
coincides with the periodic polyGACT-polyCTGA chain (recall that, accord-
ing to Eq.(6.34), the bps G:C/C:G and A:T/T:A are indistinguishable in the
renormalized chain). The following terms in the sequence describe periodic
DNA chains whose unit cell becomes progressively more complex, attaining
the quasiperiodic order characteristic of the Fibonacci sequence in the thermodynamic limit $N \to \infty$. Accordingly, the systematic study of these approximants series provides useful information regarding the progressive emergence
of quasiperiodic order in the system.

Following the approach introduced in the previous section the dispersion
relations of a successive series of approximants can be obtained from the
knowledge of the corresponding global transfer matrices, respectively given by
$Q_\alpha Q_\beta Q_\alpha$, $Q_\beta Q_\alpha Q_\alpha Q_\beta Q_\alpha$, and so on. The corresponding spectra are shown
in Fig.6.35. By inspecting this figure we see that the four bands originally
present in the energy spectrum of the polyGACT-polyCTGA chain become

progressively fragmented as we consider successive approximants. Thus, the two central bands in the energy spectrum of the $\alpha\beta$ chain split into two sub-bands in the energy spectrum of the $\alpha\beta\alpha$ approximant, into three subbands in the energy spectrum of the $\alpha\beta\alpha\beta$ approximant, and into five subbands in the energy spectrum of the $\alpha\beta\alpha\beta\alpha\beta\alpha$ approximant. As we see, this fragmentation scheme follows the series $\{1, 2, 3, 5, ...\}$ subbands. In a similar way, we see that the edge bands in the energy spectrum of the $\alpha\beta$ chain follow the fragmentation scheme $\{1, 1, 2, 3, ...\}$ subbands. In both cases the fragmentation scheme is described by the Fibonacci series $F_n = \{1, 1, 2, 3, 5, 8, ...\}$. Thus, the total number of subbands composing the spectrum of a given approximant can be expressed as $2F_{\nu-1} + 2F_{\nu-2} = 2F_\nu$, where ν is the number of Watson-Crick bps contained in the approximant unit cell.

This kind of highly fragmented energy spectrum is a typical feature of quasiperiodic systems (see Section 5.3) and gives rise to the presence of two different energy scales in the DNA spectrum. On the one hand, we have a large energy scale (within the range $5 - 14$ eV) determined by the width of the gaps among the main bands. On the other hand, due to the progressive fragmentation of these main bands, an increasing number of narrow gaps progressively appear in the spectra of higher order approximants. In this way, the emergence of the quasiperiodic order naturally introduces a specific, small energy scale in the DNA electronic structure, ranging from about $0.1 - 0.5$ eV for the low order $\alpha\beta\alpha\beta$ approximant, to values well below 100 meV for higher order approximants. The presence of these small activation energies in the electronic structure brings an additional mechanism in order to explain the anomalous absorption feature observed at low $(10 - 100 \text{ meV})$ energies in optical conductivity spectra of biological DNA samples.[143]

What is the nature of the states belonging to this highly fragmented spectra? For systems described in terms of Fibonacci on-site Hamiltonians it has been rigorously proven that the energy spectrum is singular continuous and the amplitudes of their eigenstates do not tend to zero at infinity but are bounded below throughout the system, yielding the value $\Gamma(E) = 0$ in the thermodynamic limit (see Section 5.5). This result certainly holds for the Fibonacci DNA chain as well. In Fig.6.36 we illustrate the progressive fragmentation of the energy spectrum around the energy value $E \simeq -0.4$ eV for increasing order Fibonacci DNA approximants. As we can see, a self-similar, nested structure, characteristic of the long-range quasiperiodic order present in Fibonacci systems, progressively appears in the Lyapunov coefficient overall structure as the complexity of the corresponding unit cell is increased for successive approximants. Nevertheless, the vanishing of the Lyapunov exponent should not be naively interpreted as indicating a Block-like nature for the electronic states (see Section 9.5.4). In most quasiperiodic systems we have critical wavefunctions whose amplitudes are roughly modulated by scaling exponents and one may reasonably expect their related transport properties to be more similar to those corresponding to extended states than to localized ones (see Section 5.6.1). The influence of the nature of the electronic states

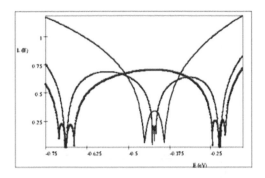

FIGURE 6.36
Lyapunov coefficient as a function of the energy for successive approximants of the Fibonacci DNA chain containing $N = 6$ bp (3 $\alpha\beta$ unit cells, solid gray), $N = 12$ bp (4 $\alpha\beta\alpha$ unit cells, dotted curve), and $N = 20$ bp (4 $\alpha\beta\alpha\alpha\beta$ unit cells, solid black). ([105] Reprinted figure with permission from Maciá E 2006 *Phys. Rev. B* **74** 245105 © 2006 by the American Physical Society.)

on the transport properties of different kinds of aperiodic sequences can be studied in terms of the energy averaged transmission coefficient [144]

$$\bar{T}(N) = \frac{\int_{W_-}^{W_+} T_N(E) \left[1 + \cosh\left(\frac{E-\mu}{k_b T}\right)\right]^{-1} dE}{\int_{W_-}^{W_+} \left[1 + \cosh\left(\frac{E-\mu}{k_b T}\right)\right]^{-1} dE}, \tag{6.75}$$

where $W_\pm$ denote the edges of the allowed energy spectrum and μ is the chemical potential. The transmission coefficient is obtained following the procedure described in Section 6.4.3 for the model Hamiltonian given by Eq.(6.44). Making use of Eq.(6.75) one can estimate the intrinsic resistivity of the DNA molecule from the expression (see Section 9.5.3)

$$\rho(N) = G_0^{-1} \frac{1 - \bar{T}(N)}{\bar{T}(N)} \frac{\pi R_0^2}{N h_0}, \tag{6.76}$$

where $R_0 = 1$ nm and $h_0 = 0.34$ nm are the B-DNA form equilibrium values. Fig.6.37 plots the dependence of the resistivity on the length for different types of periodic and aperiodic sequences. Notice that polyG has the highest conductivity with $< \rho > \sim 10^{-4}$ Ωcm, whereas the conductivity of polyGC is about one order of magnitude lower (see the inset). For a given length value the resistivity of aperiodic sequences is larger than that of periodic

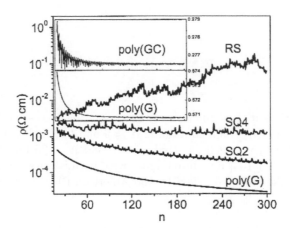

FIGURE 6.37

Dependence of the electrical resistivity with the chain length (expressed in terms of the number of base-pairs, n) for different kinds of aperiodic chains (main frame) according to the following key: Rudin-Shapiro (RS), $G \rightarrow GGC$ and $C \rightarrow CCG$ (SQ4), $G \rightarrow GGC$ and $C \rightarrow GG$ (SQ2). Inset: averaged transmission coefficient $\bar{T}(n)$ versus n for polyG and polyGC chains in the range $n = 20$ to $n = 300$ bp. The DNA chain is described in terms of the Hamiltonian given by Eq.(6.44) with model parameters $t = t_M = \tau = 1$ eV and $\varepsilon_M = \varepsilon_G = 7.75$ eV.[144] (Courtesy of Ai-Min Guo.)

ones, as expected from basic principles. (Note that the relative resistivity relation of aperiodic versus periodic chains in Fig.6.37 bears a close similarity with that exhibited by quasicrystalline versus periodic alloys in Fig.3.2, see Section 3.1.2.) The resistivity value of the aperiodic lattices based on the substitution sequences SQ2 and SQ4 progressively decreases as the chain length is increased, clearly indicating that long-range correlations favor the electrical conductance. In this regard, it is interesting to note that the $\rho(n)$ curve corresponding to the sequence SQ2 (which satisfies the Pisot property and it is therefore quasiperiodic, see Section 4.2.1) almost parallels the $\rho(n)$ curve corresponding to the polyG chain. On the other hand, the $\rho(n)$ curve corresponding to the sequence SQ4 (which is not quasiperiodic) becomes essentially independent of the system length, hence indicating that long-range correlation effects just balance backscattering effects. Finally, we observe that the resistivity of the Rudin-Shapiro sequence (which shares a diffuse Fourier spectrum with random lattices, see Section 5.2) increases as the chain length is increased. This behavior can be understood as indicating that backscattering effects are dominant over short-range correlations effects in this case.

Recent studies on the magnetic properties of biological DNA samples, indicating the existence of well defined currents (apparently on a micron scale) in $\lambda-$DNA molecules support the presence of some sort of extended states in these biopolymers.[145] Additional evidence for charge transport through double-stranded DNA oligonucleotides with a non-periodic nucleotide sequence has been obtained from current-voltage curves showing currents within the nA range (see Section 6.3). Therefore, the nature of electronic states in complex macromolecules of biological interest and their related transport properties is an interesting open topic in the field of condensed matter biophysics.

6.5.3   Biological DNA chains

The DNA sequence of the first completely sequenced human chromosome 22 contains about 33.4×10^6 nucleotides. Statistical analyses have unveiled the presence of long range power law correlations which are inferred from scale invariance properties.[140] However, at variance to common substitution aperiodic sequences (see Section 4.2.1), no construction rule allows to generate the whole chain, so that these correlations are of an intrinsically different nature. It is an important question to identify the limits of coherent charge transport in such a complex DNA sequence, that may be related with biological features.[146] Given the huge amount of nucleotides, the physically relevant task would rather be to determine to which extent charge transfer takes place through the G-HOMO, in comparison with uncorrelated chains. The results previously obtained for quasiperiodic chains will help us deepen our discussion about the relation between long range correlations and charge transport in aperiodic systems. Indeed, scale invariance characteristic of quasiperiodic sequences illustrates the way correlations among different bases at several scales give rise to a similar scaling of transmission properties. We thus start

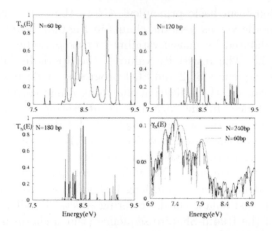

FIGURE 6.38

Transmission coefficients for chromosome 22 based sequences with increasing number of base pairs. Bottom-*right*: Energy dependent Lyapunov coefficient for two chromosome 22 based sequences. ([134] Courtesy of Stephan Roche.)

focusing on Ch22-based sequences extracted from the sequenced part entitled NT_{011520} containing 182.606 bp and retrieved from the Biotechnology Information (NCBI). The first 20nm sequence is constructed by starting from site 1.500 of the full NT_{011520} sequence and then extracting the first 60 first bp, namely GTGAAACCCC ATCTCTACTA AAAATCCAAA AAAATTAGCC GGGTGTG-GTG GCAGGCGCCT. Next sequences are constructed by adding the following next 60 bp of the sequence. Fig.6.38 presents the energy-dependent transmission coefficients and Lyapunov coefficients of chains with lengths between 20 nm and 90 nm. Lyapunov coefficients are computed for two Ch22-based finite sequences (Fig.6.38-bottom right). Compared to the quasiperiodic case (Fig.6.33), it is striking that self-similarity seems absent from the spectrum, so that the scaling in chromosome 22 relies on a totally different kind of long range correlations.

Let us now consider the case of the λ-phage DNA whose transport properties have been widely investigated experimentally (see Section 6.3). Its complete genome contains 48502 base pairs with a total length of about $16\mu m$. In Fig.6.39 we show the transmission spectra for short sequences of the chain but with increasing number of base pairs, corresponding to systems from 20 nm to 80 nm long. The starting sequence with 60bp is λ_1-GGGCGGCGAC CTCGCGGGTT TTCGCTATTT ATGAAAATTT TCCGGTTTAA GGCGTTTCCG while larger sequences are constructed from λ_1 by successively adding the next 60bp of the complete sequence. The transmission spectrum is critically length dependent. By increasing the sequence length, transmission degrades and for

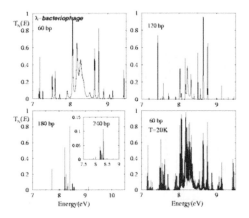

FIGURE 6.39

Energy dependent transmission coefficients for λ-bacteriophage-based sequences with increasing number of bp. The effect of thermal fluctuations on the energy spectrum structure of the genome fragment containing 60bp is shown in the bottom-right panel. (Adapted from ref.[134]. Courtesy of Stephan Roche.)

lengths above ~ 100 nm, almost all states are strongly backscattered by the energetic profile. Different parts of the λ-phage sequence have also been investigated, suggesting that the exact details of a given transmission pattern critically depend on the exact structure of the sequence.[125] In conclusion, genomic DNA sequences do manifest long range correlations which however can not be assumed to be self-similar in the usual sense.

Self-similarity as present in quasiperiodic chains has been demonstrated to induce extended states at a finite number of energies. The whole spectrum presents scale invariant features associated with the progressive partitioning of the spectrum with increasing length. To further illustrate this point in Fig.6.40 we explicitly compare the transmission coefficients of ideal quasiperiodic (Fibonacci) and biological DNAs. In both cases, $T_N(E)$ is characterized by a series of resonant peaks with high transmission. As the sequence length increases, fewer states will present good transmissivity, due to the progressive fragmentation of the spectrum, although several peaks with high transmission remain at certain energy values, and new ones may appear. For Fibonacci and Ch22-based sequences, these resonant energies are robust enough to persist against backscattering effects due to interspersed G bases along the sequence. In addition, the Lyapunov coefficient shown in Fig. 6.41 illustrates intrinsic properties of the two correlated, albeit of different nature sequences. Indeed, the series of main elliptic bumps found in the Fibonacci sequence with 60

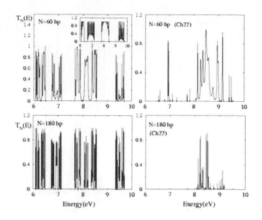

FIGURE 6.40

Transmission coefficient for Fibonacci-GC (left frames) and Ch22-based sequences (right frames). Inset: $T_N(E)$ obtained from Eq.(6.74) for a periodic approximant of length $N = 50$ bp. (From ref.[141]. Reprinted figure with permission from Roche S, Bicout D, Maciá E and Kats E 2004 *Phys. Rev. Lett.* **92** 109901 © 2004 by the American Physical Society.)

bp is reproduced in the 480 bp sequence, which present additional features associated with the partitioning of spectrum. While self-similarity fully characterizes the quasiperiodic sequence, the scaling properties in Ch22 rely on a totally different kind of long range correlations, with no hints of self-similar patterns.

In contrast, the fragmentation of the spectrum strongly affects the transmissivity of the uncorrelated random sequences (not shown here). All resonant states (when any) are evenly affected and the corresponding transmission decreases as the sequence length is increased. From a statistical analysis over many random sequences, it clearly appears that Ch22-based sequences exhibit much higher charge transfer efficiency over much longer distances in comparison with uncorrelated random sequences.

6.5.4   Comparison between different species

As shown before, genomic sequences have different long range correlations which result in more or less general trends in charge transfer. The case of Human and Pygmy Chimpanzee D1s80 DNA is particularly interesting since, despite their lack of periodicity, such sequences contain a large number of GG-pairs that have critical consequences on charge transfer. But more striking is the particularly strong efficiency of delocalization in the case of Pygmy

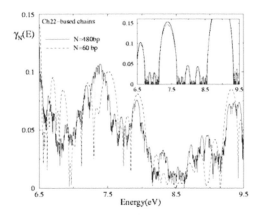

FIGURE 6.41

Lyapunov coefficient for Ch22-based (main frame) and Fibonacci- GC quasi-periodic sequences (inset). (From ref.[141]. Reprinted figure with permission from Roche S, Bicout D, Maciá E and Kats E 2004 *Phys. Rev. Lett.* **92** 109901 © 2004 by the American Physical Society.)

Chimpanzee D1s80 DNA when compared to its Human counterpart.

Fig.6.42 shows the transmission coefficient spectra for an increasing number of base pairs in the case of Pygmy Chimpanzee and Human D1s80 DNA. The charge transfer properties corresponding to relatively long chains are shown to be particularly stable, as compared to that observed in other samples of biological interest. In particular, charge transfer through a chimpanzee's sequence is much more efficient, since its transmission coefficient is enhanced by several orders of magnitude as compared to the λ-bacteriophage transmission coefficient corresponding to chains longer than 80 nm. Even more striking is the fact that the transmission corresponding to the Human D1s80 DNA sequence is more quickly damped with increasing sequence length. In fact, although it is supposed that this gene encodes similar biological functions in both Human and Chimpanzee genomes, the corresponding charge transfer properties turn out to be quite different, as deduced from their related transmission fingerprints.

In Fig.6.43 a comparison between the tunneling currents of different D1s80 sequences is shown. As found in the case of transmission coefficients, the D1s80 chimpanzee sequence appears much more conductive when compared with the D1s80 human sequences. This difference can be understood in terms of density of repeating units that are very important in such sequences.

The possible relation between electric transport properties and the gene coding/non-coding character of genomic sequences corresponding to different

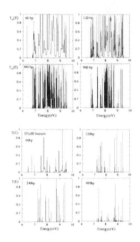

FIGURE 6.42

Transmission coefficients for D1s80-based sequences extracted from the Pygmy chimpanzee (upper diagram) and Human (lower diagram) genetic codes. (Adapted from ref.[134]. Courtesy of Stephan Roche).

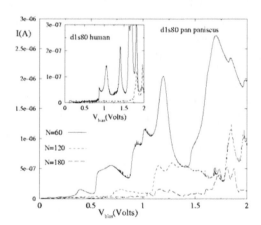

FIGURE 6.43

Tunneling currents for chains extracted from the D1s80 chimpanzee (main frame) and human (inset) sequences as a function of applied bias for $E_M = 6.75$ eV while all other hopping integrals are set to 1. ([134] Courtesy of Stephan Roche).

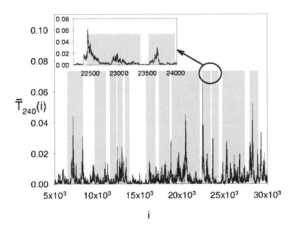

FIGURE 6.44

Comparison of the transmission coefficient corresponding to the propagation
of charge through a sequence of 240 nucleotides located around the ith one in
a fragment of the yeast chromosome III. All states comprised in the energy
interval 5.75-9.75 eV are included and the value $t = 1$ eV is adopted for the
$\pi-\pi$ coupling transfer integral. The coding regions included in the range from
5000th to 30000th nucleotides are shaded. An enlarged plot corresponding to
the 22000th to 24000th interval is shown in the inset.[147] (Courtesy of Chi-
Tin Shih).

species has been thoroughly discussed, combining the transfer matrix formalism and symbolic sequence analysis, in a number of works.[147, 148] Making use of the Hamiltonian given by Eq.(6.44), energy-averaged transmission spectra were obtained for several sequences of *complete* genomes. The statistical results reported indicate that, when all possible states contribute to conduction, most sharp peaks of the averaged transmission coefficients are mainly located in the protein coding region (Fig.6.44). At first sight, this result suggests that coding regions might exhibit larger conductance values than non-encoding ones. This correlation, however, is very sensitive to the value of the $\pi - \pi$ stacking coupling adopted in the calculations, and significantly depends on the particular genome sequence being considered. Thus, when realistic values of the transfer integral ($t \leq 0.5$ eV) are adopted, the electronic conductance becomes poorer in the coding regions, reversing the trend observed for larger transfer integral values. On the other hand, it was reported that coding regions have larger conductance values for yeast chromosomes III, VIII and X, and Ureaplasma parvum sequences, whereas for Acinetobacter sp. ADPI, Deinococcus radidurans, and Chlamidia trachomatis D/UW-3/Cx sequences the coding regions exhibit smaller conductance.[147] Subsequent work has demonstrated that the relation between the transmission efficiency and the coding nature of a given sequence is strongly dependent on the incident energy of the injected carrier as well.

The biophysical emerging scenario relates charge transfer properties and long-range oxidation mechanisms in DNA molecules with important biological processes in living organisms, as that played by the so-called base excision repair enzymes. These enzymes locate different kinds of DNA joint mutations or mismatches between complementary bases by probing the DNA-mediated charge transport. Accordingly, it has been tentatively suggested that certain cancerous mutations could shortcut the DNA repair process, hence leading to carcinogenicity.[148]

References

[1] In a lecture derived in May 1948 at the Sir Jesse Boot Foundation in England. This quotation is adapted from an extract taken from *Linus Pauling: Chemist and Molecular Biologist* by Rich A, in 1995 *DNA: The Double Helix Perspective and Prospective at Forty Years*, Ed. Chambers D A, *Ann. N. Y. Acad. Sci.* **758** 79

[2] Crick F H C *The Impact of Linus Pauling on Molecular Biology: A Reminiscence*, in 1992 *The Chemical Bond. Structure and Dynamics*, Ed. Zewail A (Academic Press, San Diego), p. 89. The text continues: "Jim and I didn't read that article until after we had discovered the double helix, but there was no doubt that Linus put forward that idea before we did".

[3] In Ref.[1], p.79

[4] In Ref.[2], p.93

[5] Astbury W T and Bell F O 1938 *Nature* **141** 747

[6] Kendrew J C *Physics, Molecular Structure and Biological Function*, in 1969 *Biology and the Physical Sciences*, Ed. Devons S (Columbia University Press, New York), p.97

[7] Furberg S 1952 *Acta Chem. Scand.* **6** 634

[8] Squires G L 2003 *Contemporary Physics* **44** 289

[9] Cochran C, Crick F H C and Vand V 1952 *Acta Cryst.* **5** 581

[10] Stokes A R 1955 *Prog. Biophys. Biophys. Chem.* **5** 140

[11] Franklin R E and Gosling R G 1953 *Nature* **171** 740; Watson J D and Crick F H C 1953 *Nature* **171** 737; Wilkins M H F, Stokes A R and Wilson H R 1953 *Nature* **171** 738

[12] Lucas A A and Lambin P 2005 *Rep. Prog. Phys.* **68** 1181

[13] Anbramowitz A and Stegun I 1993 *Bessel functions J and Y* in *Handbook of Mathematical Functions with Formulas, Graphs and Mathematical Tables* (New York, Dover)

[14] Lucas A A, Lambin Ph, Mairesse R, and Mathot M 1999 *J. Chem. Edu.* **76** 378

[15] Klug A F, Crick F H C, and Wyckoff II W 1958 *Acta Cryst.* **11** 199

[16] Kornyshev A A, Lee D J, Leikin S, and Wynveen A 2007 *Rev. Mod. Phys.* **79** 943

[17] Zimmermann S B and Pheiffer B H 1979 *Proc. Natl. Acad. Sci USA* **76** 2703

[18] In Ref.[2], p.95

[19] Franklin R and Gosling R G 1953 *Acta Cryst.* **6** 673; **6** 678

[20] Chargaff E 1950 *Experientia* **6** 201

[21] Rogers Y H and Venter J C 2005 *Nature (London)* **437** 326

[22] Zwolak M and Di Ventra M 2005 *Nano Lett* **5** 421

[23] Xu M S, Endres R G and Arakawa Y 2007 in *Charge Migration in DNA: Perspectives from Physics, Chemistry and Biology*, Ed. Chakraborty T (Nanoscience and Technology series, Springer-Verlag, Berlin) p.205

[24] Artacho E, Machado M, Sánchez-Portal D, Ordejón P, and Soler J M 2003 *Molecular Phys.* **101** 1587

[25] Gervasio F, Carloni P, and Parrinello M 2002 *Phys. Rev. Lett.* **89** 108102

[26] Taniguchi M and Kawai T 2004 *Phys. Rev. B* **70** 011913

[27] Endres R G, Cox D L, and Singh R R P 2004 *Rev. Mod. Phys.* **76** 195

[28] Maragakis P, Barnett R L, Kaxiras E, Elstner M, and Frauenheim T 2002 *Phys. Rev. B* **66** 241104(R)

[29] Starikov E B 2005 *Phil. Mag.* **85** 3435; 2002 *Phys. Chem. Chem. Phys.* **4** 4523

[30] Kato H S, Furukawa M, Kawai M, Taniguchi M, Kawai T, Hatsui T, and Kosugi N 2004 *Phys. Rev. Lett.* **93** 086403

[31] Wadati H, Okazaki K, Niimi Y, and Fujimori A 2005 *Appl. Phys. Lett.* **86** 023901

[32] Furukawa M, Kato H S, Taniguchi M, Kawai T, Hatsui T, Kosugi N, Yoshida T, and Aida M 2007 *Phys. Rev. B* **75** 045119

[33] Ye Y J, Chen R S, Martínez A, Otto P, and Ladik J 2000 *Physica B* **279** 246

[34] Ladik J 2003 *J. Molec. Struct. (Theochem)* **666-667** 1

[35] Szent-Györgyi A 1941 *Science* **93** 609; 1941 *Nature (London)* **148** 157

[36] Duchesne J, Depireux J, Bertinchamps A, Cornet N, and van der Kaa J M 1960 *Nature (London)* **188** 405

[37] Eley D D 1953 *Trans. Faraday Soc.* **49** 79; 1960 *Trans. Faraday Soc.* **56** 1432; Eley D D and Spivey D I 1962 *Trans. Faraday Soc.* **58** 411

[38] Heeger A J, Kivelson S, Schrieffer J R, and Su W P 1988 *Rev. Mod. Phys.* **60** 781

[39] Havinga E E, Tenhoeve W, and Wynberg H 1993 *Synth. Met.* **55** 299

[40] Pohl H and Kauzman W 1963 *Rev. Mod. Phys.* **36** 721

[41] Lewis F, Wu T, Zhang Y, Letsinger R L, Greenfield S R, and Wasielewski M R 1997 *Science* **277** 673; Lewis P, Cheatham Th E III, Starikov E B, Wang H, and Sankey O F 2003 *J. Phys. Chem. B* **107** 2581

[42] Kelley S O and Barton J K 1999 *Science* **283** 375

[43] Braun E, Eichen Y, Sivan U, and Ben-Yoseph G 1998 *Nature (London)* **391** 775

[44] Fink H W and Schönenberger C 1999 *Nature (London)* **398** 407

[45] Kasumov A Y, Kociak M, Gueron S, Reulet B, Volkov V T, Klinov D V, and Bouchiat H 2001 *Science* **291** 280

[46] Hwang J S, Kong K J, Ahn D, Lee G S, Ahn D J, and Hwang S W 2002 *Appl. Phys. Lett.* **81** 1134

[47] Heim T, Deresmes D, and Vuillaume D 2004 *J. Appl. Phys.* **96** 2927

[48] Porath D, Bezryadin A, de Vries S, and Dekkar C 2000 *Nature (London)* **403** 635

[49] Hartzell B, Melord B, Asare D, Chen H, Heremans J J, and Sughomonian V 2003 *Appl. Phys. Lett.* **82** 4800

[50] Zhang Y, Austin R H, Kraeft J, Cox E C, and Ong N P 2002 *Phys. Rev. Lett.* **89** 198102

[51] Storm A J, van Noort J, de Vries S, and Dekker C 2001 *Appl. Phys. Lett.* **79** 3881

[52] Tran P, Alavi B, and Gruner G 2000 *Phys. Rev. Lett.* **85** 1564

[53] de Pablo P J, Moreno-Herrero F, Colchero J, Gómez-Herrero J, Herrero P, Baró A M, Ordejón P, Soler J M, and Artacho E 2000 *Phys. Rev. Lett.* **85** 4992

[54] Ullien D, Cohen H, and Porath D 2007 *Nanotechnology* **18** 424015

[55] Cohen H, Nogues C, Naaman R, and Porath D 2005 *Proc. Natl Acad. Sci USA* **102** 11589

[56] Xu B, Zhang P, Li X, and Tao N 2004 *Nano Letters* **4** 1105

[57] Hihath J, Xu B, Zhang P, and Tao N 2005 *Proc. Natl. Acad. Sci. USA* **102** 16979; Hihath J, Chen F, Zhang P, and Tao N 2007 *J. Phys. Condens. Matter* **19** 215202

[58] Tuukkanen S, Kuzyk A, Toppari J J, Hytönen V P, Ihalainen T, and Törmä P 2005 *Appl. Phys. Lett.* **87** 183102

[59] Roy S, Vedala H, Roy A D, Kim D H, Doud M, Mathee K, Shin H K, Shimamoto N, Prasad V, and Choi W 2008 *Nano Letters* **8** 26

[60] Guo X, Gorodetsky A A, Hone J, Barton J K, and Nuckolls C 2008 *Nature Nanotechnology* **3** 163

[61] van Zalinge H, Schiffrin D J, Bates A D, Haiss W, Ulstrup J, and Nichols R J 2006 *ChemPhysChem* **7** 94

[62] Iguchi K 1994 *RIKEN Review* **6** 49

[63] Seeman N C 1999 *Acc. Chem. Res.* **30** 253

[64] Braun E, Eichen Y, Sivan U, and Ben-Yoseph G 1998 *Nature* **391** 775

[65] Pike A, Horrocks B, Connolly B, and Houlton A 2002 *Aust. J. Chem.* **55** 191

[66] Peng C K, Buldyrev S V, Goldberger A L, Havlin S, Sciotino F, Simons M, and Stanley H E 1992 *Nature* **356** 168

[67] Li W and Kaneko K 1992 *Europhys. Lett.* **17** 655

[68] Voss R F 1992 *Phys. Rev. Lett.* **68** 3805

[69] Arneodo A, Bacry E, Graves P V, and Mury J F 1995 *Phys. Rev. Lett.* **74** 3293

[70] Jerome D and Schulz H 1982 *Adv. Phys.* **31** 299

[71] Di Felice R, Calzolari A, Molinari E, and Garbesi A 2002 *Phys. Rev. B* **65** 045104

[72] Wang H, Lewis J P, and Sankey O F 2004 *Phys. Rev. Lett.* **93** 016401

[73] Berlin Y A, Burin M L, and Ratner M A 2000 *Superlattices Microstruct.* **28** 241

[74] Sugiyama H and Saito I 1996 *J. Am. Chem. Soc.* **118** 7063

[75] Voityuk A A, Jortner J, Bixon M, and Rösch N 2001 *J. Chem. Phys.* **114** 5614

[76] Duan Y, Wilkosz P, Crowley M, and Rosenberg J 1997 *J. Mol. Biol.* **272** 553

[77] Brauns E et al. 1999 *J. Am. Chem. Soc.* **121** 11644

[78] Wan C, Fiebig T, Kelley S O, Treadway C R, Barton J K, and Zewail A H 1999 *Proc. Natl. Acad. Sci. U.S.A.* **96** 6014

[79] Bruinsma R, Grüner G, D'Orsogna M R, and Rudnick J 2000 *Phys. Rev. Lett.* **85** 4393

[80] Cocco S and Monasson R 1999 *Phys Rev. Lett.* **83** 5178; 2000 *J. Chem. Phys.* **112** 10017

[81] Ratner M A 1999 *Nature (London)* **397** 480

[82] Treadway C, Hill M G, and Barton J K 2002 *Chem. Phys.* **281** 409

[83] Berlin Y A, Burin A L, and Ratner M A 2000 *Superlattices Microstruct.* **28** 241; 2002 *Chem. Phys.* **275** 61

[84] van Zalinge H, Schiffrin D J, Bates A D, Starikov E B, Wenzel W, and Nichols R J 2006 *Angew. Chem. Int. Ed.* **45** 5499

[85] Cuniberti G, Maciá E, Rodríguez A, and Römer R A 2007 in *Charge Migration in DNA: Physics, Chemistry and Biology Perspectives*, Ed. Chakraborty T (Springer, Berlin)

[86] Cuniberti G, Craco L, Porath D, and Dekker C 2002 *Phys. Rev. B* **65** 241314

[87] Iguchi K 2004 *Int. J. Mod. Phys. B* **18** 1845; 2001 *J. Phys. Soc. Jpn.* **70** 593; 1997 *Int. J. Mod. Phys. B* **11** 2405

[88] Yamada H 2004 *Int. J. Mod. Phys. B* **18** 1697

[89] Xiong G and Wang X R 2005 *Phys. Lett. A* **344** 64

[90] Xu M S, Endres R G, Tsukamoto S, Kitamura M, Ishida S, and Arakawa Y 2005 *Small* **1** 1168

[91] Klotsa D, Römer R A, and Turner M S 2005 *Biophys. J.* **89** 2187

[92] Maciá E, Triozon F, and Roche S 2005 *Phys. Rev. B* **71** 113106

[93] Sil S, Maiti S K, and Chakrabarty A 2008 cond-mat 0801.26/0v1

[94] Wang X F and Chakraborty T 2006 *Phys. Rev. Lett.* **97** 106602

[95] Díaz E, Sedrakyan A, Sedrakyan D, and Domínguez-Adame F 2007 *Phys. Rev. B* **75** 014201

[96] Caetano R A and Schulz P A 2005 *Phys. Rev. Lett.* **95** 126601; 2006 **96** 059704; Sedrakyan A and Domínguez-Adame F 2006 *Phys. Rev. Lett.* **96** 059703

[97] Yan Y J and Zhang H 2002 *J. Theor. Comput. Chem.* **1** 225

[98] Iguchi K 2004 *Int. J. Mod. Phys. B* **18** 1845

[99] Ladik J, Bende A, and Bogár F 2008 *J. Chem. Phys.* **128** 105101

[100] Fonseca Guerra C, Bickelhaupt F M, and Baerends E J 2002 *Cryst. Growth Des.* **2** 239

[101] Sugiyama H and Saito I 1996 *J. Am. Chem. Soc.* **118** 7063

[102] Natsume T, Dedachi K, Tanaka S, Higuchi T, and Kurita N 2005 *Chem. Phys. Lett.* **408** 381

[103] Voityuk A A, Jortner J, Bixon M, and Rösch N 2001 *J. Chem. Phys.* **114** 5614

[104] Maciá E and Roche S 2006 *Nanotechnology* **17** 3002

[105] E. Maciá 2006 *Phys. Rev. B* **74**, 245105; 2007 *ibid.* **75**, 035130

[106] Andresen K, Das R, Park H Y, Smith H, Kwok L W, Lamb J S, Kirkland E J, Herschlag D, Finkelstein K D, and Pollak L 2004 *Phys. Rev. Lett.* **93** 248103

[107] Das R, Mills T T, Kwok L W, Maskel G S, Millet I S, Doniach S, Finkelstein K D, Herschlag D, and Pollak L 2003 *Phys. Rev. Lett.* **90** 188103

[108] York D M, Lee T S, and Yang W 1998 *Phys. Rev. Lett.* **80** 5011

[109] Kato H S, Furukawa M, Kawai M, Taniguchi M, Kawai T, Hatsui T, and Kosugi N 2004 *Phys. Rev. Lett.* **93** 086403

[110] Wadat H, Okazaki K, Niimi Y, Fujimori A, Tabata H, Pikus J, and Lewis J P 2005 *Appl. Phys. Lett.* **86** 023901

[111] Emberly E G and Kirczenow G 1998 *Phys. Rev. B* **58** 10911; 1999 *J. Phys. Condens. Matter* **11** 6911

[112] Zhu Y, Kaun C C, and Guo H 2004 *Phys. Rev. B* **69** 245112

[113] Voityuk A A, Siriwong K, and Rösch N 2001 *Phys. Chem. Chem. Phys.* **3** 5421; Voityuk A A 2007 *Chem. Phys. Lett.* **439** 162

[114] Barbi M, Cocco S, and Peyrard M 1999 *Phys. Lett. A* **253** 358

[115] Agarwal J and Hennig D 2003 *Physica A* **323** 519

[116] Maciá E 2007 *Phys. Rev. B* **76** 245123

[117] Dauxois T, Peyrard M, and Bishop A R 1993 *Phys. Rev. E* **47** R44

[118] Dauxois T and Peyrard M 1995 *Phys. Rev. E* **51** 4027

[119] Campa A and Giansanti A 1998 *Phys. Rev. E* **58** 3585

[120] Cule D and Hwa T 1997 *Phys. Rev. Lett.* **79** 2375

[121] Troisi A and Orlandi G 2002 *J. Chem. Phys. B* **106** 2093

[122] Voityuk A A 2008 *J. Chem. Phys.* **128** 115101

[123] Voityuk A A 2008 *J. Chem. Phys.* **128** 045104

[124] Yu Z G and Song X 2001 *Phys. Rev. Lett.* **86** 6018

[125] Roche S 2003 *Phys. Rev. Lett.* **91** 108101

[126] Ren W, Wang J, Ma Z, and Guo H 2005 *Phys. Rev. B* **72** 035456; 2006 *J. Chem. Phys.* **125** 164704

[127] Guo A M and Xu H 2007 *Physica B* **391** 292

[128] Zhang W, Govorov A O, and Ulloa S E 2002 *Phys. Rev. B* **66** 060303(R)

[129] Senthilkumar K, Grozema F C, Fonseca Guerra C, Bickelhaupt F M, Lewis F D, Berlin Y A, Ratner M A, and Siebbeles L A 2005 *J. Am. Chem. Soc.* **127** 14894

[130] Conwell E M and Rakhmanova S V 2000 *Proc. Natl. Acad. Sci. U.S.A* **97** 4556

[131] Maciá E and Roche S 2006 *Nanotechnology* **17** 3002

[132] Gutiérrez R, Mandal S, and Cuniberti G 2005 *Phys. Rev. B* **71** 235116

[133] Gutiérrez R, Mohapatra S, Cohen H, Porath D, and Cuniberti G 2006 *Phys. Rev. B* **74**

[134] Roche S and Maciá E 2004 *Modern Phys. Lett. B* **18** 847

[135] Palmero F, Archilla J F R, Hennig D, and Romero F R 2004 *New J. Phys.* **6** 13

[136] Starikov E B 2003 *Phil. Mag. Lett.* **83** 699

[137] Zhu J X, Rasmussen K Ø, Balatsky A V, and Bishop A R 2007 *J. Phys. Condens. Matter* **19** 136203

[138] Hakim M B, Lindsay S M, and Powell J 1984 *Biopolymers* **23** 1185

[139] Yang I S and Anderson A C 1987 *Phys. Rev. B* **35** 9305

[140] Holste D, Grosse I, and Herzel H 2001 *Phys. Rev. E* **64** 041917

[141] Roche S, Bicout D, Maciá E, and Kats E 2003 *Phys. Rev. Lett.* **91** 228101; 2004 *Phys. Rev. Lett.* **92** 109901(E)

[142] Wang S Ch, Li P Ch, and Tseng H Ch 2008 *Physica A* **387** 5159

[143] Hübsch A, Endres R G, Cox D L, and Singh R R P 2005 *Phys. Rev. Lett.* **94** 178102

[144] Guo A M 2007 *Phys. Rev. E* **75** 061915

[145] Nakame S, Cazayous M, Sacuto A, Monod P, and Bouchiat H 2005 *Phys. Rev. Lett.* **94** 248102

[146] Carpena P, Bernaola-Galvan P, Ivanov P Ch, and Stanley H E 2002 *Nature* **418** 955; 2003 ibidem *Nature* **421** 764

[147] Shih C T 2006 *Phys. Rev. E* **74** 010903(R); 2006 *Phys. Stat. Sol. (b)* **243** 378; 2007 *Chin. J. Phys.* **45** 703

[148] Shih C T, Roche S, and Römer R A 2008 *Phys. Rev. Lett.* **100** 018105

7

Exploiting aperiodic order in technological devices

7.1 Periodic versus aperiodic

The notion of aperiodic order, that is, order without periodicity, brings in a novel paradigm in Condensed Matter Physics. The fundamental nature of this notion has been profusely illustrated in previous Chapters, where the physical properties of such diverse systems as quasicrystalline alloys, semiconductor heterostructures, metallic and dielectric multilayers, or different classes of both synthetic and biological DNA chains have been considered. In this and the next chapter we will turn our attention towards more practical issues.

It usually occurs that the development of novel, more efficient devices is often hindered by the lack of materials with the proper combination of desirable properties for some specific application. For this reason, much effort has recently been directed towards the search of materials possessing these desirable properties. The most common approach adopted for this quest is to try to synthesize new materials. In the light of the results reviewed in the preceding Chapters quasicrystalline alloys can certainly provide these raw materials for certain applications. In addition, a novel, complementary approach gradually appears: it consists in designing aperiodic structures aimed to achieve a better performance than usual periodic ones for some specific applications. In this way, wandering across the aperiodic order realm we will progressively go from the workings of nature to the workshop of craftsmen.

Clear indications on certain advantages of quasiperiodic systems over periodic ones come from the non-linear optics field, where it was shown that second harmonic generation processes (aimed at producing high energy photons – say, in the blue region of the electromagnetic spectrum – by properly combining low energy ones) were more efficient in Fibonacci dielectric multilayers (FDM, see Section 4.3) in virtue of their richer Fourier's spectrum structure. In fact, due to their higher space-group symmetry, quasiperiodic multilayers can provide *more* reciprocal vectors to the so-called quasi-phase-matching optical process, and this ultimately results in a more plentiful spectrum structure than that of a periodic multilayer.[1] The importance of the role played by the quasiperiodicity of the substrate is further highlighted when considering third harmonic generation processes, where it has been shown that the conversion efficiency in a quasiperiodic multilayer is increased by a factor of eight in comparison with the two-step process required for an usual third-harmonic generator based on periodic systems.[2, 3] In a similar way, emission enhancement effects occurring at wavelengths corresponding to multiple resonance states in light-emitting SiN/SiO_2 multilayered structures arranged according to the Thue-Morse sequence have been demonstrated as well.[4] On this basis, the possibility of designing quasiperiodic structures able to simultaneously phase match any two non-linear interactions by properly introducing an aperiodic modulation of the non-linear coefficient in ferroelectric devices has been proposed in one [5] and two dimensions.[6]

The non-linear properties of optical heterostructures can also be used to fabricate a compact-sized compressor for laser pulse. This compression is physically determined by the group velocity dispersion in the material, so that one can expect that by adding more layers to a periodic multilayer one should obtain narrower optical bands and the compression effect will be increased. However, this is inevitably accompanied by an increase of the total thickness of the structure, which is undesirable. In this context, the recourse to aperiodic structures, exhibiting a significantly larger fragmentation of their optical spectrum for similar system sizes, appears as a natural choice. Inspired by this principle, the laser compression performance of both periodic and FDMs made from high-index ZnS and low-index Na_3AlF_6 layers has been experimentally compared. As expected from theoretical considerations, the Fibonacci structure exhibits a compression enhancement due to its larger group velocity dispersion.[7]

The possible use of different kinds of photonic multilayers based on porous silicon nanotechnology has been tested for the detection of gases, liquids, and biological molecules. The sensing mechanism is based on the refractive index changes of porous silicon due to the partial substitution of the air in the pores on exposure to biochemical substances. The refraction index change is transduced, in turn, in a characteristic shift of the reflectivity spectrum. In a recent study the sensitivities of resonant optical biochemical sensors, based on both periodic and Thue-Morse porous silicon multilayers, were compared. The measurements clearly indicated that the aperiodic multilayer is *more sensitive* than the aperiodic one.[8] The physical reason for the observed improvement was traced back to the following properties: i) for a given system size Thue-Morse multilayers have less interfaces than the periodic ones, hence exhibiting a higher filling capability, and ii) multiple interference effects give rise to a great number of narrower resonance transmittance peaks in the aperiodic multilayers, hence increasing their spectral resolution for detection of vapors and liquids. On these basis, a novel approach for optical sensing, based on the excitation of critically localized modes in two-dimensional deterministic aperiodic structures generated by the Rudin-Shapiro sequence has been recently introduced.[9]

These few examples nicely illustrate the main point we will address through this Chapter, namely, that carefully designed aperiodic structures can satis-

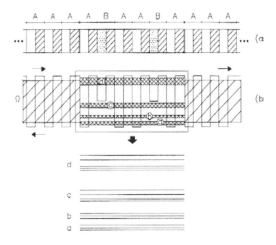

FIGURE 7.1

(a) Fibonacci heterostructure based on a sequential deposition of AsGa (A) and AsAlGa (B) semiconductor bilayers is sandwiched between AsGa periodic heterostructures. (b) The corresponding conduction band profile. The Fibonacci heterostructure is embedded between contacts (periodic potentials) whose bandwidth Ω is shadowed. Charge carriers are injected from the left contact. Due to interference effects a fraction of the carriers are backscattered and the rest propagate through the Fibonacci heterostructure and reach the contact on the right (arrows). The energy spectrum of the Fibonacci heterostructure is fragmented in four main bands (labeled a,b,c, and d), which further split themselves according to a hierarchical fractal scheme as the system size is increased (lower magnification). (From [10]. With permission from IOP Publishing Ltd.)

factorily cope with certain physical requirements necessary for the fabrication of improved devices of technological interest.

7.2 Layered systems

7.2.1 From quantum to classical waves

As we discussed in Chapter 4, the one-dimensional ordering introduced in the manufacturing process of heterostructures gives rise to novel physical properties (such as the formation of minibands) which reflect the long-range, quantum correlation present in these superstructures. In fact, the electronic properties of superlattices are determined by both the chemical nature of the constituent bulk materials and the layer thicknesses, so that these structures can be grown, in principle, to tailor their electronic properties at will. In this sense, one of the most appealing motivations for the experimental study of aperiodic heterostructures is the theoretical prediction that they should exhibit peculiar quantum states, associated to highly fragmented fractal energy spectra displaying self-similar patterns (see Sections 5.3 and 5.4).

What is the physical reason for the presence of such a fragmented band structure in quasiperiodic systems? In essence the existence of allowed bands separated by forbidden gaps stems from resonant tunneling effects between states belonging to neighboring building blocks. Now, since the structural self-similarity of the superlattice guarantees the existence of suitable resonant conditions at all scales, a fractal energy spectrum naturally arises from a delicate balance between short-range and long-range effects in these systems.[11] Evidence of a hierarchical fragmentation pattern in the electronic spectra of short heterostructures has been confirmed using different experimental techniques (see Section 4.3). Accordingly, the possibility of using Fibonacci heterostructures as efficient electronic filters was theoretically considered as well,[12] according to the basic principles sketched in Fig.7.1.

However, a number of severe limitations appear in realistic set-ups due to electron-phonon, electron-electron, or spin-orbit interaction effects, which make it a very difficult task to efficiently exploit spectrum fractal features in actual electronic applications. At this point it is important to note that multiple scattering and interference of scattered waves can be found in quantum as well as in classical physical processes. In quantum physics some representative examples are resonant tunneling of a particle through a sequence of potential barriers, development of energy bands in periodic crystals, or the progressive emergence of a fractal-like energy spectrum in quasiperiodic superlattices. In classical physics we found the development of resonant transmission bands in a Fabry-Perot interferometer, formation of stop and transmission bands in multilayered dielectric mirrors, or the scaling properties of transmission spectra in fractal multilayer structures. Accordingly, in order to fully appreciate the fingerprints of quasiperiodic order, the study of *classical* waves propagating through a quasiperiodic substrate offers a number of advantages over the study of *quantum* elementary excitations. Thus, a number of experimental studies dealing with the propagation of elastic waves,[13, 14] third sound,[15] and ultrasonic waves[16] in Fibonacci systems have been reported, confirming that characteristic self-similar features in the transmission spectra are observable at different scale lengths, up to the micrometer range.

Similarly, light transmission in aperiodic media has deserved a major attention in order to understand the interplay between optical properties and the underlying aperiodic order of the substrate. The mathematical analogy between the Schrödinger equation describing the motion of an electron with an energy ε and effective mass m, under the action of a potential $V(z)$, in one

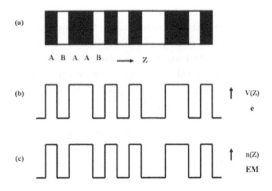

FIGURE 7.2
a) Sketch of a FDM grown along the z direction, b) electronic potential profile $V(z)$ for an electron propagation along the growth direction, c) refractive index profile $n(z)$ for a classical (electromagnetic) wave propagating through the structure).

dimension

$$\frac{\hbar^2}{2m}\frac{d^2\psi}{dz^2} + [\varepsilon - V(z)]\psi = 0, \tag{7.1}$$

and the Helmholtz equation describing a monochromatic electromagnetic wave of frequency ω propagating in a lossless, dispersionless medium with a variable index of refraction profile $n(z)$

$$\frac{d^2E}{dz^2} + \left[\frac{\omega^2}{c^2}n^2(z) - k_{\parallel}^2\right]E = 0, \tag{7.2}$$

where E is the transversal component of the electric field, $k_{\parallel}$ is the wave vector in the XY plane (perpendicular to the propagation direction z), and c is the vacuum speed of light, provides a powerful tool to relate electron motion in superlattices to electromagnetic waves propagating in multilayers (Fig.7.2). In the case of periodic media a general description, which borrows some basic notions like Bloch waves or Brillouin zones from condensed matter theory, has been developed,[17] and successfully applied to numerous optical devices, such as dielectric multilayer mirrors, beamsplitters, filters and so on, which are of common use in nowadays optoelectronics and optical communication applications.[18, 19] In this way, the so-called *photonic crystal* concept has been introduced to describe optical systems which exhibit large frequency stop bands due to interference effects, in close analogy with the presence of a band structure in conventional atomic lattices, or the formation of energy minibands in superlattices. Naturally, the very notion of photonic crystal can be extended to describe the properties of aperiodic photonic structures as well. To this end, one can consider that the optical properties of the medium are modulated by a quasiperiodic refraction index function, $n(z)$, in Eq.(7.2). The resulting structure could be properly regarded as a one-dimensional photonic quasicrystal. Similarly, one could construct a photonic fractal with a self-similar refraction index profile by arranging the structure according to a certain fractal pattern.[20]

Fibonacci dielectric multilayers have been intensively studied during the last decade,[21, 22, 23, 24, 25, 26, 27, 28, 29] since they provide a canonical example of quasiperiodic structures for optical applications. Nevertheless, other classes of self-similar structures also exhibit interesting photonic properties. For instance, the optical transmission spectra of multilayers based on the metallic mean sequences (which do not have the Pisot property, see Sections 4.2.1 and 4.2.2) were compared with those of multilayered structures based on the precious means (which exhibit the Pisot property). The observed differences nicely illustrate the fact that both kinds of aperiodic systems belong to quite different aperiodic order domains.[25, 30] The optical properties of multilayers based on the Thue-Morse sequence, characterized by a singular continuous Fourier spectrum (see Section 5.2), have also received a considerable attention during the last two decades.[29, 31, 32, 33, 34] The presence of band gap scaling, similar to that previously observed in Fibonacci multilayers,[23] along with the presence of an omnidirectional band gap, has been experimentally confirmed for representatives of this class, based on the stacking of either Si/SiO_2,[35] or TiO_2/SiO_2 layers,[36] according to the Thue-Morse sequence. The physical origin of the fundamental band gaps in structures with different Fourier measures but sharing the self-similar property can then be properly attributed to the presence of short-range correlations among certain basic building blocks interspersed through the whole multilayer.

On the other hand, the possible localization of light by effect of the quasiperiodic order has been explored both theoretically and experimentally. Recent experiments have studied the propagation of light through porous silicon Fibonacci multilayers by means of ultrashort pulse interferometry. A strongly suppressed group velocity has been observed for frequencies close to a main band gap.[37] Calculations show that the optical path drift naturally occurring during the growth of the Fibonacci multilayer triggers the localization of the first few band edge modes, although preserving some fingerprints of their characteristic self-similar pattern.[38]

Finally, it should be noted that most basic results obtained from the study of aperiodic dielectric multilayers can be readily extended to the case of acoustic waves propagating through aperiodic superlattices, where discrete dips in the transmission coefficient have been reported both theoretically [39] and experimentally.[40]

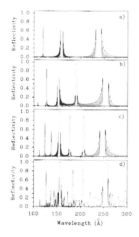

FIGURE 7.3

X-ray reflectivity of multilayer structures with refractive indices $n_A = 0.9200$ and $n_B = 0.9995$, each layer thickness being $d = 5$ nm. Results for periodic approximants of the form a) $(ABAAB)^m$ with $m = 47$ ($N = 235$), b) $(ABAABABA)^m$ with $m = 29$ ($N = 232$), c) $(ABAABABAABAAB)^m$ with $m = 18$ ($N = 234$), and d) a Fibonacci multilayer with $N = F_{12} = 233$ are shown. (From ref.[41]. With permission from Elsevier.)

7.2.2 Optical engineering

As previously mentioned, the isomorphism of Shrödinger and Helmholtz equations provides a helpful interplay of basic concepts in modern optoelectronics. On one hand, basic features known for electrons in periodic, quasiperiodic, and random potentials have been successfully transferred to classical electromagnetic waves, introducing challenging concepts in photonic engineering (photonic band gaps design and light localization). On the other hand, knowledge gained from the study of optical properties of complex structures can be extended to understand some electronic properties of complex nanostructures with characteristic length scale on the order of electron de Broglie wavelength $(1 - 10$ nm) instead of optical wavelength $(10 - 10^3$ nm).

It is clear that in any multilayer stack, characterized by a given dielectric function $n(z)$ in Eq.(7.2), a number of resonant modes exist corresponding to transmission of waves throughout the structure without reflection. These modes are properly described in terms of their related transmission $T_N(\omega)$, or reflection $R_N(\omega)$, spectra, where N measures the number of layers in the stack. In this way, by considering different kinds of refraction index profiles (i.e., periodic, quasiperiodic, self-similar, random) a characteristic "spectral portrait" could be assigned to every multilayer structure. For the sake of illustration in Fig.7.3 we show the x-ray reflectivity spectra for a series of periodic

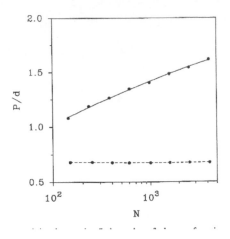

FIGURE 7.4
Overall reflectivity in units of d as a function of the number of layers N in periodic approximants (dashed line) and Fibonacci multilayers of similar size (solid line). (From ref.[41]. With permission from Elsevier.)

approximants of the Fibonacci multilayer. In all cases, two main reflection peaks are observed at about $\lambda = 150\,\text{Å}$ and $\lambda = 250\,\text{Å}$, along with a set of subsidiary peaks, whose number increases as the order of the approximant increases. As a consequence, the envelope of the reflectivity spectrum, $R_N(\lambda)$, becomes more and more spiky as the considered multilayer approaches the quasiperiodic limit. This property can be quantified by means of the expression

$$P_N = \frac{\left[\int_{\lambda_{min}}^{\lambda_{max}} R_N(\lambda)\, d\lambda \right]^2}{\int_{\lambda_{min}}^{\lambda_{max}} R_N^2(\lambda)\, d\lambda}, \tag{7.3}$$

which gives an estimation of the overall reflectivity of the sample as a function of the number of layers. The amplitude of different reflectivity peaks can be characterized by the scaling of P_N with the number of layers, the higher its value the higher the whole reflectivity. Figure 7.4 shows the results obtained in both periodic approximants and in the Fibonacci multilayer as a function of the number of layers. Periodic multilayers present an almost constant value of P_N for different number of layers. On the contrary, the value of P_N for Fibonacci multilayers increases monotonically with N and it is always larger than the corresponding value for periodic approximants of the same size. In fact, whereas the number of high reflectance peaks in periodic multilayers changes only slightly on increasing the number of layers, the fragmentation

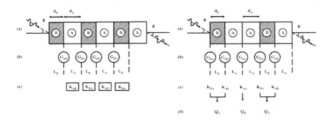

FIGURE 7.5

Sketch illustrating a) the propagation of light through a periodic dielectric multilayer (on the left) and a FDM (on the right), b) the sequence of $G_{i+1,i}$ and L_i auxiliary matrices, c) the sequence of light transfer matrices $K_{i+1,i}$, and d) the renormalization scheme adopted for the FDM (on the right).

process characteristic of the energy spectrum in quasiperiodic multilayers gives rise to the occurrence of new reflectivity peaks. This leads to an overall increase of the reflectivity, allowing for a fine tuning of different narrow lines at the same time. In this way, Fibonacci multilayers based on high-Z/low-Z metallic layers, like those grown by alternating deposition of V/Mo, W/Ti, or Ta/Al thin films[42] (see Section 4.2.1), may be efficiently used as selective filters of soft x-rays.[41]

From the considered example two basic questions naturally arise. Is it possible to predict the transmission spectrum main features from the knowledge of the $n(z)$ modulating function for a given multilayer stack? Is there any fundamental difference in the transmission properties among different classes of aperiodically arranged optical multilayers? In this Section we will discuss the basic properties of optical devices based on quasiperiodically ordered structural designs, comparing their optical performance with that corresponding to the more usual periodic arrangements.

7.2.2.1 Periodic dielectric multilayers

In order to properly compare the optical response of both periodic and quasiperiodic systems we will rely on the transfer matrix technique (see Section 9.2). This approach is particularly well suited to our purposes, since it describes the optical response of the global system in terms of the light behavior in contiguous layers (Fig.7.5). We will focus on light propagating through layered aperiodic multilayers composed of linear, homogeneous, lossless materials with no optical activity, under general incidence conditions. When studying the propagation of light in multilayered systems one must consider the propagation across the interface separating two neighboring layers (described by the matrices $G_{i+1,i}$ in Fig. 7.5(b)), along with the light propagation within each layer (described by the matrices L_i in Fig. 7.5(b)). For electromagnetic waves with electric field amplitude perpendicular to the plane of light path (TE waves), one has[21]

$$G_{i+1,i} \equiv \begin{pmatrix} 1 & 0 \\ 0 & \frac{n_i \cos \theta_i}{n_{i+1} \cos \theta_{i+1}} \end{pmatrix}, \quad L_i \equiv \begin{pmatrix} \cos \delta_i & -\sin \delta_i \\ \sin \delta_i & \cos \delta_i \end{pmatrix}, \quad (7.4)$$

where n_i are the refractive indices, θ_i are the refraction angles determined by Snell's law, $\delta_i \equiv n_i d_i k \cos \theta_i$ are the layer phase thickness, d_i are the widths of the layers, and $k = 2\pi/\lambda$ is the wave vector in vacuum. The transmission of light through a binary multilayer composed of two different materials can be properly described in terms of a product involving the matrices [see Fig. 7.5(c)]

$$K_{AA} \equiv G_{AA}L_A = \begin{pmatrix} \cos \delta_A & -\sin \delta_A \\ \sin \delta_A & \cos \delta_A \end{pmatrix},$$

$$K_{AB} \equiv G_{AB}L_B = \begin{pmatrix} \cos \delta_B & -\sin \delta_B \\ u^{-1} \sin \delta_B & u^{-1} \cos \delta_B \end{pmatrix},$$

$$K_{BA} \equiv G_{BA}L_A = \begin{pmatrix} \cos \delta_A & -\sin \delta_A \\ u \sin \delta_A & u \cos \delta_A \end{pmatrix},$$

$$K_{BB} \equiv G_{BB}L_B = \begin{pmatrix} \cos \delta_B & -\sin \delta_B \\ \sin \delta_B & \cos \delta_B \end{pmatrix}, \quad (7.5)$$

where $u \equiv R\beta$, with $R \equiv n_A/n_B$ and $\beta \equiv \cos \theta_A / \cos \theta_B$. Let us consider a multilayer made of ν bilayers AB which repeat periodically (Fig.7.5, left panel). In this case the global transfer matrix can be straightforwardly expressed in terms of the auxiliary matrices (7.5) in the form $\mathcal{M}_N \equiv Q^\nu$, where $N = 2\nu$ is the total number of layers and

$$Q = \begin{pmatrix} \cos \delta_A \cos \delta_B - u \sin \delta_A \sin \delta_B & -\sin \delta_A \cos \delta_B - u \cos \delta_A \sin \delta_B \\ \sin \delta_A \cos \delta_B + u^{-1} \cos \delta_A \sin \delta_B & \cos \delta_A \cos \delta_B - u^{-1} \sin \delta_A \sin \delta_B \end{pmatrix}. \quad (7.6)$$

Since Q is unimodular (i. e., its determinant is unity) we can make use of the Cayley-Hamilton theorem for unimodular matrices in order to explicitly evaluate the matrix $\mathcal{M}_N$ in the closed form (see Section 9.2)

$$\mathcal{M}_N = \begin{pmatrix} T_\nu + \frac{(u^{-1}-u)U_{\nu-1}}{2} \sin \delta_A \sin \delta_B & -\Upsilon_+ U_{\nu-1} \\ \Upsilon_- U_{\nu-1} & T_\nu - \frac{(u^{-1}-u)U_{\nu-1}}{2} \sin \delta_A \sin \delta_B \end{pmatrix}, \quad (7.7)$$

where $\Upsilon_\pm \equiv \sin\delta_A \cos\delta_B + u^{\pm 1} \cos\delta_A \sin\delta_B$, $T_\nu(w)$ and $U_{\nu-1}(w)$ are Chebyshev polynomials of the first and second kinds, respectively, and

$$w \equiv \frac{1}{2}\mathrm{tr}Q = \cos\delta_A \cos\delta_B - \frac{u + u^{-1}}{2}\sin\delta_A \sin\delta_B. \qquad (7.8)$$

From Eq.(7.7) the dispersion relation for the periodic multilayer can be easily obtained from the condition $\cos[q\nu(d_A + d_B)] = \mathrm{tr}\mathcal{M}_N/2$, which leads to the well-known expression[43]

$$\cos[q(d_A + d_B)] = \cos\delta_A \cos\delta_B - \frac{u^2 + 1}{2u}\sin\delta_A \sin\delta_B. \qquad (7.9)$$

The transmission coefficient can be obtained from the standard expression

$$T_N = \frac{4}{||\mathcal{M}_N|| + 2}, \qquad (7.10)$$

where $||\mathcal{M}_N||$ denotes the sum of the squares of the four elements of the global transfer matrix. From Eq.(7.7) one obtains

$$T_N^{QP} = \frac{1}{1 + \left[\frac{u^2-1}{2u}\right]^2 U_{\frac{N}{2}-1}^2(w)\sin^2\delta_B}. \qquad (7.11)$$

Expressions given by Eqs.(7.9) and (7.11) hold for any arbitrary wavelength electromagnetic wave impinging onto the system at any arbitrary incidence angle.

7.2.2.2 Quasiperiodicity effects in the light transmission

Let us now consider a FDM consisting of two kinds of layers, labelled A and B, which are arranged according to the Fibonacci sequence, obeying the concatenation rule $S_{j+1} = S_{j-1}S_j$, for $j \geq 1$, with $S_0 = B$ and $S_1 = A$.[21] The number of layers is given by $N = F_j$, where F_j is a Fibonacci number (see Section 2.4.3). The resulting structure for S_4 is shown in Fig. 7.5 (right panel). Therefore, to obtain the global transfer matrix we must evaluate a matrix product involving three different types of transfer matrices (K_{AA}, K_{BA}, and K_{AB}) which, in addition, are quasiperiodically ordered. At this point, we shall take advantage of the transfer matrix renormalization technique (see Section 9.4) to obtain an analytical expression for the transmission coefficient of light propagating in a FDM. The key point consists in renormalizing the set of transfer matrices $K_{i+1,i}$ according to the blocking scheme $Q_B \equiv K_{AA}$ and $Q_A \equiv K_{AB}K_{BA} = Q$. Note that the renormalized transfer matrix sequence is also arranged according to the Fibonacci sequence and, consequently, the topological order present in the original FDM is preserved by the renormalization process. Now, we realize that the Q_i matrices commute under certain circumstances. In fact, after some algebra one gets

$$[Q_A, Q_B] = \frac{u^2 - 1}{u}\sin\delta_A \sin\delta_B \begin{pmatrix} -\cos\delta_A & \sin\delta_A \\ \sin\delta_A & \cos\delta_A \end{pmatrix}. \qquad (7.12)$$

The commutator (7.12) vanishes in three different cases ($u > 0$): i) The choice $u = 1$ is satisfied for the special case $n_A\beta = n_B$ (which reduces to the trivial periodic case $n_A = n_B$ under normal incidence conditions), and the choices ii) $\delta_A = n\pi$, and iii) $\delta_B = n\pi$, with $n = 1, 2 \ldots$. Therefore, in order to satisfy the commutation condition (7.12), it is *not* necessary to impose restrictive conditions onto both kinds of layers *simultaneously*. For those wavelengths verifying the condition $[Q_A, Q_B] = 0$, we can express the global transfer matrix of the system as $\mathcal{M}_N \equiv Q_A^p Q_B^q$, where $p = F_{j-2}$ and $q = F_{j-3}$. Note that for a FDM of length N, p indicates the number of B layers present in the system. Since the matrices Q_A and Q_B are unimodular for any choice of the system parameters and for any value of the light wavelength one can exploit the Cayley-Hamilton theorem in order to explicitly evaluate $\mathcal{M}_N$ in terms of Chebyshev polynomials, as we did in the study of the periodic case.

Taking into account the commutation conditions ii) and iii) we are led to consider three different possible situations:

- The case $\delta_A \equiv \delta_B = n\pi$, which implies $n_A d_A \beta = n_B d_B$. In this case the half-wavelength condition is satisfied at every layer. Therefore, $Q_A = I$ and $Q_B = (-1)^n I$, where I denotes the identity matrix, and the transparency condition, $T_N = 1$, is trivially obtained.

- The case $\delta_B = n\pi$, for which the global transfer matrix has the form

$$\mathcal{M}_N = (-1)^{np} \begin{pmatrix} \cos\gamma & -\sin\gamma \\ \sin\gamma & \cos\gamma \end{pmatrix}, \tag{7.13}$$

where $\gamma \equiv n\pi F_{j-1} R\beta/\eta$, and $\eta \equiv d_B/d_A$ measures the filling factor. In this case, we get $T_N = 1$ for any β as well. Physically this result can be easily interpreted if we keep in mind that, when the B layers satisfy the half-wavelength condition, the transmission properties of the FDM will depend entirely on the interaction of light with the layers of material A. Now, since the optical behavior of the double layers AA is completely equivalent to that of single A layers ($G_{AA} = I$) and, according to the Fibonacci sequence, B layers always appear flanked by A layers, those wavelengths satisfying the resonance condition $n\lambda = 2n_B d_B \cos\theta_B$ will effectively *see* a *periodic* distribution of A layers separated by fully transparent slabs of constant width d_B. In this case the two-component FDM will behave like an equivalent homogeneous *periodic* medium, characterized by an effective thickness $d' \equiv F_{j-1} d_A$, and a refraction index n_A. Note that, if instead of imposing the half-wavelength condition onto B layers, we make the substitution $B \rightarrow A$, replacing all the B type layers originally present in the FDM by type A ones, we should obtain a homogenous system of length $N d_A > d'$. Consequently, the FDM will exhibit an effective *optical phase shrinkage* for those wavelengths satisfying the resonance condition. It should be noted, however, that this

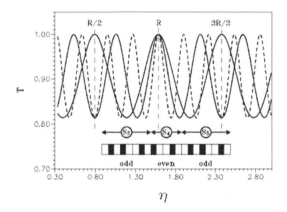

FIGURE 7.6
Dependence of the transmission coefficient with the design parameter η, for $n = 1$ and $\beta = 1$, corresponding to S_4 (thick solid line), S_5 (thin solid line), and S_6 (dashed line) FDMs. We have considered that A (B) layers are composed of SiO_2 (TiO_2) whose indices of refraction (at 700 nm) are $n_A = 1.45$ and $n_B = 2.30$, respectively. The inset shows a scheme for a FDM resonating microcavity. ([45] Reprinted figure with permission from Maciá E 1998 *Appl. Phys. Lett.* **73** 3330 © 1998 by the American Institute of Physics.)

effect is a generic property of any system exhibiting short-range correlations, and, therefore, it is not specific of self-similar arrangements. In fact, a similar behavior was reported in photonic crystals with correlated disorder.[44]

- This physical scenario changes substantially if we impose the condition $\delta_A = n\pi$. In this case, the layers of material A become fully transparent to the incoming light and, consequently, the transmission properties of the FDM will depend on the interaction of light with the layers of material B. The key point now is to realize that these layers are spaced by *two different* distances, d_A and $d_{AA} = 2d_A$, arranged according to a quasiperiodic sequence. Hence, those wavelengths satisfying the resonance condition $n\lambda = 2n_A d_A \cos\theta_A$ will effectively see a quasiperiodic distribution of B layers, instead of a periodic one. In this case, the global transfer matrix for the FDM can be expressed in the closed form

$$\mathcal{M}_N = (-1)^{nF_{j-1}} \begin{pmatrix} \cos\gamma' & -u\sin\gamma' \\ u^{-1}\sin\gamma' & \cos\gamma' \end{pmatrix}, \qquad (7.14)$$

where $\gamma' \equiv p\delta_B$. The explicit evaluation of the transmission coefficient leads to[45]

$$T_{QP} = \frac{1}{1 + \left[\frac{u^2-1}{2u}\right]^2 \sin^2(p\varphi)},\qquad(7.15)$$

where $\varphi = n\pi\eta(R\beta)^{-1}$.

In Fig. 7.6 we show the dependence of the transmission coefficient with η for three FDMs containing $N = 5$, $N = 8$, and $N = 13$ layers, under normal incidence conditions. By inspecting this figure we observe that the curves $T(\eta)$ are symmetrical with respect to the axis $\eta = R = 1.5862\ldots$, and they exhibit a series of maxima ($T_{max} = 1$) and minima ($T_{min} \simeq 0.814$). The number of both maxima and minima increases as the FDM size increases, and their relative positions shift with respect to the symmetry axis. Physically the origin for such oscillatory patterns can be understood as follows. Under the A layer resonance condition, which fixes the value for the incoming light wavelength, the constructive or destructive nature of the interferences, arising from the interaction of light with the quasiperiodic distribution of B layers, will be strongly dependent on the precise relationship between λ and the structural parameters d_A and d_B. Thus, when $\eta = R$ we get the well known relationship $n_A d_A \beta = n_B d_B$, corresponding to the particular case where both A and B layers simultaneously satisfy the resonance condition. Interestingly, Fig. 7.6 shows that there exists a significant number of *additional* values of η for which transparency condition is satisfied in FDMs of variable length. Conversely, if we choose the value of η in such a way that the wavelengths satisfying the resonant condition at the A layers verify the quarter-wavelength condition at the B layers, the quasiperiodic distribution of B layers efficiently backscatter the incoming light, resulting in a significant reduction of the transmission coefficient value. For any choice of the parameter η other than those just described, the transmission coefficient will adopt different values in the interval $T_{min} \le T \le 1$. Accordingly, η can be regarded as a *control design parameter* able to determine the overall FDM optical behavior, varying from that corresponding to a selective filter ($T = 1$) to that proper of a reflective coating (T_{min}).

On the other hand, plugging $n = \beta = 1$ in Eq.(7.15) we see that, when $\eta = R/2$ or $\eta = 3R/2$, the transmission coefficient attains an extreme value, which should be a minimum or a maximum depending on the *parity* of the integer p: when p is even we get $T = 1$, while for odd p we have $T = T_{min}$. Since the parity of the Fibonacci numbers exhibits the recurrence odd-odd-even, the transmission coefficients corresponding to consecutive S_j FDMs should alternate accordingly, as it is illustrated in Fig. 7.6 for the FDMs corresponding to $p = 2$, $p = 3$, and $p = 5$. This interesting property could be used to construct resonating optical cavities where an "even" FDM (for instance S_4), exhibiting full transmission, is sandwiched between two "odd" FDMs (for instance S_5), behaving as optical mirrors. A sketch of such a device is shown, as an inset, in Fig. 7.6.

As discussed above, in order to compare the relative performance of quasiperiodic and periodic dielectric multilayers it is convenient to focus on the case $\delta_A = n\pi$. In so doing, Eqs.(7.11) and (7.15) are conveniently rewritten in the form,[46]

$$T_{QP} = \frac{1}{1 + a(R,x)\sin^2\left[p\, b_n(R,x,\eta)\right]}, \tag{7.16}$$

and

$$T_P = \frac{1}{1 + a(R,x)\sin^2\left[\nu\, b_n(R,x,\eta)\right]}, \tag{7.17}$$

where we made use of Snell's law, $R\sin\theta = \sin\theta_B$, to introduce the auxiliary variable $x \equiv \sin^2\theta$, describing the light incidence geometry $(\theta_A \equiv \theta)$, and we have also introduced the auxiliary functions

$$a(R,x) \equiv \left[\frac{R^2 - 1}{2R}\right]^2 (1-x)^{-1}(1 - R^2 x)^{-1}, \tag{7.18}$$

and

$$b_n(R,x,\eta) \equiv n\pi y\sqrt{\frac{1 - R^2 x}{1 - x}}, \tag{7.19}$$

where $y \equiv \eta/R = n_B d_B/n_A d_A$ measures the phase ratio between both dielectric layers. In Fig. 7.7 we show the dependence of the transmission coefficient with the phase ratio y for a periodic multilayer with $\nu = 4$ (dashed line) and a FDM with $p = 3$ (solid line), both of them containing $N = 8$ layers, under normal incidence conditions. This figure exhibits a series of maxima and minima similar to those plotted in Fig.7.6. Nevertheless, the physical origin of such an oscillatory pattern is somewhat different in this case. Thus, when $y = 1$ ($n_A d_A = n_B d_B$) both A and B layers simultaneously satisfy the resonance condition, and the long-range order of the layers sequence becomes irrelevant. However, for $y \neq 1$ the optical response of periodic and quasiperiodic multilayers progressively differs as the phase ratio is progressively increased (or decreased), in the way displayed in Fig. 7.7.

A particularly interesting situation occurs if we choose the phase ratio value in such a way that the wavelengths satisfying the half-wavelength condition at the A layers of the periodic multilayer verify the quarter-wavelength condition at the B layers of the FDM. In this case, the quasiperiodic distribution of B layers efficiently backscatter the incoming light, resulting in a significant reduction of the transmission coefficient value. Conversely, the periodic multilayer exhibits full transmission for the same wavelength. From Eqs. (7.18) and (7.19) we obtain that this condition is satisfied when $y = k/2n|p-\nu|$, with $k = 1, 3, 5 \ldots$. The cases corresponding to $k = 1$ and $k = 3$ are indicated by vertical lines in Fig. 7.7. This behavior opens the possibility of constructing a mixed device composed of both periodically and quasiperiodically arranged multilayers, so that the refractive index contrast and the layers thicknesses

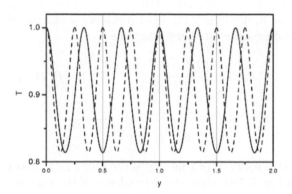

FIGURE 7.7

Dependence of the transmission coefficient with the phase ratio y for a periodic multilayer with $\nu = 4$ (dashed line) and a Fibonacci dielectric multilayer with $p = 3$ (solid line) containing $N = 8$ layers, under normal incidence geometry and $n = 1$. For the sake of illustration we have taken $n_A = 1.45$ and $n_B = 2.30$ as suitable representative values. The vertical lines indicate the phase ratios satisfying the condition $y = 1/2$ and $y = 3/2$, respectively. ([46] Reprinted figure with permission from Maciá E 2001 *Phys. Rev. B* **63** 205421 © 2001 by the American Physical Society.)

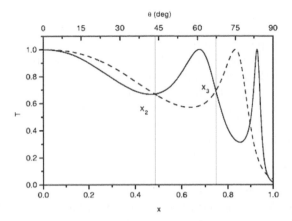

FIGURE 7.8

Dependence of the transmission coefficient with the angle of incidence θ for a periodic multilayer with $\nu = 4$ (dashed line) and a Fibonacci dielectric multilayer with $p = 3$ (solid line) containing $N = 8$ layers and $n = 1$, for the case $y = 1$. We have taken $n_A = 1.45$ and $n_B = 2.30$. ([46] Reprinted figure with permission from Maciá E 2001 *Phys. Rev.* B **63** 205421 © 2001 by the American Physical Society.)

determining the phase ratio value can act as *control design parameters* able to determine the optical response of their different constitutive substructures from that corresponding to a selective filter ($T = 1$) to that proper of a reflective coating (T_{min}). It is worth to highlight, at this point, that such a mixed device can be viewed as an *hybrid order device* made of two different kinds of subunits, each one exhibiting a *different kind of topological ordering* which respectively gives rise to a reversal in the value of the corresponding transmission coefficients. The key point here is that such a *complementary behavior* can be obtained by just changing the kind of topological order in the stacking sequence of layers composing each subunit, so that both the chemical nature of the different layers and the wavelength of the incoming light remains unchanged. This is a quite remarkable result, since the pertinent codes to alternate between periodic and quasiperiodic orderings in the sequence of the different layers can be easily implemented in current state-of-the-art deposition processes.

Such a complementary behavior can be further exploited under oblique angle incidence conditions. In Fig. 7.8 we show the dependence of the transmission coefficient with the angle of incidence, θ, for the periodic multilayer (dashed line) and the FDM (solid line) previously considered in Fig. 7.7, for the case $\eta = R$. Both transmission curves exhibit a series of maxima, whose number increases with the length of the system. By inspecting Fig. 7.8 we can distinguish two different regimes. At low incidence angles the optical response of both periodic and quasiperiodic multilayers is quite similar, although the transmission curve for the FDM systematically departs from that corresponding to the periodic multilayer, exhibiting lower transmission values. The second regime starts at the critical angle value $x_2 \simeq 0.48$ where a crossing point between both transmission curves occurs. Afterwards, the transmission curve for the FDM suddenly grows as the incidence angle is increased, reaching a broad peak at $x \simeq 0.67$ and it rapidly decreases again to reach another crossing point with the periodic multilayer transmission curve at $x_3 \simeq 0.75$. A similar oscillatory pattern repeats itself as the incidence angle is further increased, determining the subsequent crossing points.

In summary, these results suggest the possibility of exploiting the quasiperiodic order of the system by properly matching the wavelength of the incoming light with some of the different characteristic lengths present in the aperiodic substrate. Thus, by properly choosing the different design parameters, as determined by the refractive index contrast, R, and the multilayer filling factor, η, we can select specific resonance conditions, directly associated to the quasiperiodic order of the substrate. In addition, we can also play with the incidence angle geometry of the incoming light in order to further exploit the capabilities previously determined by the architecture of the multilayer. In fact, the most relevant feature of the transmission curves shown in Fig. 7.8 is the existence of broad incidence angle intervals where the periodic and quasiperiodic multilayers respectively exhibit *complementary optical responses*, in the sense that when one of them exhibits high T values the other one exhibits low T values (and vice versa). This complementary behavior suggests the possibility of constructing resonating optical devices that exploit the novel interference possibilities associated to the coexistence of different kinds of topological order (i.e., periodic and quasiperiodic order) in a given arrangement of the substrate. In this way, plentiful possibilities for new tailored materials should appear. This appealing possibility will be further analyzed in Section 8.2.1.

7.2.2.3 Number recognition

Within the framework described in previous sections propagation of classical waves in aperiodic multilayers can be usefully exploited for number recognition and data storage purposes.[47] To this end, one should consider a multilayer composed of K different types of materials as a number written in base K. Thus, layered structures built of two different substances will correspond to binary numbers, where layers with low (high) refraction index are labelled with symbol "0" ("1"), respectively (Fig.7.9). For instance, periodic multilayers consisting of alternating high/low refraction indices of the form 10, 1010, 101010, ... correspond to decimal numbers 2, 10, 42,..., respectively. In Table 7.1 we list the binary sequence equivalence of the shorter Fibonacci

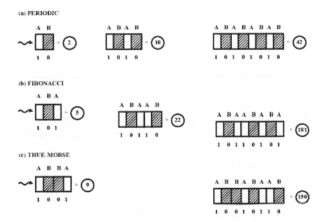

FIGURE 7.9

Sketch illustrating the different code number capabilities of different dielectric multilayers depending on the type of order (periodic, Fibonacci sequence, or Thue-Morse sequence) in the arrangement of their layers. We note (see the diagrams on the right) that depending on the stacking order (which determines the binary representation of the refraction index profile) three multilayers with the same number of layers ($N = 8$) will code for three diferent decimal numbers values.

multilayers (containing up to 21 layers) along with their associated decimal numbers and their prime factors.

TABLE 7.1
Binary number representation of Fibonacci multilayers of increasing size. The prime factor decomposition of the associated decimal numbers is given in the column on the right.

FIBONACCI MULTILAYER	$N = F_n$	CODE NUMBER	PRIME FACTORS
0	1	0	–
1	1	1	1
10	2	2	2
101	3	5	5
10110	5	22	2×11
10110101	8	181	181
1011010110110	13	5814	$2 \times 3^2 \times 17 \times 19$
101101011011010110101	21	1488 565	$5 \times 13 \times 22\,901$

The key point of this approach is that due to multiple scattering and interference effects the propagation of an electromagnetic wave through the multilayer results in the formation of a characteristic transmission (or reflection) pattern, which could be properly used to identify the number associated to each multilayer. In this sense, transmission coefficients could be regarded as spectral portraits allowing to identify the value of the number associated to a given multilayer by means of a suitable physical process (light propagation). Due to the sensitivity of interference processes to minor modifications in the arrangement of the layers this method allows for an efficient detection of possible errors in the layer distribution in a non-invasive way. This property could be used in optical data recording and read-out, though it is clear that not only the identification of certain specific numbers (i.e., those related to a given substitution sequence) but also the possibility to code and identify any given number is necessary for information coding. In this regard, the very possibility of performing operations with numbers using wave propagation through multilayer based Fabry-Perot interferometers was illustrated.[47] In the same vein, a systematic analysis of progressive sequential splitting of some peaks in the spectral portraits of Cantor dielectric multilayers (see Section 4.2.3) revealed a direct correlation between their associated number value and its prime factors (Fig.7.10).[47, 48]

7.2.3   Thermal emission control

The multilayered structures considered up to now are made of lossless, linear optical materials. Some interesting results are obtained when one considers more realistic materials, able to couple with the electromagnetic field. For instance, one may consider that some of the layers are able to absorb a fraction

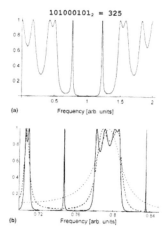

(a)

(b)

FIGURE 7.10
(a) Spectral portrait of a triadic Cantor dielectric multilayer of second generation corresponding to the decimal number 325. (b) Progressive splitting of the transmission spectrum as the generation of the Cantor multilayer is progressively increased: 2nd generation (short dash), 3rd generation (long dash), and 4th generation (solid line). (Courtesy of Sergey V. Gaponenko.[47])

of the incident radiation. In that case, the intensity of light propagating through the system decays exponentially according to the expression $I(\omega) = I_0 e^{-\alpha z}$, where

$$\alpha = \frac{2n''\omega}{c} \qquad (7.20)$$

measures the absorption per unit length for an electromagnetic wave of frequency ω propagating in a medium of complex index $n = n' + in''$. Thus, the absorptance increases linearly with increasing frequency. Now, even in the case of highly absorptive materials, not all incident radiation is usually absorbed by a given material. This is because the material reflects some radiation at its surface, with a reflection coefficient given by

$$R_S = \left| \frac{n_0 - n}{n_0 + n} \right|^2, \qquad (7.21)$$

where n_0 is the refraction index of the surrounding medium. Now, according to the Kirchoff's second law, the ratio of the thermal emittance to the absorptance is a constant, independent of the nature of the material, and that constant is unity only when the material is a perfect blackbody. Therefore, most absorbing materials behave as the so-called gray-body when heated, exhibiting emittance values lower than unity. An ingenious way to circumvent this shortcoming, enhancing the thermal emissivity of a given material above

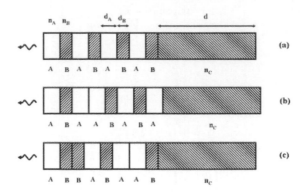

FIGURE 7.11

A multilayered stack (thin film) made of alternating layers of refractive indices n_A (real) and n_B (complex) coats a thick absorbing substrate of refraction index n_C. The entire structure is embedded in a medium, taken to be air ($n = 1$). Different working regimes are obtained depending on the adopted absorbers distribution (shadowed layers). In a) the absorbers are periodically distributed, whereas in b) and c) they are respectively arranged according to the Fibonacci and Thue-Morse sequences.

that corresponding to its gray-body value *at certain* frequencies was proposed on the basis of the structural design depicted in Fig.7.11. In order to fully understand the physical principle of the proposed device we must recall some basic concepts.

Thermal radiation is just spontaneous emission that is thermally pumped and that has a blackbody spectrum, which is in thermal equilibrium with its surroundings. The radiant power per unit area emitted by the surface of a perfect blackbody is given by $\mathcal{P} = cu/4$, where c is the vacuum speed of light and u is the energy density per unit frequency ω at a given temperature T, which can be written as

$$u(\omega, T) = N(\omega)\frac{\hbar\omega}{e^{\frac{\hbar\omega}{k_B T}} - 1}, \qquad (7.22)$$

where $N(\omega)$ is the electromagnetic density of modes. For free-space boundary conditions one has

$$N(\omega) = \frac{2\omega^2}{\pi^2 c^3}, \qquad (7.23)$$

so that one obtains the well-known Planck law [49]

$$P(\omega, T) = \frac{\hbar \omega^3}{2\pi^2 c^2} \frac{1}{e^{\frac{\hbar \omega}{k_B T}} - 1}. \tag{7.24}$$

Now, it has been observed that quantum confinement effects modify the spontaneous emission processes in photonic band gap multilayered structures. Accordingly, one should expect a modification of the thermal radiation in such systems. A convenient way to deal with this problem within the framework of the transfer matrix technique relies on the evaluation of the so-called thermal optical power

$$\mathcal{E}(\omega) = 1 - R(\omega) - T(\omega), \tag{7.25}$$

where $R(\omega)$ and $T(\omega)$ are the reflection and transmission coefficients, respectively. Eq.(7.25) is physically interpreted as the ratio of the optical power emitted at frequency ω into a spherical angle element $d\Omega$ by a unit surface area of the thin film, to the power emitted by a blackbody with the same area at the same temperature. In this way, the power spectrum of the heated multilayered system located in front of an emitting hot surface (Fig.7.11) is given by the expression

$$\tilde{P}(\omega, T) = \mathcal{E}(\omega) P(\omega, T), \tag{7.26}$$

which properly modifies Planck's law given by Eq.(7.24).[50, 51]

In order to evaluate the thermal optical power some care must be taken into account for complex indices of refraction used to model the presence of absorbing dielectric layers (which will emit upon thermal excitation). Some illustrative results are shown in Fig.7.12. Let us first consider the case where a periodic film coating sits atop the heated substrate (Fig.7.12a). As we see the film significantly blocks heat radiation emitted by the substrate at the frequencies corresponding to the photonic crystal bandgap ($\omega/\omega_0 = 1$) as expected, but we also observe that the substrate's emission is enhanced from the gray-body level all the way up to the perfect blackbody rate at a number of frequencies corresponding to the pass-band transmission resonances of the multilayered film. This occurs because the thin film acts as an antireflective coating at these resonances, hence removing the impedance mismatch given by Eq.(7.21). In this way, all the radiation incident from the left tunnels through the multilayer structure into the substrate for these selected frequencies, so that the substrate effectively behaves as a perfectly absorbing blackbody in that case.[51] A similar enhancement of the substrate's thermal emittance at certain resonance frequencies accompanied by the corresponding inhibition at the stop-bands is observed in the aperiodically arranged thin film coatings as well. In the case of the Fibonacci coating (Fig.7.12b) a characteristic trifurcation splitting can be clearly appreciated around a number of frequencies, and one finds a strong emittance within the spectral range corresponding to the midgap in the periodic case. In fact, the presence of allowed bands in

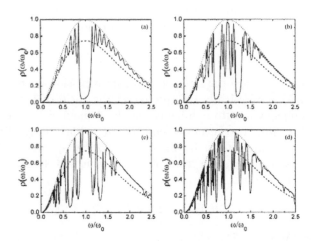

FIGURE 7.12

Thermal radiation spectra as a function of the reduced frequency ($\lambda_0 = 700$ nm) under normal incidence conditions for a device as that shown in Fig.7.11 with $n_A = 1.45$ (SiO$_2$) and $n_B = 1.0 + 0.01i$, where the thin film coat layers are respectively arranged according to the following sequences (a) periodic, (b) Fibonacci ($N = 377$), (c) Thue-Morse ($N = 512$), and period-doubling ($N = 512$). The perfect blackbody thermal spectrum is given by the dotted curve, whereas the dashed curve gives the thermal spectrum for the substrate, with refractive index $n_C = 3 + 0.03i$. The temperature is chosen so that the blackbody (Wien) peak is aligned with the midgap frequency $\omega_0 = 2\pi c/\lambda_0$. All the curves are properly normalized by this peak power. ([52] Courtesy of Eudenilson L. Albuquerque. With permission from IOP Publishing Ltd.)

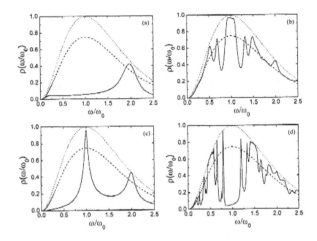

FIGURE 7.13

The same as in Fig.7.12, but considering that B layers are composed of a metamaterial with refraction index $n_B = -1.0 + 0.01i$. ([52] Courtesy of Eudenilson L. Albuquerque. With permission from IOP Publishing Ltd.)

certain forbidden regions of the periodic system is a characteristic feature of quasiperiodic systems (see Fig.5.11 for an illustration of this effect in the phonon spectrum of Fibonacci lattices). An analogous pattern is observed in the thermal spectrum corresponding to the Thue-Morse thin film, whereas that corresponding to the period-doubling sequences is more similar to the periodic one around the $\omega/\omega_0 = 1$ spectral range. In all the aperiodic arrangements, however, one has a richer thermal emission spectrum, reflecting in a conspicuous way the highly fragmented nature of their transmission profiles. Quite interestingly, these spiky thermal emission spectra can be substantially smoothed (hence obtaining broader spectral ranges with enhanced emittance) by using metamaterials in the composition of the multilayer coat (Fig.7.13).

Metamaterials are artificially constructed composites exhibiting a negative electrical permittivity ϵ together with a negative magnetic permeability μ in the same frequency range. This yields a negative refractive index (i.e., $n = i^2 \sqrt{|\epsilon|} \sqrt{|\mu|} = -\sqrt{\epsilon\mu}$). One of the main features of negative refraction index materials is that the electric field $\mathbf{E}$, the magnetic field $\mathbf{H}$, and the wave vector $\mathbf{k}$ form a left-handed triplet. Due to that, these materials are also referred to as left-handed materials, because they support waves with the phase velocity opposite to the direction of the energy flow. Moreover, their phase and group velocity are antiparallel, since usually the group velocity has the same direction as the energy flow (Poynting vector). It has been shown that multilayered structures made of alternating layers of metamaterials and ordinary dielectrics can exhibit a new type of photonic bandgaps which are not based on interference effects. The first kind of the so-called non-Bragg gap arises naturally when the volume average refractive index of the multilayer, $\bar{n}$, equals zero. The second kind appears in dispersive materials at frequencies where either μ or ϵ vanishes in the metamaterial.[54] The average refraction indices for nth order Fibonacci and Thue-Morse multilayers are given by[55, 56]

$$\bar{n}_F(n) = \frac{n_A d_A F_{n-1} + n_B d_B F_{n-2}}{d_A F_{n-1} + d_B F_{n-2}}, \qquad \bar{n}_{TM} = \frac{n_A d_A + n_B d_B}{d_A + d_B}, \qquad (7.27)$$

where we note that $\bar{n}_{TM}$ does not depend on the system size, since the number of A layers and B layers is just the same in a Thue-Morse sequence of any order. On the other hand, making use of Eq.(2.5) one gets the quasiperiodic limit

$$\lim_{n\to\infty} \bar{n}_F(n) = \frac{n_A d_A \tau + n_B d_B}{d_A \tau + d_B}. \qquad (7.28)$$

The non-Bragg zero-$\bar{n}$ gap condition for these aperiodic multilayers then reads $n_A = -n_B \eta$ and $n_A = -n_B \eta \tau^{-1}$, for Thue-Morse and Fibonacci multilayers, respectively, where $\eta = d_B/d_A$ is the filling ratio and the indices of refraction generally will depend on the frequency. Therefore, the precise location of the zero-$\bar{n}$ gap can be finely tuned by properly choosing the optical parameters of the system. From Eqs.(7.27) we see that the central frequency of the zero-$\bar{n}$ gap depends on the kind of order present in the structure. Thus, the location of this gap in the Thue-Morse multilayer coincides with that present in the periodic sequence generated from the substitution rule $A \to AB$, $B \to AB$, which shares the same substitution matrix (see Section 4.2.1).

The possibility of tailoring the thermal emittance of a substrate by coating it with a film composed of alternating negative and positive refraction index materials arranged according to a triadic Cantor set has been numerically analyzed as well.[57] As it can be appreciated from Fig.7.14, the sequential splitting characteristic of self-similar Cantor sets gives rise to the presence of narrow emittance peaks located at the midgap frequency region. Thus, while a full refection band appears in the emittance spectra of periodic multilayers, Cantor-type ones exhibit sharp and narrow resonances throughout the band even for low generation numbers.

7.2.4   Photovoltaic cells

A photovoltaic cell is a device that converts solar energy into electricity by means of the so-called photovoltaic effect. This effects consists in the generation of an electron-hole pair from the absorption of a photon with energy enough to promote an electron from the valence band to the conduction band in a semiconducting material. The theoretical efficiency of ideal solar cells significantly depends on their electronic band structure, since the photon-induced electrical current originates from the difference between the rates of

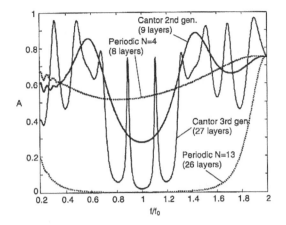

FIGURE 7.14
Thermal emittance (absorptance) spectra for periodic films based on alternating negative/positive refraction index materials are compared to those corresponding to triadic Cantor type ones of similar size. The adopted optical parameters are $n_A = 1.41$, $n_B = -2.0$, and $n_C = 3 + 0.03i$. (From ref.[57]. With permission from IOP Publishing Ltd.)

photon absorption and photon emission due to radiative recombination of the electron-hole pairs. Thus, though the theoretical thermodynamic limit is as high as 93%, standard devices based on silicon p-n junctions yield a sunlight-to-electricity efficiency of about 33%. This low efficiency spurred a flurry of research aimed at improving this figure. Among the different proposals reported to date, we will focus on those based on photon-induced transition at intermediate energy levels,[58] since this approach significantly benefits from the highly fragmented structure of the energy spectrum in quasiperiodic systems.[59] In this case, the photovoltaic efficiency is enhanced due to the presence of a number of new possible excitation channels between the valence (conduction) bands and the intermediate bands, which allow for the absorption of additional, lower energy photons (Fig.7.15). In usual, crystalline materials an intermediate band can be introduced in several ways: impurities, lone pair bands, or superlattice stacking (leading to the formation of minibands). Now, it is reasonable to expect that different intermediate band structures give different efficiency limits. In addition, for a fixed band gap value (the most usual situation in practical applications), the photovoltaic efficiency limit progressively increases from the predicted ideal efficiency of 63% for one intermediate band up to 76% and 80% for two and three intermediate bands, respectively (Fig.7.16). These results strongly suggest that the trench between the thermodynamic limit and reported efficiencies in actual devices

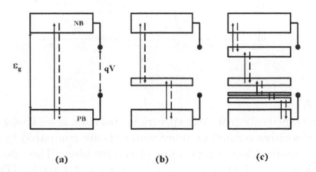

FIGURE 7.15

Band diagrams of solar cells based on (a) a typical p-n semiconductor, (b) an intermediate band structure, and (c) a highly fragmented multiband structure as that sketched for a semicoductor based Fibonacci superlattice in Fig.7.1. The model solar cell combines a negative-contact band (NB), a positive-contact band (PB), and a number of intermediate narrower bands. It is assumed that all incident photons with energy equal or greater than the band gap ε_g are absorbed, and they generate one electron-hole pair, whereas radiative recombination annihilates those pairs. The balance of electrons gives the current, which is delivered to an external load by the positive and negative contacts, yielding an electrical work qV. Non-radiative transitions (v.g. electron-phonon interactions) between any two bands are forbidden in ideal conditions.

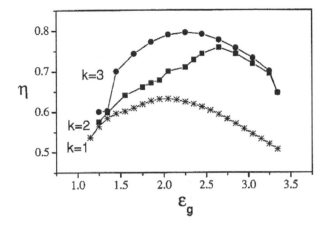

FIGURE 7.16
Limiting efficiency for solar cells with one ($k = 1$), two ($k = 2$), and three ($k = 3$) intermediate bands as a function of the band gap ε_g. The lines are a guide to the eye. ([59] Courtesy of Ru-wen Peng).

may gradually be filled by considering materials having highly fragmented energy spectra in photovoltaic cells.

The quest for the optimum band structures to be considered to this end opens promising avenues in semiconductor based materials science engineering which naturally lead one to the consideration of aperiodic materials. In fact, one reasonably expects that devices based on aperiodic arrangements (including Fibonacci, Thue-Morse, or Cantor superlattices as representative examples, Fig.7.17) are a quite natural choice, for they properly combine the presence of highly fragmented miniband electronic structures, with a self-similar distribution of energy levels (see Sections 5.3 and 5.7), hence further contributing to improve the photovoltaic efficiency by adding up the contributions coming from electronic transition involving lower and lower energy scales.

In summary, in this Section we have analyzed on the possibility of exploiting aperiodic order in devices based on aperiodic multilayered structures. As we have seen, in many instances the recourse to Fibonacci, Thue-Morse, or fractal arrangements of layers enhances the response obtained from periodic arrangements. One of the main lessons to be extracted from the different experimental realization described in this Section is that aperiodic order is able by itself to endow a previously existing device with novel properties, opening new avenues for innovative applications on the basis of well-known physical mechanisms. Such a feature has been illustrated in some detail in the case

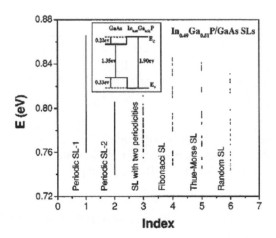

FIGURE 7.17

Electronic miniband structures for several periodic and aperiodic $In_{0.49}Ga_{0.51}P/GaAs$ superlattices whose band-edge diagram is shown in the inset. The structural parameters of the superlattices labelled by the index number read as follows: 1) $N = 21$, $a = b = 2.5$ nm, 2) $N = 21$, $a = b = 3.5$ nm, 3) is composed of two parts, one with $N = 10$, $a = b = 2.5$ nm, and the other with $N = 11$, $a = b = 3.5$ nm, 4) $N = 21$, $a = 2.5$ nm, $b = 3.5$ nm; 5) $N = 16$, $a = 2.5$ nm, $b = 3.5$ nm; 6) $N = 21$, $a = 2.5$ nm, $b = 3.5$ nm; where N is the number of layers, and a (b) are the thicknesses of blocks A (B). ([59] Courtesy of Ru-wen Peng. With permission from IOP Publishing Ltd.)

of resonant optical cavities, multiband filters, x-ray mirrors, thermal emissivity control, and photovoltaic cells. Many more applications are currently being considered in these[60, 61, 62] and other related fields of technological interest as, for instance, the propagation of acoustic waves in phononic circuits,[39] or the physical properties of ferromagnetic,[63] piezoelectric,[64] or piezomagnetic[64] superlattices based on aperiodic sequences.

7.3 Photonic and phononic quasicrystals

In the previous Section we have discussed how aperiodic order affects the physical properties of quasiperiodic multilayered systems in one-dimension. In this Section we will address a further extension of quasiperiodicity effects by considering bulk aperiodic systems in two and three dimensions. In so doing, we will usually deal with composite materials which do not exist in nature and provide a remarkable example of materials by design.[65]

For the sake of illustration, let us start by introducing a nice experiment which was originally designed in order to study the nature of the frequency spectrum of an acoustic system analogous to a two-dimensional Schrödinger equation with a quasiperiodic potential. The system consists of a number of tuning forks (natural frequency 440 Hz) glued into a heavy aluminum plate patterned according to a standard Penrose tile (Fig.7.18a). The tuning forks are mounted at the centers of the rhombuses, with the two tines oriented in line with the shorter diagonal. The tuning forks are coupled to each other by means of 1-mm-diameter steel wire arcs which are spot welded from one tine of a tuning fork to that of a nearest neighbor. Using the four sides of each rhombus as a reference, four nearest neighbors are identified, and each tine of a tuning fork is coupled to the two nearest tines of the adjacent tuning forks. In this way, four different lengths (3.7, 5.8, 7.26 and 7.77 cm) are distributed through the system forming a quasiperiodic network of coupled resonators. An electromagnet is positioned near one tine of the array, and an ac current is passed through it in order to drive the system, and its response is monitored with four electric guitar pickups positioned next to random tines in the array. The resulting resonant frequencies correspond to the eigenvalues of the two-dimensional quasiperiodic system. The obtained frequency spectrum is shown in Fig.7.18, along with its related DOS. In analogy with the phonon spectra of general Fibonacci lattices, the frequency spectrum of the Penrose tile exhibits four characteristic bands (labeled A, B, C, and D) separated by the corresponding gaps α, β, and γ. In addition, a series of subsidiary gaps appear at different locations within the four main bands, giving the overall spectrum the typical appearance of a code-bar pattern. Quite remarkably, the widths of these bands and gaps are in ratios involving the golden mean,

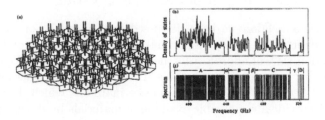

FIGURE 7.18

(a) Schematic drawing of a tuning fork quasicrystal based on the Penrose tile (for the sake of drafting simplicity the tuning forks are not drawn with their actual orientation in the experimental setup). The frequency spectrum (c) and the DOS (b) of the tuning fork quasicrystal were determined as a composite of the resonant spectra obtained from twenty different positions in the Penrose lattice. The DOS is obtained as the inverse of the difference in frequency for neighboring eigenvalues in the frequency spectrum. ([66] Reprinted figure with permission from He S and Maynard J D 1989 *Phys. Rev. Lett.* **62** 1888 © 1989 by the American Physical Society.)

namely, $C/B = \tau$, $A/B = \tau^2$, $\beta/\alpha = \gamma/\beta = \tau$.[66] The presence of frequency stop bands allows one to extend the photonic crystal notion to its analogous acoustic version, the so-called *phononic* crystal. Accordingly, most characteristic features of energy spectra in one-dimensional quasiperiodic systems, namely, hierarchical fragmentation and scalability, are also present in the spectra of two-dimensional ones.

7.3.1 Optical devices in two dimensions

Let us consider a periodic photonic crystal in two or three dimensions exhibiting a characteristic photonic gap over a certain frequency range. What happens to the photonic gap if one replaces the periodic arrangement of the constitutive elements by an aperiodic one? This question was addressed by several groups both theoretically and experimentally, concluding that two-dimensional photonic band-gaps can be obtained in systems comprised of quasiperiodic arrangements of dielectric materials. This property is illustrated in Fig.7.19 for two octogonal arrangements of dielectric cylinders in air. The frequency spectrum corresponding to the arrangement shown in Fig.7.19a exhibits a number of relatively broad gaps inside which TM mode propagation (with the magnetic field parallel to an axis perpendicular to the plane) is forbidden (Fig.7.19c), whereas the connected work shown in Fig.7.19b is required in order to obtain a sizable spectral gap for TE mode propagation (with electric field parallel to the cylinders). The fact that isolated cylinders are good in order to obtain band gaps for the TM modes while a connected network is necessary for the TE modes is not a specific feature of quasiperiodic order, since it was previously reported for periodic systems.[67, 68]

Point defects (missing cylinders) or line defects (missing rows of cylinders) can be respectively used to create highly localized defect modes or to form waveguides in photonic band-gap systems in both periodic and aperiodic systems alike (Fig.7.20). In the case of quasiperiodic arrangements the absence of translational symmetry leads to some specific differences, namely, the point defect properties become much richer and the light guides derived from quasiperiodic systems become more frequency selective.[69] To further check these properties an octogonal photonic crystal was constructed as a 23×23 array of alumina cylinders ($\epsilon = 8.9$) of diameter 6.12 mm, with $L = 9$ mm, embedded in a Styrofoam template. Its wave guiding properties were measured in the microwave region (7.5-12.5 GHz) for the TM modes. It was found that both the position and width of the predicted band-gap do not significantly depend on the light propagation direction in the plane, that is, one obtains a complete gap for the TM. When compared with similar gaps in periodic photonic crystals it was found that complete gaps in photonic quasicrystals allow for a more uniform light reflection. This property stems from the fact that the first Brillouin zone has more symmetries in quasiperiodic crystals, so that these structures are able to support much more isotropic band-gaps as compared to conventional photonic crystals based on periodic square or

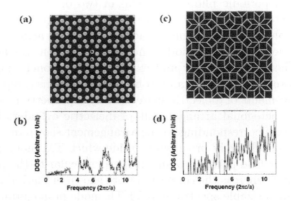

FIGURE 7.19

Two related quasiperiodic photonic band-gap systems are formed placing 164 dielectric cylinders ($\epsilon = 10$) in the vertices (a), or along the sides (c), of a two-dimensional quasiperiodic lattice with octogonal symmetry. The cylinders occupy a volume fraction of 30% (a) or 25% (c) in an air ($\epsilon = 1$) background. The octogonal pattern is a tile composed of squares and rhombuses (acute angle of 45°) of equal side L. The quasi-lattice constant is then given by $a = (6 + 4\sqrt{2})L$. The DOS for TM modes and TE modes are shown in (b) and (d), respectively. (Adapted from ref.[67]. Reprinted figure with permission from Chan Y S, Chan C T and Liu Z Y 1998 *Phys. Rev. Lett.* **80** 956 © 1998 by the American Physical Society.)

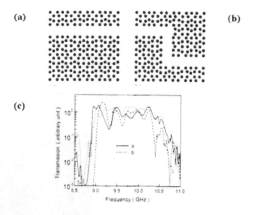

FIGURE 7.20
Schematic of straight (a) and bending (b) waveguide designs in the octogonal quasicrystal. The corresponding transmission spectra are shown in (c). (Adapted from ref.[72]. Reprinted with permission. © 1999, American Institute of Physics.)

hexagonal lattices, hence favoring the possible appearance of a complete gap in these systems. Nevertheless, one should keep in mind that quite isotropic photonic band-gaps are also obtained for Archimedean tilings, which are periodic structures displaying high order local rotational symmetry (see Fig.1.2). [70, 71]

Two types of waveguides (a straight guide and a bending guide with two 90° shape corners (Fig.7.20a and b, respectively)) were fabricated by properly removing three rows of cylinders in the octogonal lattice. The measured transmission spectra indicate a very clear guiding effect in the gap region. However, the transmission curve is not so smooth as that corresponding to periodic crystals. In the octogonal crystal a number of small peaks and dips appear in the transmission profile, specially in the case of the bending waveguide (Fig.7.20b). These ripples appear due to the presence of a more complex resonance pattern in the quasiperiodic system, so that the presence of aperiodic order is not so beneficial in this case.[72] The existence of a complete bandgap in a two-dimensional Penrose lattice made of dielectric alumina rods was experimentally reported in the microwave regime and similar waveguiding properties (straight and 90° sharp bend) were observed by removing one row of rods from the perfect Penrose lattice.[73]

To open a complete gap in two-dimensional photonic crystals the refractive index needs to be larger than 2. This condition is easily obtained in systems like those shown in Fig.7.20, but it rules out the possibility of using polymer

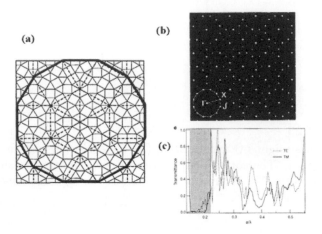

FIGURE 7.21

(a) The Sampfli inflation tiling starts with a big parent square-triangle tiling (thick, dashed lines) from which one generates an offspring square-triangle tiling (thin, solid lines).[78] (b) Representation of the dodecagonal crystal reciprocal lattice, showing 12-fold symmetry. The Brillouin zone (white dodecagon) corresponds to the cell highlighted in (a) with a thick line. (c) The corresponding experimental transmission along the $\Gamma - J$ direction. The illuminating laser wavelength $\lambda = 633$ nm compared to the photonic quasicrystal pitch of $a = 260$ nm with a filling factor of 28%. The shadowed area highlights the gap for both TE (dashed line) and TM (solid line) wave polarizations. (Adapted from refs.[78] and [80]. Reprinted figure with permission from Oxborrow N and Henley C L 1993 *Phys. Rev. B* **48** 6966 © 1993 by the American Physical Society.)

materials to fabricate periodic photonic crystals because their refractive indices are typically lower than 1.7. Compared with periodic photonic crystals, photonic quasicrystals have a smaller threshold value of refractive index to open a complete gap. For example, the threshold value is $n = 1.26$ for an octogonal quasilattice, which indicates that optoelectronic components based on photonic quasicrystals may be realized in low refractive index materials such as silica, glasses, and polymers. Motivated by this possibility a high-quality octogonal photonic crystal was fabricated by using a laser induced microexplosion method. The resulting structure consisted of air cylinders embedded in a polymer material dielectric matrix arranged according to the pattern shown in Fig.7.19a. Thus, in this case one has a complementary design where the holes (low dielectric constant) play the role of cylinders and the air substrate is replaced by a suitable resin with a higher value of ϵ. Fabry-Pérot cavities were then fabricated by removing the central layer of holes in these systems.[74]

As we have seen, the anisotropy of a photonic bandgap depends on the symmetry of the photonic crystal lattice: the higher the degree of local symmetry the more isotropic the resulting band-gap. Accordingly, as the order of the symmetry increases, the Brillouin zone becomes more circular, resulting in the formation of a full band-gap (Fig.7.21b). Thus, it was natural to investigate aperiodic structures exhibiting progressively higher-order rotational and mirror symmetries, such as those found in 10-fold,[75] 12-fold, 14-fold, 18-fold,[76] or even 40-fold and 120-fold quasiperiodic lattices.[77] Experimental evidence for complete bandgaps for both TE and TM polarizations was reported for dodecagonal quasicrystals based on a square-triangle tiling obeying the Stampfli inflation rule (Fig.7.21a).[79] The photonic QCs were composed of 150-nm air pores (filling fractions 28-30%) separated by a distance $a = 260$ nm in a 260 nm thick silicon nitride slab ($n = 2.02$) which is capped above and below by silicon dioxide to confine the light in two-dimensions. In contrast to periodic lattices, the TM bandgap lies exactly in the middle of the TE bandgap, leading to an efficient overlapping for both polarization states in the resulting complete photonic bandgap (Fig.7.21c). According to numerical simulations this bandgap remains open for very low refraction index materials such as glass ($n = 1.45$) with similar filling fractions.[80] This property makes photonic QCs very promising for a number of optical applications. For instance, anomalous refraction and focusing properties of electromagnetic waves in the microwave region (resembling that reported for superlenses based on negative refraction index materials) were demonstrated for dodecagonal QCs consisting of dielectric cylinders ($\epsilon = 8.6$) and radii 3.0 mm embedded in a Styrofoam template according to the arrangement shown in Fig.7.21a with a lattice constant of $a = 10$ mm.[81] Recent numerical studies suggest that these anomalous properties arise from complex scattering effects and short-range interactions associated with certain symmetry points in the photonic QCs, which are not present in periodic photonic crystals.[82] This photonic QC is composed of pure dielectric materials and therefore is subject to far

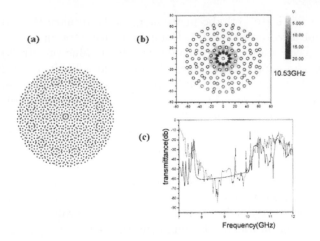

FIGURE 7.22

(a) Schematics of a dodecagonal Penrose lattice. (b) Distribution of the electric field at the frequency $\nu = 10.53$ GHz corresponding to a sharp peak within the bandgap. (c) Measured (dotted line) and calculated (solid line) transmittance of a defect-free Penrose dodecagonal crystal. (Adapted from ref.[83]. Reprinted figure with permission from Wang Y, Hu X, Xu X, Chang B and Zhang D 2003 *Phys. Rev. B* **68** 165106 © 2003 by the American Physical Society.)

less absorption loss than the usual metal-based metamaterials. Since the increased absorption in metals prevents the scaling of these metamaterials to the optical wavelengths, the effective negative refractive index obtained in this QC structure in the microwave regime could be reasonably expected to hold in the optical wavelengths as well.

The occurrence of localized modes in defect-free photonic QCs arising from the competition of self-similarity and aperiodicity was reported in a different dodecagonal structure (the so-called dodecagonal Penrose lattice) based on the tiling of squares and two kinds of rhombi, whose acute angles are 30° and 60°, respectively (Fig.7.22a). In the experimental setup 229 alumina cylinders ($\epsilon = 8.9$) are inserted in a polystyrene foam template ($\epsilon = 1.04$). The radius of the cylinders is 3.0 mm and the side length of the square and the rhombi is 12.1 mm. The measured transmittance shows a first gap located from 8.06 to 11.38 GHz which exhibits three sharp peaks at certain specific frequencies (Fig.7.22c). This feature is quite different from the results obtained for both perfect photonic crystals and defect free octogonal and decagonal photonic QCs, where the transmission spectra are smooth inside the gaps. Therefore, the experiment proves the existence of some electromagnetic waves which are localized in the dodecagonal sample only, hence indicating some peculiar feature of 12-fold symmetry as compared to either 8-fold and 10-fold ones. A plausible physical scenario accounting for this difference can be introduced by considering the delicate balance between delocalization effects due to long-range self-similarity (measured in terms of the local isomorphism prescribed by the Conway's theorem, see Section 2.2.2) and confinement effects related to the absence of periodicity in this system.[83] In fact, numerical calculations showed that the distribution of the electric field at the three frequencies corresponding to peaks within the bandgap was essentially confined in the two concentric dodecagonal rings in the central part of the structure (Fig.7.22b). The existence of such localized modes can be fruitfully exploited in order to design single-cell resonators with high quality factors,[84] or coupled-resonator optical waveguides.[85]

Most basic results reported for photonic quasicrystals can be readily extended to phononic quasicrystals as well. The existence of gaps in such materials provides an opportunity to confine and control the propagation of acoustic waves in a completely analogous way. Thus the possibility of designing flat lenses by using quasiperiodic arrays of steel cylinders embedded in air background has been theoretically discussed, comparing the behavior of some acoustic QCs and superlenses based on a periodic distribution of metamaterials. In this case, the motivation for using the high-symmetry QC is to maintain an efficient interference of waves (long-range order) while reducing the orientational order of the system (crystallographical restriction theorem is relaxed) to get a more isotropic propagation.[86]

7.3.2 Three-dimensional arrays

As we have seen the complete photonic band gap appears due to the overlapping of band gaps in all the possible directions. To this end, a high degree of symmetry is an advantage as it makes the optical properties more isotropic and the operation less dependent on the angle of the incident light. Now, for a three-dimensional periodic distribution of the dielectric function one generally has different periodicities associated to different directions in a solid and, correspondingly, different frequencies for the band gap centers. In that case, a relatively large size is required for these stop bands to properly overlap to each other in order to give rise to a complete band gap. This condition is achieved by constructing dielectric-air lattices with materials having a high refractive index contrast. Such a requirement could be certainly relaxed to some extent if the considered solid exhibits a more isotropic periodicity, so that the Brillouin zone is close to being spherical. Motivated by this basic principle, several theoretical studies were performed, and it was concluded that lattices with diamond symmetry have the minimum permittivity modulation contrast necessary for the appearance of a complete band gap among the materials based on periodic order. Now, it is well known that quasicrystals have higher point group rotational symmetries and long-range quasiperiodic order, so photons can be diffracted as efficiently as they are by periodic structures. Therefore, their band structure can be almost isotropic and such structures are reasonably preferable for the appearance of a complete photonic band gap.[87] In fact, in photonic QCs the complex wave scattering processes can confine light by interference effects, giving rise to the phenomenon of light localization even in the absence of disorder. Another advantage of photonic quasicrystals is that they are best suited to phase matching for nonlinear optical effects such as frequency conversion.

Following this guiding idea a photonic quasicrystal based on a generalized version of Penrose tiling in three dimensions was created by stereolithography method. The 3D-Penrose tiling consists of a quasiperiodic arrangement of a prolate and an oblate rhombohedron similar to those considered by Ammann (Fig.2.7). This sample exhibited large stop bands in certain directions in the microwave range.[88] Three-dimensional photonic quasicrystals were subsequently obtained for the infrared,[89] and visible ranges,[90] by using laser writing and optical interference holography, respectively. The ability of 3D-Penrose tiling to form complete band gaps has been recently illustrated in the case of phononic quasicrystals as well.[91]

7.4   Complex metallic alloys

7.4.1   The basic strategy: Smart materials

In order to be of technological relevance a novel material should present challenging properties, but it should also be easy to produce in the desired shape, stable in operating conditions, of low cost, and non-toxic. In addition, it must benchmark already existing materials currently used in existing devices. It is clear that it is by no means easy to fulfill these criteria in full, so that materials engineering research is progressively adopting new strategies. Metals-based industries nowadays rely largely on materials based on elementary metals and binary metallic alloys to which a number of additional elements are added in smaller quantities to tailor them for particular properties. These metallic alloys are mostly either periodic crystals slowly grown from the melt, like those samples obtained by means of Czochralski, Bridgman, or flux growth techniques, or amorphous alloys obtained by rapid quenching. On the other hand, different classes of quasicrystals and their related approximant phases (see Section 2.5.3) have been synthesized by conventional metallurgy techniques. These samples resist thermal treatment without transforming into usual periodic crystals upon annealing (see Table 2.1). Most of these stable quasicrystals consist of aluminum mixed with transition metals, like Fe, Ni, or Co; or normal metals, like Mg and Zn (see Section 3.1.3). Stable quasicrystals were also discovered in titanium based alloys,[92] and in the Mg-Zn-rare earth system.[93] In addition, a binary quasicrystalline phase was discovered in Cd(Ca,Yb) system.[94] Among all of them the icosahedral AlPd(Re,Mn) and AlCuFe phases have been widely considered for practical applications. In particular, the AlCuFe icosahedral alloy has a higher commercial interest because it is made of inexpensive components.

In Chapter 3 we reviewed most relevant physical properties of QCs, and we realized that these solids exhibit a plethora of anomalous behaviors (as compared to common metallic systems), resembling those of either semiconducting or ionic materials in most cases. These anomalous behaviors are shared by their approximants and, to a lesser extent, are also observed in a number of metallic phases located near the QC-forming region in the phase diagrams. For instance, we recall that the electrical resistivity of the ξ' phase of the Al-Pd-Mn alloys system showed an almost negligible temperature dependence between 300 and 4 K (see Section 3.4.3).[95] While weakly temperature-dependent resistivities are not uncommon for both amorphous alloys and bulk metallic glasses lacking long-range ordered crystalline lattices,[96] the temperature-independent resistivity of ξ'-Al-Pd-Mn was observed on monocrystalline samples of good lattice perfection and structural homogeneity. It is interesting that ξ'-Al-Pd-Mn possesses another unusual physical property—an anomalously low thermal conductivity, which is as low as that of amorphous SiO_2, a known thermal (and electrical) insulator. This combination of a temperature independent electrical resistivity of moderate value along with a remarkably low thermal conductivity supports the potential interest of ξ'-Al-Pd-Mn and related complex metallic alloys for technological applications as "smart" materials (e.g., temperature independent electrical resistors with low heat dissipation).

In alloys consisting of three or more components, including not only metals but also metalloids or rare earths, phases are quite frequently formed whose crystal structure is based on giant unit-cells containing up to hundreds or even thousands of atoms. All these unit cells have a substructure based on polyhedral atom arrangements or clusters that partially overlap or are linked by bridging elements, which are making up the unit cells (Fig.7.23). Since the complex structure puts restrictions on the stoichiometry, the structurally complex alloy phases exist only within fairly narrow ranges of composition. Accordingly, it is natural to group together all these odd metallic phases by taking into account their most characteristic distinctive feature: all of them possess giant unit cells. As a result these materials, referred to as complex metallic alloys, can offer unique combinations of properties which are excluded in conventional materials. Among these we find (see Section 3.1.2):

- Combination of high hardness with low thermal conductivity for thermal barriers design

- Low friction coefficients combined with corrosion and oxidation resistant properties, reduced solid-solid adhesion, and wetting by polar liquids for hard coating applications

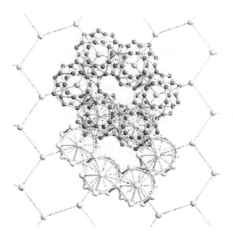

FIGURE 7.23

A view of the ξ'-AlPdMn skeleton structure along the [010] direction. Mn atoms form a planar flattened-hexagon lattice and are located in the centres of pseudo-Mackay icosahedra.[97] The two interpenetrating polyhedra that form the outer shell of the pseudo-Mackay cluster (a 12-atom Pd icosahedron, white atoms) and a 30-atom Al icosidodecahedron (black atoms) are shown. ([98] Courtesy of Janez Dolinšek).

- Thermoelectric figure of merit tunable by composition variation for thermoelectric applications

- Combination of good infrared light absorption properties with high-temperature stability for heat conversion

- Enhanced hydrogen storage capacity under reversible conditions for batteries industry.

Another possible strategy in order to obtain novel materials with improved properties for specific applications is to combine two or more constituent materials with significantly different physical or chemical properties which remain separate on a macroscopic level within the whole structure (the so-called composites). Among composites, a quasicrystalline precipitation-strengthened steel is marketed by Sandvik Steel in Sweden. The steel is strong, relatively ductile, corrosion-resistant, and resistant to overaging. The steel is attractive for tools in the health industry (surgery, dentistry, acupuncture) and comprises key components of a 'wet' electric shaver currently marketed by Philips. Another type of bulk composite is an Al-based alloy which can be formed by rapid solidification and powder processing. The quasicrystals form by precipitation, yielding nanoscale particles surrounded by an Al matrix.[99, 100]

New materials opening a wide variety of adjustable properties for optical applications are based on composites consisting of inhomogeneous media with metal particles in a dielectric matrix (the so-called cermets). In Section 3.1.2.4 we saw that quasicrystals of high structural quality show optical properties which are quite different compared to those of usual insulators, semiconductors, or metals. In particular, they exhibit a relatively high absorption at visible wavelengths combined with a significant thermal stability, which can be exploited in order to design solar selective absorbers for heat production.[101, 102]

A type of composite which is promising involves polymers with quasicrystalline fillers. Tests with certain high-performance thermoplastic resins have shown that quasicrystalline AlCuFe particles significantly enhance the wear resistance of the polymers. This effect is not understood, but it may be related to a combination of the hardness, low friction, and low thermal conductivity intrinsic to the quasicrystal. At the same time, key thermochemical characteristics of the polymer, that is, its glass and melting transitions, are not degraded, and this indicates that the quasicrystal does not catalyze cross-linking or other disadvantageous reactions in the resin, so that it may be potentially used as a prosthetic biomaterial.[103]

7.4.2 Hard coatings and thermal barriers

Quasicrystals are very brittle at low and intermediate temperatures. The toughness of single grains is less than 0.5 MPa m$^{\frac{1}{2}}$. In contrast, metallic crystals like aluminum are characterized by a toughness value about two orders of magnitude higher. As a result quasicrystals have little resistance to crack propagation below 450 °C. Hence, they are not very useful as bulk materials but may be more fruitfully employed as surface coatings or as composites. In such a case, the mechanical integrity is supplied by the substrate or the matrix, while the quasicrystal offers supplemental functions. For instance, quasicrystals have gained most attention as coatings which combine wear resistance with low friction and/or low adhesion. One product has been marketed which exploits these properties–the quasicrystal-coated frying pan, sold by Sitram under the trademark Cybernox^{TM}. The quasicrystal is not directly competitive with Teflon$^{®}$ in terms of its so-called non-stick property, but adhesion for many foods is reduced relative to a metal pan, and of course the coating is much harder than Teflon$^{®}$. The hardness means that normal cutlery can be used with the quasicrystal coating, an attractive feature relative to Teflon$^{®}$. Furthermore, the low thermal conductivity also presents an advantage in the cookware, since it leads to even surface heating.[99]

The brittleness of quasicrystals is overcome at high temperatures. Above about 450 °C, quasicrystals become plastic and deform more and more with applied stress. Deformation is facilitated by dislocations, as in conventional crystals. However, in contrast to conventional crystals, there is no work hardening. This proves the absence of pinning centers and indicates a viscous flow regime that might be due to some internal friction between supposedly undeformable atomic entities.[104]

The oxidation resistance of quasicrystals does not significantly depart from that of aluminum alloys or compounds. When in contact with pure O_2 or air, the quasicrystal surface is soon covered by a thin layer of pure alumina. In good metals like aluminum, the energy of the clean-surface is typically 1 Jm^{-2} or greater. Very low values in the range of $0.40 - 0.60$ Jm^{-2} have been reported recently for Al-based quasicrystals, assuming a certain form of the analytic dependence of the friction coefficient on surface energy.[105] Other, more indirect evidence that the surface energy of quasicrystals is indeed small can be deduced from a comparison of contact angles measured by depositing small liquid droplets on the surface of a quasicrystalline sample as well as on reference samples such as Teflon$^{®}$, metals, or a bulk oxide such as alumina. By making certain assumptions, the contact angle can be related fairly simply to the reversible adhesion energy W of a given liquid onto the surface, larger angles and poorer wetting corresponding to smaller values of W. Poor wetting is an appropriate description of the interaction between carefully polished Al-rich quasicrystals and water. Correspondingly, their reversible adhesion energy with water is only 25 % larger than that of Teflon (55 and 44 mJm^{-2}, respectively) and one-third of that of window glass. The fundamental origin of this behavior is not yet understood, but a natural question is whether it is controlled by the properties of the surface of the quasicrystal, or those of the bulk.[106] Quasicrystals always carry a thin surface oxide layer, consisting mainly of aluminum oxide.[107] However, bulk alumina is comparable to window glass in that it is completely wet by water, indicating a high value of

W. Thus, wetting of the quasicrystals is probably not controlled, at least to zeroth order, by the surface layer, but rather in some way by the underlying atomic and electronic structure of the bulk. (Note that this discussion is only demonstrably valid for water; common organic solvents wet quasicrystals well, and this indicates that the wetting may depend on the polarity of the liquid).

Surface friction is a further important issue for quasicrystals. Numerous experiments using a spherical diamond pin have demonstrated that friction is lower on a quasicrystal ($\mu = 0.04 - 0.05$) than on steel ($\mu = 0.07$) of comparable hardness. With more conventional riders such as mild steel or WC – Co, the comparison is less favorable due to the transfer of material from the indenter to the surface. During such friction measurements, a major drawback arises from the brittleness of the contacting bodies and hence formation of wear particles. However, scratch tracks indicate that quasicrystals subjected to repetitive and severe shear develop some ductility during this process, and consequently they acquire an ability to self-repair. This brittle-to-ductile behavior might be associated with the nucleation of crystalline nanometer-sized grains within the quasicrystalline matrix.[108]

Thermal barriers based on quasicrystalline alloys can also be manufactured as thick coatings by thermal spraying techniques or magnetron sputtering. This application relies upon bulk properties of quasicrystals, namely, low thermal conductivity and plasticity at high temperature. As such, the quasicrystal is a poor heat conductor which is additionally able to resist the shear stress, if any, generated at the interface with the substrate due to the difference in thermal expansion coefficients. In contrast to zirconia, such a difference is small, since expansion coefficients for quasicrystals, in the range of $(13 - 16) \times 10^{-6}$ K^{-1}, are close to those of metallic alloys. In comparison to metals, quasicrystals also resist oxidation and corrosion by sulfur at high temperature. The main drawback to using quasicrystals comes from their rather low melting temperature and significant atomic mobility. Intercalation of a thin diffusion barrier, typically made of an oxide such as Y_2O_3, allows the latter difficulty to be easily overcome. A specific composition with nearly congruent melting at 1170 °C was developed to enhance the temperature range in which quasicrystalline thermal barriers may be useful. Although not competitive with doped zirconia above 1050 – 1100 °C, such barriers were successfully tested at 950 °C in a real-time ground test of an aircraft engine. Application to other combustion engines, to power generators or, in general, to heat insulation of fast-moving mechanical parts is also of interest.[100, 109]

7.4.3 Hydrogen storage

A hydrogen molecule approaching a metal can be dissociated at the interface, absorbed at appropriate surface and near surface sites, and dissolved at interstitial sites of the host metal. The ability to absorb hydrogen in metals or alloys is greatly dependent upon chemical affinity between hydrogen and host atoms, type of interstitial sites, and the total number and actual size of interstitial sites in the crystal. There are basically two types of interstitial sites in crystal lattices, namely, octahedral and tetrahedral interstices. The sites that the hydrogen atoms occupy first depend on the physical size of the interstices as well as the chemical affinities between hydrogen and the metal atoms surrounding the sites.

Following the discovery of the second largest group of thermodynamically stable QCs in the TiZrNi system, studies on the hydrogenation properties of such QCs started, searching for suitable samples for their possible application as hydrogen-storage materials.[92] In fact, although the local structure of icosahedral phases is not yet completely known in all detail (see Section 3.2), diffraction studies of the i-TiZrNi phase indicated that it is based on Bergman atomic clusters as main building blocks (see Fig.2.1).[110]

TABLE 7.2
Distribution of interstices in different types of lattices.

LATTICE	TETRAHEDRAL	OCTAHEDRAL	TOTAL
fcc	8	13	21
bcc	24	18	42
QC	140	0	140

This cluster contains a significantly large number of tetrahedral interstitials and no octahedral interstitials, clearly outnumbering the amount of interstices available in the case of common cubic lattices (see Table 7.2). Thus, one can reasonably expect that the more complex structure of quasicrystalline phases may provide a lot of suitable sites for hydrogen absorption. The hydrogen absorption properties of icosahedral phase powders has been compared to those of both amorphous and crystalline samples of interest in hydrogen energy industry (such as $LaNi_5$ or TiFe hydrides) in high pressure vessels experiments as well as by using electrochemical hydrogenation at room temperature.[111] The maximum hydrogen concentrations obtained for $Ti_x Zr_{83-x} Ni_{17}$ QCs, with $41 \leq x \leq 61$ ranged within $1.50 - 1.74$ hydrogen-to-metal ratio. These figures are not as good as those reported for the most efficient materials discovered for hydrogen storage purposes (i.e., Mg and V, with an hydrogen-to-metal ratio of 2.0), but are certainly better than those obtained for the samples TiFe (0.98), $LaNi_5$ (1.00), and Mg_2Ni (1.33) of common use in H-batteries research. In addition, hydrogen in the icosahedral phase can be desorbed comparatively more easily than in the amorphous phase. After hydrogen desorption by heating the hydrogenated icosahedral phase powder to 800 K, the quasilattice constant of the icosahedral phase completely returned to that before hydrogenation, showing good reversibility for the absorption and desorption of hydrogen. It was also observed that after a second gas-phase loading of hydrogen, the icosahedral phase is still stable, though it now coexists with a $(Ti,Zr)H_2$ hydride phase.

In summary, the maximum hydrogen concentration for the TiZrNi QC phase powders reach to about 60% either by gas-phase or electrochemical loading of hydrogen. They perform better than crystalline materials of comparable specific mass because the occurrence of icosahedral quasiperiodic order enforces the formation of a greater number of interstitial sites that can be occupied by protons. This level exceeds that of commonly used hydrides, promising new hydrogen-storage material for fuel-storage and battery applications. [111]

7.4.4 Surface catalysis

The potential application of QCs for catalysis was first investigated regarding the catalytic activity of various ultra-fine particles extracted from AlPd, AlPdMn, AlCuFe, or AlCuCo alloys by spark evaporation. It was observed that the generation of hydrogen from the temperature decomposition of methanol ($CH_3OH \rightarrow CO + 2H_2$) started at significantly lower temperatures than with the conventional crystalline catalysts like Pd or Cu. Results also showed that the highest quantity of hydrogen was formed on the catalysts under quasicrystalline form.[112] Subsequently, the catalytic properties of i-AlCuFe QCs with various compositions were investigated for the steam reforming reaction of methanol. This endothermic process takes place above 500 K between methanol and water steam, according to the formula $CH_3OH + H_2O \rightarrow CO_2 + 3H_2$.[113, 114] This reaction is an effective way to produce hydrogen under mild reaction conditions, e.g., for fuel cells. The high brittleness of QCs allows one to prepare large amounts of ultra-fine powders through grinding, without losing chemical homogeneity. The quasicrystal is prepared as a powder, then crushed to maximize surface area, and finally leached in NaOH or Na_2CO_3. The leaching selectively removes Al and Al oxides, leaving each particle consisting of concentric layers. At its core the solid is quasicrystal, and at its perimeter it consists of porous Al-OH that supports dispersed nanoparticles of Cu and Fe. The Cu nanoparticles are far more resistant to sintering than Cu particles in the Cu-based catalysts that are the current industrial standard. Therefore, i-AlCuFe QCs shows a unique combination of properties that makes it an efficient catalyst, namely, low cost of the raw elements, brittleness to warrant large specific areas after grinding, and optimum Fe/Cu concentration ratio.[100]

7.4.5 Thermoelectric materials

During the last few years we have witnessed a growing interest in searching for novel, high performance thermoelectric materials for energy conversion. The efficiency of thermoelectric devices depends on the transport coefficients of the constituent materials and it can be properly expressed in terms of the figure of merit given by the dimensionless expression

$$ZT = \frac{T\sigma S^2}{\kappa_e + \kappa_{ph}},$$ (7.29)

where T is the temperature, $\sigma(T)$ is the electrical conductivity, $S(T)$ is the Seebeck coefficient, and $\kappa_e(T)$ and $\kappa_{ph}(T)$ are the thermal conductivities due to the electrons and lattice phonons, respectively. The appealing question regarding what electronic structure provides the largest possible figure of merit was addressed some time ago, concluding that (i) the best thermoelectric material is likely to be found among materials exhibiting a sharp singularity (Dirac delta function) in the density of states close to the Fermi level, and (ii), in that case, the effect of the DOS background contribution onto the figure of merit value may be quite dramatic, the figure of merit value being inversely proportional (in a marked non-linear way) to the DOS value near the singularity.[115]

Quite interestingly the electronic structure of quasicrystalline alloys satisfies these requirements in a natural way, since it exhibits a pronounced pseudo-gap at the Fermi level as well as some narrow features on the DOS close to the Fermi level (see Section 3.4.1). At first sight it may seem surprising to propose a metallic alloy as a suitable thermoelectric material, since it is well known that metallic compound usually exhibit very low $ZT \sim 10^{-3}$. However, such a proposal makes sense due to the peculiar transport properties of QCs (see Section 3.1.2).[116] In fact, their electrical conductivity (i) is remarkably low (ranging from 100 to 5000 Ω^{-1} cm^{-1} at room temperature), (ii) it steadily *increases* as the temperature increases up to the highest temperatures of measurement ($T \simeq 900$ K), and (iii) it is extremely sensitive to minor variations in the sample composition. This sensitivity to the sample stoichiometry is also observed in other transport parameters, like the Hall or Seebeck coefficients, and resembles doping effects in semiconductors. In addition, the temperature dependence of the Seebeck coefficient: (i) is clearly non-linear, exhibiting well-defined extrema in most instances, (ii) small variations in the chemical composition give rise to sign reversals in the $S(T)$ value, and (iii) for a given sample stoichiometry it shows a strong dependence on the heat treatments applied to the sample.[117] Therefore, the electronic transport properties of quasicrystalline alloys exhibit unusual composition and temperature dependences, resembling more semiconductor-like than metallic character. Furthermore, the thermal conductivity of QCs is unusually low for a metallic alloy and it is mainly determined by the lattice phonons (rather than the charge carriers) over a wide temperature range. Thus, for most icosahedral phases the thermal conductivity at room temperature is comparable to that of zirconia ($\sim 1 - 2$ W m^{-1} K^{-1}). This low thermal conductivity of QCs is particularly remarkable in the light of Slack's phonon-glass/electron-crystal proposal for promising thermoelectric materials,[118] and it has considerably spurred the interest on the potential application of QCs as thermoelectric materials from an experimental viewpoint.

Consequently, according to their transport properties, quasicrystalline alloys are marginally metallic and should be properly located at the *border line* between metals and semiconductors. Thus, QCs bridge the gap between metallic materials and semiconducting ones, occupying a very promising position in the quest for novel thermoelectric materials. In fact, it has been recently pointed out that the electrical conductivity of i-AlPd(Mn,Re) QCs may be strongly dependent on the bonding nature of their constitutive icosahedral clusters, so that small changes in the cluster structure may induce a metallic-covalent bonding conversion (see Section 3.1.3).[122] Thus, one of the main advantages of QCs over other competing thermoelectric materials is that one can try to modify both the electrical conductivity and the thermoelectric power, without losing the low thermal conductivity, by properly varying the sample stoichiometry.[117]

TABLE 7.3

Room temperature transport coefficients values for different quasicrystalline families, after: a) Ref. [117], b) Ref. [119], c) Ref. [120], d) estimated, e) Ref. [122], and f) Ref. [121].

Sample	$\sigma(\Omega^{-1}\text{cm}^{-1})$	S $(\mu\text{V K}^{-1})$	$\kappa(\text{Wm}^{-1}\text{K}^{-1})$	ZT
AlCuRu	250^b	27^b	1.8^d	0.003
AlCuFe	310^b	44^b	1.8^c	0.01
CdYb	7000^f	16^f	4.7^f	0.01
AlCuRuSi	390^b	50^b	1.8^d	0.02
AlPdRe	175^e	95^e	0.7^e	0.07
AlPdMn	640^a	85^a	1.6^a	0.08

According to their chemical composition QCs can be grouped into several families. In Table 7.3 we list the transport coefficients values for those representatives yielding the best figure of merit values at room temperature. From the listed data we appreciate a progressive trend towards larger values of ZT resulting from the synthesis of suitable QCs. Furthermore, significantly enhanced figure of merit values are obtained at higher temperatures. Thus we have $ZT = 0.25$ for i-AlPdMn samples at $T = 550$ K;[123] $ZT = 0.11$ for i-$\text{Al}_{71}\text{Pd}_{20}\text{Re}_9$ samples at $T = 660$ K, and $ZT = 0.15$ for i-$\text{Al}_{71}\text{Pd}_{20}(\text{Re}_{0.45}\text{Ru}_{0.55})_9$ samples at $T = 700$ K.[122]

As we have seen in Section 3.4.1 the $\sigma(T)$, $S(T)$, and $\kappa_e(T)$ curves of several quasicrystals can be consistently described in terms of the two-Lorentzian spectral conductivity functions given by Eq.(3.10). Then plugging Eqs.(3.16)-(3.18) into Eq.(7.29) one gets[124, 125]

$$ZT = \frac{J_1^2}{J_0 J_2 - J_1^2 + c^2 J_0 \varphi},\tag{7.30}$$

where $c \equiv 2e/k_B$ and $\varphi(T) \equiv \kappa_{ph}(T)/T$. As we explained in Section 3.4.3,

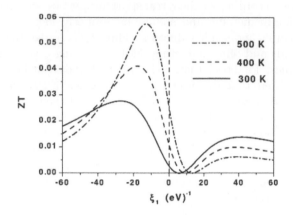

FIGURE 7.24

Dependence of the thermoelectric figure of merit as a function of the phenomenological coefficient ξ_1 at $T = 300$ K (solid line); $T = 400$ K (dashed line), and $T = 500$ K (dot-dashed line). ([124] Reprinted figure with permission from Maciá E 2004 *Phys. Rev. B* **69** 184202 © 2004 by the American Physical Society.)

the reduced kinetic coefficients, J_n, can be expressed in terms of series of phenomenological coefficients, ξ_i, which, in turn, can be related to experimental transport curves.

In Fig.7.24 we plot the ZT curve as a function of the phenomenological coefficient ξ_1 value at different temperatures, as derived from Eq.(7.30) for a suitable choice of the model parameters. The $ZT(\xi_1)$ curve exhibits a deep minimum, where ZT almost vanishes, flanked by two maxima at about $\xi_1 \simeq -25$ (eV)$^{-1}$ and $\xi_1 \simeq +40$ (eV)$^{-1}$. Considering this curve in the light of Eq.(3.33) several conclusions can be drawn concerning the relationship between the ZT curves and the sample's electronic structure. First, when the Fermi level is located close to the pseudogap minimum, Eq.(3.33) yields very small ξ_1 values. In that case, we will obtain small figures of ZT at room temperature, in agreement with the experimental results. Note, however, that the minimum of the $ZT(\xi_1)$ curve does not coincide with $\xi_1 = 0$, so that as the temperature is increased we obtain progressively larger values of ZT, in agreement with experimental findings as well. Second, as the Fermi level progressively shifts from the pseudogap's minimum (due to a systematic change in the sample stoichiometry, for instance), the ZT values progressively increase attaining well defined maxima. Note that these maxima reach different peak values depending on the sign of ξ_1. Accordingly, we conclude that best thermoelectric performances will be expected for those stoichiometries

able to locate the Fermi level below the minimum of the pseudogap, and that
the deeper the pseudogap the larger the resulting figure of merit at a given
temperature. Finally, we observe that as temperature increases the $ZT(\xi_1)$
maximum below (above) the Fermi level progressively increases (decreases)
and shifts towards (away from) the pseudogap's minimum located at $\xi_1 = 0$.
This behavior indicates that the precise stoichiometry to get an optimal thermoelectric performance will depend in general on the working temperature
for the considered sample.

From this study one concludes that band structure effects play a significant
role in the thermoelectric performance of i-AlPdRe QCs (and likely in other
quasicrystalline phases containing transition-metals as well). In particular
we find that samples whose Fermi level is located at the pseudogap's minimum exhibit very small ZT values. This condition occurs for stoichiometries
yielding $e/a \simeq 1.74$. Thus, the most stable samples (e.g., $Al_{68.5}Pd_{22.9}Re_{8.6}$,
$Al_{69.4}Pd_{21.2}Re_{9.4}$, $Al_{70}Pd_{20}Re_{10}$) are also the worse ones for thermoelectric
applications. On the the hand, large figure of merit values are expected for
those samples exhibiting narrow features in the DOS close to the Fermi level,
like i-$Al_{70.5}Pd_{21}Re_{8.5}$.[126, 127] Consequently, it seems reasonable to expect
that relatively high values of the figure of merit may be obtained by a judicious choice of sample composition, working temperature, and Peltier cell
structural design.[125]

7.5   DNA-based nanoelectronics

7.5.1   Peltier nanocells

The experimental way to the possible use of organic molecules in the design
of nanoscale thermoelectric devices was opened up by the measurement of an
appreciable thermoelectric power ($+18$ μV K^{-1} at room temperature) over
guanine molecules adsorbed on a graphite substrate using a STM tip.[128]
Subsequently, the thermoelectric response of phenyldithiol organic molecules
chemisorbed on gold surfaces was theoretically analyzed, and Seebeck coefficient values comparable to those obtained in previous experiments were
reported.[129] Similar values ($+22$ μV K^{-1} at room temperature) have been
recently reported on a sample of FeCl$_3$-doped polythiophene.[130] Although
these figures are too small to be of interest for most current thermoelectric applications, it is reasonable to expect that they may be significantly enhanced
by a proper choice of the materials composing the thermoelectric nano-cell.
Thus, the thermoelectric potential of some conducting polymers, like polythiophene and polyaminosquarine, has been recently reviewed on the basis
of their electronic band structures.[130] Also, the thermoelectric properties
of nanocontacts made of single-wall carbon nanotubes have been numerically

studied, concluding that doped semiconducting nanotubes may exhibit very high figures of thermoelectric merit.[131].

In fact, the extreme sensitivity of thermopower to finer details in the electronic structure suggests that one could optimize the device's thermoelectric performance by properly engineering its electronic structure. With the aim of exploring such a possibility, a systematic theoretical study on the thermoelectric properties of DNA nucleobases guanine (G), cytosine (C), adenine (A), and thymine (T) – either as single units or forming dimers or trimers – connected to metallic leads at different temperatures was performed.[132, 133, 134] The obtained results showed that relatively large thermopower values can indeed be obtained by properly locating the system's Fermi level. In addition, the thermoelectric response of trimer nucleobases exhibits two resonant features where the Seebeck coefficient attains large values ($200 - 400$ μV K^{-1} at room temperature), closely resembling recently reported thermopower curves of silicon based atomic junctions.[135] Since both the location and the magnitude of these peaks sensitively depend on the energetics of the considered trimer, one may think of introducing a thermoelectric signature for different codons of biological interest,[134] in close analogy with the transversal electronic signature recently proposed for single-stranded DNA chains.[136, 137]

In this Section, we will analyze the thermoelectric response of more realistic double-stranded DNA (dsDNA) chains, in order to estimate the potential of synthetic DNA chains as thermoelectric materials. From an applied viewpoint the convenience of synthetic versus biological DNA based thermoelectric devices is twofold: (i) synthetic DNA strands can be polymerized at will in order to fit any prescribed design; and (ii) quantum chemical calculations show the existence of convenient charge channels in periodic dsDNA chains. Thus, charge transfer mainly proceeds via hole (electron) propagation through the purine (pyrimidine) bases, where the HOMO (LUMO) carriers are respectively located in polyG-polyC (polyA-polyT) chains (see Section 6.2). In fact, experimental current-voltage curves show that double-stranded poly(dA)-poly(dT) chains behave as n-type semiconductors, whereas poly(dG)-poly(dC) ones behave as p-type semiconductors.[138] Accordingly, these synthetic DNAs may provide the basic building blocks necessary to construct a nanoscale thermoelectric cell, where the DNA chains will play the role of semiconducting legs in standard Peltier cells, as it is illustrated in Fig.7.25.

In order to substantiate this proposal, one must consider the energy dependence of Seebeck coefficient, S, and thermoelectric power factor ($S^2\sigma$, where σ is the electrical conductivity) of polyG-polyC and polyA-polyT chains at room temperature. The duplex DNA molecules are modeled in terms of the renormalized one-dimensional effective Hamiltonian given by Eq.(6.35) and Eq.(6.44), and the transmission coefficient is derived by embedding the chain between two semi-infinite leads (see Sections 6.4.2 and 6.4.3). Within the transfer matrix framework, considering nearest-neighbors interactions only, the corresponding Schrödinger equation can be expressed in the form

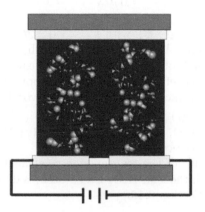

FIGURE 7.25

Sketch illustrating the basic features of a nanoscale DNA based Peltier cell. A polyA-polyT (polyG-polyC) oligonucleotide, playing the role of n-type, left (p-type, right) semiconductor legs, are connected to organic wires (light boxes) deposited onto ceramic heat sinks (dark boxes). ([141] Reprinted figure with permission from Maciá E 2007 *Phys. Rev. B* **75** 035130 © 2007 by the American Physical Society.)

$$\begin{pmatrix} \psi_{N+1} \\ \psi_N \end{pmatrix} = \begin{pmatrix} 2x\lambda^{-1} & -\lambda^{-1} \\ 1 & 0 \end{pmatrix} \begin{pmatrix} 2x & -1 \\ 1 & 0 \end{pmatrix}^{N-2} \begin{pmatrix} 2x & -\lambda \\ 1 & 0 \end{pmatrix} \begin{pmatrix} \psi_1 \\ \psi_0 \end{pmatrix}, \qquad (7.31)$$

where ψ_n is the wavefunction amplitude for the energy E at site n, $2x \equiv (E - \alpha)/t_0$ describes the DNA energetics, and the ratio $\lambda \equiv \tau/t_0$ measures the DNA-lead coupling strength (the Schrödinger equation for polyA-polyT is obtained by just replacing $\alpha \leftrightarrow \beta$). The transmission coefficient at zero bias as a function of energy is given by [139]

$$T_N(E) = \left\{ 1 + W^{-1} \left[(E - \varepsilon_M)U_{N-1} - \Omega(U_N + \lambda^2 U_{N-2}) \right]^2 \right\}^{-1}, \qquad (7.32)$$

where $W \equiv (E - E_-)(E_+ - E)$, with $E_\pm \equiv \varepsilon_M \pm 2t_M$, define the allowed spectral window determined by the lead's bandwidth, $\Omega \equiv t_M/\lambda$, and $U_k(x)$ are Chebyshev polynomials of the second kind. By inspecting Eq.(7.32) we realize that the transmission coefficient in general does not reach the full transmission condition $T_N = 1$. In fact, even in the most favorable conditions for charge transport (i.e., $E = \varepsilon_M$) we get $T_N(\varepsilon_M) = \left[1 + (\lambda U_{N-2}^* + \lambda^{-1} U_N^*)^2/4 \right]^{-1} < 1$, where $U_k^* \equiv U_k(x_M)$, and $2x_M = (\varepsilon_M - \alpha(\varepsilon_M))/t_0$. This transmission degradation stems from contact effects.[140]

From the knowledge of the transmission coefficient given by Eq.(7.32) the conductance through the lead-DNA-lead system is determined using the Landauer formula (see Section 9.5.3)

$$G_N(E_F) = G_0 T_N(E_F), \qquad (7.33)$$

where $G_0 = 2e^2/h \simeq 1/12906 \ \Omega^{-1}$, and E_F denotes the Fermi level. On the other hand, the Seebeck coefficient is obtained from the expression [129]

$$S_N(E_F, T) = -|e|L_0 \left(\frac{\partial \ln T_N(E)}{\partial E} \right)_{E_F} T, \qquad (7.34)$$

where e is the electron charge, $L_0 \equiv \pi^2 k_B^2/3e^2 = 2.44 \times 10^{-8} \ \text{V}^2\text{K}^{-2}$ is the Lorenz number, and T is the temperature. Making use of Eqs.(7.32) and (7.33) into Eq.(7.34) one gets,[141]

$$S_N(E_F, T) = \tilde{S}_0(T)\Delta G \left\{ B(E_F) + \left(\frac{\partial \ln h(E)}{\partial E} \right)_{E_F} \right\}, \qquad (7.35)$$

where $\tilde{S}_0(T) = 2|e|L_0 T$, $\Delta G \equiv 1 - G_N/G_0$, $B(E_F) \equiv (E_F - \varepsilon_M)/W(E_F)$, and $h(E) \equiv (E - \varepsilon_M)U_{N-1} - \Omega(U_N + \lambda^2 U_{N-2})$. The Seebeck coefficient is then expressed as a product involving three contributions. The factor $\tilde{S}_0$ sets the thermovoltage scale (in $\mu\text{VK}^{-1}\text{eV}$ units) and accounts for the linear temperature dependence of S_N. The factor ΔG links the thermopower magnitude to the conductance properties of the chain, so that the Seebeck coefficient progressively decreases (increases) as the conductance increases (decreases), vanishing when $T_N = 1$, as expected from basic transport theory. The last factor in Eq.(7.35) depends on two additive contributions in turn. The value of $B(E_F)$ depends on the relative position of the Fermi level with respect to both the band center, ε_M, and the band edges, $E_\pm$, of the contacts. Thus, its contribution vanishes when $E_F \to \varepsilon_M$, whereas B (and consequently S_N) asymptotically diverges as the Fermi level approaches the spectral window edges (i.e., $E_F \to E_\pm$). Finally, the logarithmic derivative term in Eq.(7.35) contains most physically relevant information, accounting for (i) contact effects (related to the coupling constants λ and Ω), (ii) size effects (described by the N parameter dependence), and (iii) resonance effects related to the DNA energetics by means of the Chebyshev polynomials' argument

$$x(E_F) \equiv x_0 = -\frac{1}{2t_0} \left(b + (a_1 - 1)E_F + \frac{2t^2}{E_F - \gamma} \right). \qquad (7.36)$$

Since we are mainly interested in the study of the *intrinsic* transport properties of DNA chains, we will minimize contact effects by adopting $t_m = t_0 = \tau$ henceforth, so that $\lambda = 1$ and $\Omega = t_0$. Thus, we can rewrite Eqs.(7.33) and (7.35) in the form

$$G_N(E_F) = \frac{G_0}{1 + C(E_F)U_{N-1}^2}, \qquad (7.37)$$

where $C(E_F) \equiv [\alpha(E_F) - \varepsilon_M]^2 / W(E_F)$, and

$$S_N(E_F, T) = \tilde{S}_0(T) [1 - T_N(E_F)] \left[B(E_F) + \frac{P_2(F_F)}{E_F - \gamma} + \left(\frac{\partial \ln U_{N-1}}{\partial E} \right)_{E_F} \right],$$
$$\qquad (7.38)$$

where

$$P_2(E_F) \equiv \frac{a_1(E_F - \gamma)^2 - 2t^2}{a_1(E_F - \gamma)^2 + (a_0 - \varepsilon_M)(E_F - \gamma) + 2t^2}. \qquad (7.39)$$

By comparing Eqs.(7.35) and (7.38) we see that the logarithmic derivative in Eq.(7.35) has been split into two separate contributions. The first one includes sugar-phosphate backbone effects through the γ parameter dependence. In particular, since $P_2(\gamma) = -1$, we realize that S_N asymptotically diverges as the Fermi level approaches the backbone on-site energy (i.e., $E_F \to \gamma$). In general, the γ value will depend on the chemical nature of the nucleotides, as well as the possible presence of water molecules and/or counterions attached to the backbone.[142] Accordingly, this resonant enhancement of thermoelectric power strongly depends on environmental conditions affecting the DNA electronic structure (see Fig.6.20). Finally, the Chebyshev polynomial logarithmic derivative appearing in Eq.(7.38) describes size effects in the thermoelectric response for DNA chains of different length.

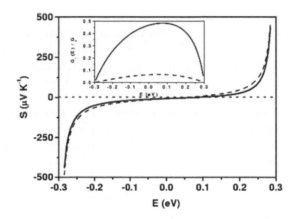

FIGURE 7.26

Room temperature dependence of the Seebeck coefficient as a function of the Fermi level energy for a G-C (solid curve) and A-T (dashed curve) Watson-Crick base pairs. Inset: The Landauer conductance as a function of the Fermi level energy for the same base pairs. The origin of energies is set at the guanine contact level $\varepsilon_M = \varepsilon_G \equiv 0$ eV. ([141] Reprinted figure with permission from Maciá E 2007 *Phys. Rev. B* **75** 035130 © 2007 by the American Physical Society.)

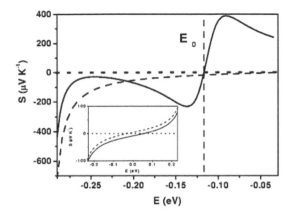

FIGURE 7.27

Seebeck coefficient as a function of the Fermi level energy for a polyG-polyC (solid curve) and a polyA-polyT (dashed curve) oligomer with $N = 5$ base pairs. The vertical dashed line separates the energy regions exhibiting n-type and p-type thermopowers, respectively. Inset: The Seebeck coefficient as a function of the Fermi level energy for an A-T Watson-Crick base pair (solid line) is compared to that corresponding to a polyA-polyT oligomer with $N = 5$ (dashed line). ([141] Reprinted figure with permission from Maciá E 2007 *Phys. Rev. B* **75** 035130 © 2007 by the American Physical Society.)

In Fig.7.26 we plot the thermopower and electrical conductance curves as a function of the Fermi energy obtained from Eqs.(7.37) and (7.38) for both G-C and A-T complementary pairs ($N = 1$) connected to guanine wires at both ends. The $S(E)$ curves exhibit typically metallic values ($1 - 10 \ \mu VK^{-1}$) over a broad energy interval around the guanine energy level and then suddenly grow (in absolute value) as E_F approaches the band edges (due to the $B(E_F)$ contribution). As we can see, the thermoelectric response is very similar for both kinds of Watson-Crick pairs, though the Seebeck coefficient is somewhat larger for the A-T one, due to its smaller conductance value (shown in the inset). In this case ($U_0 = 1$) the transmission coefficient reduces to $T_1 = (1 + C)^{-1}$ and the corresponding conductance curves attain the maximum $G_1 \simeq 3.8 \times 10^{-5}$ ($G_1 = 5.1827 \times 10^{-6}$) Ω^{-1} at the resonance energy $E_1^* = 8.64 \times 10^{-2}$ ($E_1^* = 5.50 \times 10^{-2}$) eV for G-C (A-T) bp, respectively. These conductance values are remarkably large (in particular, the G-C base pair value is about one order of magnitude larger than the values usually reported for organic molecular junctions [143]) accounting for the small values of the Seebeck coefficient in the energy interval $-0.2 \lesssim E \lesssim 0.2$, as prescribed by the ΔG factor in Eq.(7.35).

As the number of base pairs composing the DNA chain is progressively increased several topological features (i.e, maxima, minima, and crossing points) appear in the thermopower curves, as it is illustrated in Figs.7.27 for the case $N = 5$. As we see, the Seebeck coefficient is characterized by the presence of two peaks around a crossing point located at the energy $E_0 = -0.116$ eV. The thermopower values attained at the peaks are significantly high, and compare well with the values reported for benchmark thermoelectric materials. Nevertheless, as the Fermi level shifts away from the resonance energy, the Seebeck coefficient significantly decreases, clearly illustrating the fine tuning capabilities of thermopower measurements. On the contrary, the thermoelectric response of the polyA-polyT chain is rather insensitive to the chain length. This is illustrated in the inset of Fig.7.27, where we compare the thermoelectric curves of a single A-T base pair and a $N = 5$ polyA-polyT oligomer. This property is related to the fact that both adenine and thymine energy levels are far above the contact Fermi level; meanwhile the guanine level is just aligned to the contact one in the polyG-polyC chain. In that case, a pronounced resonance peak (saturating at the quantum conductance value G_0) appears in the conductance curve, as it is shown in the inset of Fig.7.28. On the other hand, according to Eq.(7.34) the main features of the polyG-polyC Seebeck coefficient shown in Figs.7.26 and 7.27 can be properly accounted for in terms of the conductance curve shown in this inset. In fact, when the Fermi level is located at the left (right) of the conductance peak the slope of the transmission coefficient curve $T_N(E)$ is positive (negative) leading to n-type (p-type) thermopower, respectively. In addition, the steeper the conductance curve the higher the thermopower value close to the resonance energy. Finally, we note that the crossover energy E_0 defines two different regimes where the polyG-polyC oligomer alternatively exhibits n-type or p-type thermopower. In this regard it is worth mentioning that when the Fermi level is located above E_0, the Seebeck coefficient of each DNA chain exhibits contrary signs, so that the polyG-polyC chain behaves as a p-type material, while the polyA-polyT chain behaves like a n-type one, in agreement with previous experimental results.[138]

By properly combining the previous results, making use of the typical values $L_N = 0.34 \times N$ nm for the length, and $R = 1$ nm for the radius of B-form DNA, we can determine the magnitude of the thermoelectric power factor $P_N = \sigma_N S_N^2 = G_N L_N S_N^2 / (\pi R^2)$ for the considered samples. The overall shape of the power factor is mainly determined by the energy dependence of the Seebeck coefficient. In fact, in the case $N = 1$ the power factor takes on relatively small values over a broad range of energies located around the conductance peak, but it significantly increases as the Fermi level approaches the band edges, as it was previously discussed. In the case $N = 5$, in addition to this general behavior the power factor also attains significantly large values close to the resonance energy of the polyG-polyC chain due to the presence of the above mentioned Seebeck coefficient peaks. The values of the power factor maxima attained in this case ($P_5 = 1.5 - 3 \times 10^{-3}$ Wm^{-1}K^{-2}) nicely fit with

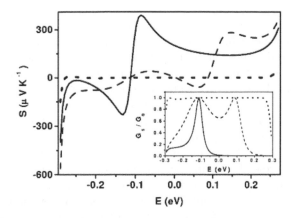

FIGURE 7.28
Seebeck coefficient as a function of the Fermi level energy for a polyG-polyC oligomer with $N = 5$ base pairs and $\gamma = 4.5$ eV (solid curve), $\gamma = 4.0$ eV (dashed curve), and $\gamma = 3.0$ eV (dotted curve) with $\tau = t_M = 0.15$ eV, and $\varepsilon_M = 0$ eV. Inset: Landauer conductance as a function of the Fermi level energy for the same samples shown in the main frame. ([141] Reprinted figure with permission from Maciá E 2007 *Phys. Rev. B* **75** 035130 © 2007 by the American Physical Society.)

those reported for benchmark thermoelectric materials ($P = 2.5 - 3.5 \times 10^{-3}$ Wm^{-1}K^{-2}) at high temperatures.[144] On the contrary, the power factor is completely negligible for polyA-polyT oligonucleotides.

Up to now, we have neglected the possible influence of environmental effects, keeping a fixed value for the backbone related on-site energy γ. However, the sensitivity of thermopower to possible backbone effects should be considered in any realistic treatment, for the presence of a number of counterions located along the DNA sugar-phosphate backbone (mainly in the vicinity of negatively charged phosphates) as well as the grooves of the DNA helix (mainly near the nitrogen electronegative atoms of guanine and adenine) is expected. In Fig.7.28 we compare the Seebeck coefficient as a function of the energy for different γ values for a polyG-polyC chain with $N = 5$. By inspecting this plot we realize the remarkable role played by environmental effects on thermopower. In fact, by systematically varying the on-site energy parameter from $\gamma = 4.5$ eV (no environmental effects) to $\gamma = 3.0$ eV, the thermoelectric response of the DNA chain can be modulated from typically semiconducting values to typically metallic ones. As expected from basic theory (see Eq.(7.35)), the degradation of the thermopower is related to a progressive enhancement of the DNA conductance. This result is shown in the inset of

Fig.7.28, where we plot the systematic variation of the polyG-polyC oligomer conductance as γ is progressively decreased.

In summary we conclude that the thermoelectric response of short dsDNA chains strongly depends on (i) the chemical nature of the considered DNA chain and (ii) the relative position between the contacts Fermi level and the DNA molecular levels. Thus, while the thermoelectric power of polyA-polyT oligomers is quite insensitive to the number of base pairs composing the chain, polyG-polyC oligomers exhibit a strong dependence on the chain length. Accordingly, we can efficiently optimize the power factor of polyG-polyC chains by properly shifting the Fermi level position close to the resonance energy, which plays the role of a tuning parameter. On the other hand, depending on the E_F position, n-type and p-type thermoelectric responses can be simultaneously obtained for polyA-polyT and polyG-polyC DNA chains, respectively. This is a very convenient feature in order to design DNA based thermoelectric devices, where both oligomers would play the role that semiconducting materials legs usually play in standard Peltier cells (Fig.7.25). To this end, the relatively low value of the polyA-polyT chain Seebeck coefficient could be significantly improved by connecting it to adenine wires, rather than guanine ones, in order to get a proper alignment between the contacts Fermi level and the DNA molecular levels.

The thermoelectric quality of a material is expressed in terms of the dimensionless figure of merit given by Eq.(7.29). Therefore, the potential of DNA oligomers as thermoelectric materials will ultimately depend on their thermal transport properties. We can make a rough estimation of ZT by assuming that the thermal transport properties recently reported for a series of simple organic semiconductors (e.g., pentacene) are representative of more complex biomolecules as well. In particular, it seems reasonable to expect that the thermal conduction is dominated by phonon transport in most organic compounds, leading to small thermal conductivities in general. In fact, room temperature thermal conductivity values in the range $\kappa = 0.25 - 0.50$ $Wm^{-1}K^{-1}$ were measured for different organic films.[145] On the other hand, it is well known that the thermal conductivity of low dimensional systems is usually lower than the bulk, accounting for the higher thermoelectric performance reported for multilayers and nanowires.[146] Accordingly, bulk values provide an upper limit to the expected thermal conductivity.

A suitable estimation of thermal conductivity for ideal coupling between a ballistic thermal conductor and the reservoirs relies on the quantum of thermal conductance $g_0 = \pi^2 k_B^2 T/(3h) = 9.46 \times 10^{-13} T$ WK^{-1}, which represents the maximum possible value of energy transported per phonon mode.[147] In the regime of low temperatures four main modes, arising from dilatational, torsional, and flexural degrees of freedom, are expected for a quantum wire.[148] Therefore, the thermal conductivity of a DNA oligomer of length $L_N = 0.34N$ nm will be given by $\kappa_N \simeq 4g_0 L_N/(\pi R^2) = 0.02$ $Wm^{-1}K^{-1}$ (at $T = 10$ K) and $\kappa_N \simeq 0.6$ $Wm^{-1}K^{-1}$ (at room temperature) in optimal conditions. By taking the value $\kappa \simeq 0.1$ $Wm^{-1}K^{-1}$ as a suitable reference value, along with the power factors values previously obtained, we get $ZT \simeq 4.5 - 9.0$ for polyG-polyC chains with five base pairs at room temperature (well above the usual highest $ZT \simeq 1 - 2$ for conventional bulk materials).[149] These remarkably high figure of merit values (comparable to those exhibited by best thermoelectric materials,[150]) must be properly balanced with the significant role played by unavoidable environmental effects, stemming from the presence of a cation/water molecules atmosphere around the DNA chain, on the actual thermoelectric efficiency of DNA based nano-cells. In addition, the role of polarons (whose formation is a very common process for organic polymers with a flexible backbone such as DNA) in the electrical transport efficiency will deserve a closer scrutiny.[151, 152, 153, 154] Broadly speaking, the on-site interaction of the charge carrier with phonon modes tends to localize it, leading to charge transfer rates within the range $\tau = 5 - 75$ ps, as reported by experiments.[155] These values are much larger than the charge transfer rates related to coherent tunneling, which are given by $\tau \simeq t_0/h \simeq 0.03$ ps. Accordingly, one reasonably expects that the presence of polarons will give rise to a degradation of the charge transfer efficiency, as compared to that corresponding to coherent transport conditions. According to Eq.(7.35) a decrease in the charge transfer efficiency is generally accompanied by an enhancement of the Seebeck coefficient. On this basis, one could then expect that polaronic effects would lead to further improvement in the thermoelectric properties of DNA chains.

In summary, prospective studies on the thermoelectric properties of synthetic DNA oligonucleotides suggest that these materials are suitable candidates to be considered in the design of highly-performing, nano-scale sized thermoelectric cells. Experimental work aimed to test the actual capabilities of DNA based thermoelectric devices under different environmental conditions as well as to accurately determine the thermal transport properties of synthetic DNA samples would be then very appealing.

7.5.2 Codon sequencing

As it was mentioned at the end of Section 6.1, there currently exists a growing interest in the search for new sequencing methods entirely based on physical principles able to allow for non-invasive analysis of a huge number of nucleotides along the DNA strands. In this regard, scanning tunnel spectroscopy, which directly detects the molecular levels of single DNA bases, has been exploited during the last few years. In fact, nucleobase-modified tip STM measurements demonstrate the ability to identify the different DNA nucleobases due to selective chemical interactions, although it remains a chemically based rather than a purely physically based technique.[156]

In this Section, we will consider the possibility of looking for coding sequence regions (introns) in long DNA fragments by employing thermoelectric measurements, hence complementing the dc conductivity based approach introduced in Section 6.5. To this end, we shall analyze the thermoelectric performance of short DNA chains connected between metallic contacts at different temperatures in order to estimate the possibility of directly sensing triplet nucleobases associations (including codons in codifying regions) via their *thermoelectric signature*. We shall consider the single-strand DNA Hamiltonian given by Eq.(6.19). In the case of a trimer oligonucleotide we have three nucleobases of energies $\varepsilon_1^v, \varepsilon_2^v$, and ε_3^v, respectively, coupled with hopping terms t and ηt. Depending on the DNA sequence composition, its length, and temperature the effective value of the hopping integral t can vary over a relatively broad range. We will adopt the values $t_{CC} = 0.3$ eV, $t_{GG} = 0.25$ eV, $t_{TT} = 0.13$ eV, and $t_{AA} = 0.035$ eV in the case of homopolymers, and $t_{GT} = 0.083$ eV, $t_{TG} = 0.26$ eV, $t_{AC} = 0.11$ eV, and $t_{CA} = 0.37$ eV in the study of heteropolymers. These values were derived from *ab initio* calculations for 5'-XY-3' intrastrand stacked pairs.[157] In this way, a more realistic description for codon triplets of biological interest is provided. We will consider values within the range $\tau = 0.1 - 0.5$ eV for the base-metal electronic coupling. Finally, we have considered two different contact parameters of technological interest. On the one hand, one corresponds to a molecule connected to an open edge of a graphene sheet at both sides. In this way, the spectral window is approximately given by the graphite π bandwidth $[-6.8, 0]$ eV, corresponds to the tight-binding parameters $t_M = 1.7$ eV, and $\varepsilon_M = -3.4$ eV. On the other hand, we consider platinum contacts corresponding to the tight-binding parameters $t_M = 2.2$ eV, and $\varepsilon_M = -5.4$ eV, determining the allowed spectral window $[-9.8, -1.0]$ eV.

In the first place we shall consider the transport properties corresponding to GGG, AAA, CCC, and TTT codon trimers, respectively, codifying for glycine, lysine, proline, and phenylalanine amino acids in the homo sapiens genetic code. In this case we have $\eta = 1$. In the main frame of Fig.7.29 we show the CCC thermopower curve for $\tau = 0.5$ eV. This curve is characterized by three peaks and two crossing points E_0 and E^*, respectively defining three different regimes exhibiting p-type or n-type thermopower alternatively. In the upper inset of Fig.7.29, we compare the conductance of a C monomer, a CC dimer, and CCC trimer as a function of the Fermi energy. While the dimer conductance is degraded as a consequence of dimerization effects, we observe that the trimer conductance is significantly enhanced as compared to that corresponding to both C and CC bases over a broad energy range. In addition, a well defined, narrow resonance peak is located at about $E^* = -3.35$ eV, which is flanked by a shallow conductance minimum at about $E_0 = -4.57$ eV. Taking into account Eq.(7.34) the origin of the crossing points in the thermopower curve can be properly traced back to these topological features.

Now we consider the transport properties corresponding to TGT, CAC, and TTG codon trimers, respectively codifying for cysteine, histidine, and leucine amino acids in the homo sapiens genetic code. The aim is to compare their properties with those previously obtained for the TTT and AAA trimers in order to see the effects stemming from the change of one of their

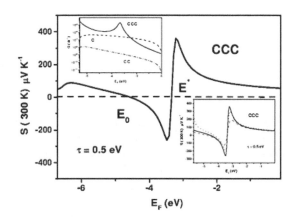

FIGURE 7.29
Room temperature dependence of the Seebeck coefficient as a function of the Fermi level energy for a CCC trimer with $\tau = 0.5$ eV, $t_M = 1.7$ eV, $\varepsilon_G = -7.75$ eV, $\varepsilon_A = -7.95$ eV and $\varepsilon_T = \varepsilon_C = -8.30$ eV, and $\varepsilon_M = -3.4$ eV. (Upper inset) Landauer conductance as a function of the Fermi level energy for CCC (solid line), CC (dot-dashed line), and C (dashed line) nucleobases. (Lower inset) Environmental effects in the thermopower due to the presence of backbone counterions giving rise to nucleobase energy shifts within the range $\Delta E = -1$ eV (dotted line) and $\Delta E = +1$ eV (dashed line). (From ref.[139]. With permission from IOP Publishing Ltd.)

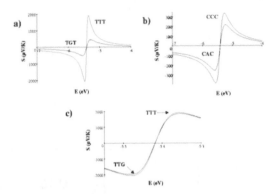

FIGURE 7.30

Dependence of the room temperature thermopower as a function of the Fermi level energy for a) TTT and TGT codons; b) CCC and CAC codons; and c) TTT and TTG codons connected to platinum leads with $\tau = 0.5$ eV, $t_M = 2.2$ eV, and $\varepsilon_M = -5.4$ eV. (From ref.[134]. With permission.)

original nucleobases ($\eta \neq 1$ in this case). In Fig.7.30a and b we respectively compare the thermopower curves for the TTT/TGT and CCC/CAC trimers as a function of the Fermi energy. We can see that the thermopower is substantially reduced upon the substitution of the central nucleobases in both cases. This is a general feature of the codons obeying the formula XYX. This result strikingly contrasts with the small effect associated to the substitution of an end nucleobase instead, as it is illustrated in Fig.7.30c. Accordingly, the thermoelectric properties of codons obeying the general formula XXY are quite similar to those observed for the corresponding homonucleotides.

In summary, the presence of resonance effects among electronic levels in oligonucleotides composed of three nucleobases leads to a significant enhancement of the thermoelectric signal. By comparing the transport curves corresponding to different types of trimers we see that a characteristic thermoelectric signature can be used to identify the XYX type codons from XXX homonucleotide ones on the basis of their different thermoelectric responses. Since the coding properties of DNA introns are closely related to codon triplet associations this preliminary result may enclose some biological relevance well deserving a more detailed study by means of more realistic modeling of both the electronic structure of nucleotides and the codon-lead bonding geometry.

References

[1] Zhu S N, Zhu Y Y, Qin Y Q, Wang H F, Ge C Z, and Ming N B 1997 *Phys. Rev. Lett.* **78** 2752

[2] Zhu S N, Zhu Y Y, and Ming N B 1997 *Science* **278** 843

[3] Chen Y B, Zhang C, Zhu Y Y, Wang H T, and Ming N B 2001 *Appl. Phys. Lett.* **78** 577

[4] Dal Negro L, Yi J H, Nguyen V, Yi Y, Michel J, and Kimerling L C 2005 *Appl Phys. Lett.* **86** 261905

[5] Fradkin-Kashi K and Arie A 1999 *IEEE J. Quantum Electron.* **35** 1649; Fradkin-Kashi K, Arie A, Urenski P, and Rosenman G 2002 *Phys. Rev. Lett.* **88** 023903

[6] Lifshitz R, Arie A, and Bahabad A 2005 *Phys. Rev. Lett.* **95** 133901

[7] Makarova L N, Nazarov M M, Ozheredov I A, Shkurinov A P, Smirnov A G and Zhukovsky S V 2007 *Phys. Rev. E* **75** 036609

[8] Moretti L, Rea I, De Stefano L, and Rendina I 2007 *Appl. Phys. Lett.* **90** 191112

[9] Boriskina S V and Dal Negro L 2008 *Optics Express* **16** 12511

[10] Maciá E 2006 *Rep. Prog. Phys.* **69** 397

[11] Maciá E and Domínguez-Adame F 2000 *Electrons, Phonons, and Excitons in Low Dimensional Aperiodic Systems* (Editorial Complutense, Madrid)

[12] Diez E, Domínguez-Adame F, Maciá E, and Sánchez A 1996 *Phys. Rev. B* **54** 16792

[13] Tamura S and Wolfe J P 1987 *Phys. Rev. B* **36** 3491; Hurley D C, Tamura S, Wolfe J P, Ploog K, and Nagle J 1988 *ibid.* **37** 8829; Tamura S and Nori F 1990 *ibid.* **40** 9790

[14] Fernández-Alvarez L and Velasco V R 1998 *Phys. Rev. B* **57** 14141; Zárate J E, Fernández-Alvarez L and Velasco V R 1999 *Superlatt. Microstruct.* **25** 519

[15] Kono K, Nakada S, Narahara Y, and Ootuka Y J 1991 *J. Phys. Soc. Jpn.* **60** 368; Komoro T, Shirahama K, and Kono K 1995 *Phys. Rev. Lett.* **75** 3106

[16] Zhu Y, Ming N, and Jiang W 1989 *Phys. Rev. B* **40** 8536

[17] Mishra S and Satpathy S 2003 *Phys. Rev. B* **68** 045121

[18] Yablonovitch E 1987 *Phys. Rev. Lett.* **58** 2059

[19] Joannopoulos J D, Meade R D, and Winn J N 1995 *Photonic Crystals: Molding the Flow of Light* (Princeton Univ. Press, New Jersey)

[20] Sibilia C, Tropea F and Bertolotti M 1998 *J. Mod. Opt.* **45** 2255

[21] Kohmoto M, Sutherland B, and Iguchi K 1987 *Phys. Rev. Lett.* **58** 2436

[22] Schwartz C 1988 *Appl. Opt.* **27** 1232

[23] Gellermann W, Kohmoto M, Sutherland B, and Taylor P C 1994 *Phys. Rev. Lett.* **72** 633

[24] Hattori T, Tsurumachi N, Kawato S, and Nakatsuka H 1994 *Phys. Rev. B* **50** 4220

[25] Dulea M, Severin M and Riklund R 1990 *Phys. Rev. B* **42** 3680

[26] Miyazaki H and Inoue M 1990 *J. Phys. Soc. Jpn.* **59** 2536

[27] Liu Y and Riklund R 1987 *Phys. Rev. B* **35** 6034

[28] Latgé A and Claro F 1992 *Opt. Commun.* **94** 389

[29] Vasconcelos M S, Albuquerque E L, and Mariz A M 1998 *J. Phys. Condens. Matter* **10** 5839

[30] Huang X Q, Jiang S S, Peng R W, and Hu A 2001 *Phys. Rev. B* **63** 245104

[31] Riklund R and Severin M 1988 *J. Phys. C* **21**, 3217; Dulea M, Severin M, and Riklund R 1990 *Phys. Rev. B* **42** 3680

[32] Liu N H 1997 *Phys. Rev. B* **55** 3543

[33] Vasconcelos M S and Albuquerque E L 1999 *Phys. Rev. B* **59** 11 128

[34] Albuquerque E L and Cottam M G 2003 *Phys. Rep.* **376** 225

[35] Dal Negro L, Stolfi M, Yi Y, Michel J, Duan X, Kimerling L C, LeBlanc J, and Haavisto J 2004 *Appl. Phys. Lett.* **84** 5186

[36] Qiu F, Peng R W, Huang X Q, Hu X F, Wang M, Hu A, Jiang S S, and Feng D 2004 *Europhys. Lett.* **68** 658

[37] Dal Negro L, Oton C J, Gaburro Z, Pavesi L, Johnson P, Lagendijk A, Righini R, Colocci M, and Wiersma D S 2003 *Phys. Rev. Lett.* **90** 055501

[38] Ghulinyan M, Oton C J, Dal Negro L, Pavesi L, Sapienza R, Colocci M, and Wiersma D S 2005 *Phys. Rev. B* **71** 094204

[39] Aynaou H, el Boudouti E H, Djafari-Rouhani B, Akjouj A, and Velasco V R 2005 *J. Phys. Condens. Matter* **17** 4245

[40] King P D C and Cox T J 2007 *J. Appl. Phys.* **102** 014902

[41] Domínguez-Adame F and Maciá E 1995 *Phys. Lett. A* **200** 69

[42] Pan F M, Jin G J, Wu X S, Wu X L, Hu A, and Jiang S S 1997 *Phys. Lett. A* **228** 301; 1996 *J. Appl. Phys.* **80** 4063

[43] Zi J, Wan J, and Zhang C 1998 *Appl. Phys. Lett.* **73** 2084

[44] Lei X Y, Li H, Ding F, Zhang W, and Ming N B 1997 *Appl. Phys. Lett.* **71** 2889

[45] Maciá E 1998 *Appl. Phys. Lett.* **73** 3330

[46] Maciá E 2001 *Phys. Rev. B* **63** 205421

[47] Gaponenko S V, Zhukovsky S V, Lavrinenko A V, and Sandomirskii K S 2002 *Optics Commun.* **205** 49; Chigrin D N, Lavrinenko A V, Yarotsky D A, and Gaponenko S V 1999 *Appl. Phys. A* **68** 25

[48] Lavrinenko A V, Zhukovsky S V, Sandomirski K S, and Gaponenko S V 2002 *Phys. Rev. E* **65** 036621; Zhukovsky S V, Lavrinenko A V, and Gaponenko S V 2004 *Europhys. Lett.* **66** 455

[49] Planck M 1959 *The Theory of Heat Radiation* (Dover, New York)

[50] John S and Wang J 1990 *Phys. Rev. Lett.* **64** 2418; 1991 *Phys. Rev. B* **43** 12772

[51] Cornelius C M and Dowling J P 1999 *Phys. Rev. B* **59** 4736

[52] de Medeiros F F, Albuquerque E L, Vasconcelos M S, and Mauriz P W 2007 *J. Phys. Condens. Matter* **19** 496212

[53] Smith D, Pendry J and Wiltshire M 2004 *Science* **305** 788

[54] Li J, Zhou L, Chan C T, and Sheng P 2003 *Phys. Rev. Lett.* **90** 083901

[55] Monsoriu J A, Depine R A, and Silvestre E 2007 *J. Eur. Opt. Soc.* **2** 07002

[56] Vasconcelos M S, Mauriz P W, de Medeiros F F, and Albuquerque E L 2007 *Phys. Rev. B* **76** 165117

[57] Maksimović M and Jakšić Z 2006 *J. Opt. A: Pure Appl. Opt.* **8** 355

[58] Luque A and Martí A 1997 *Phys. Rev. Lett.* **78** 5014

[59] Peng R W, Mazzer M, and Barnham K W J 2003 *Appl. Phys. Lett.* **83** 770

[60] Golmohammadi S, Moravvej Farshi M K, Rostami A, and Zarifkar A 2007 *Optics Express* **15** 10520

[61] Wang L, Liu Z, and Zhang W 2008 *J. Phys. Soc. Jpn.* **77** 034701

[62] Barriuso A G, Monzón J J, Sánchez-Soto L L, and Felipe A 2005 *Optics Express* **13** 3913

[63] Pantic M, Pavkov-Hrvojevic, Rutonjski M, Skrijar M, Kapor D, Rado-
sevic S, and Budincevic M 2007 *Eur. Phys. J B* **59** 367

[64] Liu Z and Zhang W 2007 *Phys. Rev. B* **75** 064207; 2005 *ibid.* **72** 134304

[65] Steurer W and Sutter-Widmer D 2007 *J. Phys. D: Appl. Phys.* **40** R229

[66] He S and Maynard J D 1989 *Phys. Rev. Lett.* **62** 1888

[67] Chan Y S, Chan C T, and Liu Z Y 1998 *Phys. Rev. Lett.* **80** 956

[68] Sigalas M M, Economou E N, and Kafesaki M 1994 *Phys. Rev. B* **50**
3393

[69] Cheng S S M, Li L M, Chan C T, and Zhang Z Q 1999 *Phys. Rev. B*
59 4091

[70] Ueda K, Dotera T, and Gemma T 2007 *Phys. Rev. B* **70** 195122

[71] Rattier M, Benisty H, Schwoob E, Weisbuch C, Krauss T F, Smith C J
M, Houdré R, and Oesterle U 2003 *Appl. Phys. Lett.* **83** 1283

[72] Jin C, Cheng B, Man B, Li Z, Zhang D, Ban S, and Sun B 1999 *Appl.
Phys. Lett.* **75** 1848

[73] Bayindir M, Cubukcu E, Bulu I, and Ozbay E 2001 *Phys. Rev. B* **63**
161104(R)

[74] Zhou G and Gu M 2007 *Appl. Phys. Lett.* **90** 201111

[75] Shchesnovich V S 2007 *Phys. Rev. B* **76** 115130

[76] Lee T D M, Parker G J, Zoorob M J, Cox S J, and Charlton M D B
2005 *Nanotechnology* **16** 2703

[77] Matsui T, Agrawal A, Nahata A, and Vardeny Z V 2007 *Nature (London)*
446 517

[78] Oxborrow N and Henley C L 1993 *Phys. Rev. B* **48** 6966

[79] Stampfli P 1986 *Helv. Phys. Acta* **159** 1260

[80] Zoorob M E, Charlton M D B, Parker G J, Vaumberg J J, and Netti M
C 2000 *Nature (London)* **404** 740

[81] Feng Z, Zhang X, Wang Y, Li Z Y, Cheng B, and Zhang D Z 2005 *Phys.
Rev. Lett.* **94** 247402

[82] Di Gennaro E, Miletto C, Savos S, Andreone A, Morello D, Galdi V,
Castaldi G, and Pierro V 2008 *Phys. Rev. B* **77** 193104

[83] Wang Y, Hu X, Xu X, Chang B, and Zhang D 2003 *Phys. Rev. B* **68**
165106

[84] Kim S K, Lee J H, Kim S H, Hwang I K, and Lee Y H 2005 *Appl. Phys.
Lett.* **86** 031101

[85] Wang Y, Liu J, Zhang B, Feng S, and Li Z Y 2006 *Phys. Rev. B* **73** 155107

[86] Zhang X 2007 *Phys. Rev. B* **75** 024209

[87] Dyachenko P N, Miklayev Yu. V, and Dmitrienko V E 2007 *JETP Letters* **86** 240

[88] Man W, Megens M, Steinhardt P J, and Chaikin P M 2005 *Nature (London)* **436** 993

[89] Liderman A, Cademartiti M, Hermatschweiler M, Toninelli C, Ozin G A, Wiersma D S, Wegener M, and Von Freymann G 2006 *Nature Mater.* **5** 942

[90] Xu J, Ma X, Wang X, and Tam W Y 2007 *Opt. Express* **15** 4287

[91] Sutter-Widmer D, Neves P, Itten P, Sainidou R, and Steurer W 2008 *Appl. Phys. Lett.* **92** 073308

[92] Kelton K F, Kim Y J, and Stroud R M 1997 *Appl. Phys. Lett.* **70** 3230

[93] Tsai A P, Niikura A, Inoue A, and Masumoto T 1994 *Phil. Mag. Lett.* **70** 169

[94] Tsai A P, Guo J Q, Abe E, Takakura H, and Sato T 2000 *Nature (London)* **408** 537

[95] Dolinšek J, Jeglič P, McGuiness P J, Jagličić Z, Bilušić A, Bihar Ž, Smontara A, Landauro C V, Feuerbacher M, Grushko B, and Urban K 2005 *Phys. Rev. B* **72** 064208

[96] Mizutani U 2001 *Introduction to the Electron Theory of Metals* (Cambridge University Press, Cambridge)

[97] Boudard M, Klein H, de Boissieu M, Audier M, and Vincent H 1996 *Phil. Mag. A* **74** 939

[98] Maciá E and Dolinšek J 2007 *J. Phys.: Condens. Matter* **19** 176212

[99] Dubois J M 1997 *New Horizons in Quasicrystals: Research and Applications*, Eds. Goldman A I, Sordelet D J, Thiel P A, and Dubois J M (World Scientific, Singapore) p 208

[100] Dubois J M 2005 *Useful Quasicrystals* (World Scientific, Singapore)

[101] Eisenhammer T 1995 *Thin Solid Films* **270** 1

[102] Eisenhammer T and Lazarov M 1994 German Patent No. 4425140

[103] Anderson B C, Bloom P D, Baikerikar K G, Sheares V V, and Mallapragada S K 2002 *Biomaterials* **23** 1761

[104] Dubois J M, Brunet P, and Belin-Ferré E 2000 *Quasicrystals: Current Topics* edited by Belin-Ferré E, Berger C, Quiquandon M, and Sadoc A (World Scientific, Singapore) p 498

[105] Dubois J M, Brunet P, Costin W, and Merstallinger A 2004 *J. Non-Cryst. Solids* **334** & **335** 475-480

[106] Belin-Ferré E and Dubois J M 2006 *Int. J. Mat. Res.* **97** 1

[107] Thiel P A 2004 *Prog. Surf. Sci.* **75** 69; 2004 *Prog. Surf. Sci.* **75** 191

[108] Wu S, Brien V, Brunet P, Dong C, and Dubois J M 2000 *Phil. Mag. A* **80** 1645 – 1655.

[109] Maciá E, Dubois J M, and Thiel P A 2008 *Quasicrystals* entry in *Ullmann's Encyclopedia of Industrial Chemistry*, 7th Edition, 2008 (Wiley-VCH, Winheim)

[110] Henmig R G, Majzoub E H, and Kelton K F 2006 *Phys. Rev. B* **73** 184205

[111] Tasaki A and Kelton K F 2006 *Int. J. Hydrogen Energy* **31** 183

[112] Nosaki K, Masumoto T, Inoue K, and Yamaguchi T 1998 US patent No. 5.800638

[113] Yoshimura M and Tsai A P 2002 *J. Alloys and Compounds* **342** 451; 2001 *Appl. Cat. A: General* **214** 237

[114] Tanabe T, Kameoka S, and Tsai A P 2006 *Catalysis Today* **111** 153;.Kameoka S, Tanabe T, and Tsai A P 2004 *Catalysis Today* **93-95** 23

[115] Mahan G D and Sofo J O 1996 *Proc. Natl. Acad. Sci. USA* **93** 7436

[116] Maciá E 2001 *Phys. Rev B* **64**, 094206

[117] Pope A L, Tritt T M, Chernikov M A, and Feuerbacher M 1999 *Appl. Phys. Lett.* **75** 1854

[118] Slack G A 1995 *CRC Handbook of Thermoelectrics*, Ed. Rowe D M (CRC Press, Boca Raton, FL)

[119] Biggs B D, Poon S J, and Munirathnam N R 1990 *Phys. Rev. Lett.* **65** 2700; Pierce F S, Poon S J, and Biggs B D 1993 *Phys. Rev. Lett.* **70** 3919; Biggs B D, Li Y, and Poon S J 1991 *Phys. Rev. B* **43** 8747; Pierce F S, Bancel P A, Biggs B D, Guo Q, and Poon S J 1993 *Phys. Rev. B* **47** 5670

[120] Dubois J M, Kang S S, Archembault P, and Colleret B 1993 *J. Mat. Res.* **8** 38; Archembault P and Janot C 1997 *MRS Bulletin* p. 48, November 1997

[121] Pope A L, Tritt T M, Gagnon R, and Strom-Olsen J 2001 *Appl. Phys. Lett.* **79** 2345

[122] Nagata T, Kirihara K, and Kimura K 2003 *J. Appl. Phys.* **94** 6560; Kirihara K and Kimura K 2002 *J. Appl. Phys.* **92** 979

[123] Pope A L, Zawilski B M, Gagnon R, Tritt T M, Ström-Olsen J, Schneidmiller R, and Kolis J W 2001 *Mat. Res. Soc. Symp. Proc.* **643** K14.4.1

[124] Maciá E 2004 *Phys. Rev. B* **69** 184202

[125] Maciá E 2004 *Phys. Rev. B* **70** 100201(R)

[126] Escudero R, Lasjaunias J C, Calvayrac Y, and Boudard M 1999 *J. Phys. Condens. Matter* **11** 383

[127] Maciá E 2004 *Phys. Rev. B* **69** 132201

[128] Poler J C, Zimmermann R M, and Cox E C 1995 *Langmuir* **11** 2689

[129] Paulsson N and Datta S 2003 *Phys. Rev. B* **67** 241403(R)

[130] Gao X, Uehara K, Klug D D, Patchkovskii S, Tse J S, and Tritt T M 2005 *Phys. Rev. B* **72** 125202

[131] Esfarjani K, Zebarjadi M, and Kawazoe Y 2006 *Phys. Rev. B* **73** 085406

[132] Roche S and Maciá E 2004 *Mod. Phys. Lett. B* **18** 847

[133] Maciá E 2005 *Nanotechnology* **16** S254

[134] Maciá E 2005 *Rev. Adv. Mater. Sci.* **10** 166

[135] Zheng X, Zheng W, Wei Y, Zeng Z, and Wang J 2004 *J. Chem Phys.* **121** 8537

[136] Lagerqvist J, Zwolak M, and Di Ventra M 2006 *Nano Lett.* **6** 779

[137] Zwolak M and Di Ventra M 2005 *Nano Lett.* **5** 421

[138] Yoo K H, Ha D H, Lee J O, Park J W, Kim J, Kim J J, Lee H Y, Kawai T, and Choi H Y 2001 *Phys. Rev. Lett.* **87** 198102

[139] Maciá E and Roche S 2006 *Nanotechnology* **17** 3002

[140] Maciá E, Triozon F, and Roche S 2005 *Phys. Rev. B* **71** 113106

[141] Maciá E 2007 *Phys. Rev. B* **75** 035130

[142] Endres R G, Cox D L, and Singh R R P 2004 *Rev. Mod. Phys.* **76** 195

[143] Kergueris C, Bourgoin J P, Palacin S, Esteve D, Urbina C, Magoga M, and Joachim C 1999 *Phys. Rev. B* **59** 12505

[144] Yamaguchi S, Nagawa Y, Kaiwa N, and Yamamoto A 2005 *Appl. Phys. Lett.* **86** 153504

[145] Kim N, Domercq B, Yoo S, Christensen A, Kippelen B, and Grahm S 2005 *Appl. Phys. Lett.* **87** 241908

[146] Nolas G S, Sharp J, and Goldsmid H J in 2001 *Thermoelectrics Basic Principles and New Materials Developments* (Springer Series in Materials Science **45**, Springer, Berlin), p. 235

[147] Schwab K, Henriksen E A, Worlock J M, and Roukes M L 2000 *Nature (London)* **404** 974

[148] Nishiguchi N 1995 *Phys. Rev. B* **52** 5279; Nishiguchi N, Ando Y, and Wybourne M N 1997 *J. Phys. Condens. Matter* **9** 5751

[149] Harman T C, Taylor P J, Walsh M P, and LaForge B E 2002 *Science* **297** 2229

[150] Dresselhaus M S et al. 2001 *Semiconductors and Semimetals: Recent Trends in Thermoelectric Materials Research III,* Ed. Tritt T M (Academic, San Diego, CA)

[151] Conwell E M, and Rakhmanova S V 2000 *Proc. Natl. Acad. Sci. U.S.A.* **97** 4556

[152] Zhang W, Govorov A O, and Ulloa S E 2002 *Phys. Rev. B* **66** 060303(R).

[153] Alexandre S S, Artacho E, Soler J M, and Chacham H 2003 *Phys. Rev. Lett.* **91** 108105

[154] Yamada H, Starikov E B, Hennig D, and Archilla J F R 2005 *Eur. Phys. J. E* **17** 149

[155] Wan C, Fiebig T, Kelley S O, Treadway C R, Barton J K, and Zewail A H 1999 *Proc. Natl. Acad. Sci. U.S.A* **96** 6014

[156] Xu M S, Endres R G, and Arakawa Y 2007 in *Charge Migration in DNA: Perspectives from Physics, Chemistry and Biology*, Ed. Chakraborty T (Nanoscience and Technology series, Springer-Verlag, Berlin) p.205

[157] Yan Y J and Zhang H 2002 *J. Theor. Comp. Chem.* **1** 225

8

Novel designs based on the aperiodic order

8.1 Order and design: The technological viewpoint

In the previous Chapter we have seen that aperiodic order can be fruitfully exploited in a number of structures with potential practical applications. For instance, one can grow layered structures consisting of a large number of aperiodically arranged films. The simplest example of such nanostructured materials is a two-component Fibonacci heterostructure, where layers of two different materials (metallic, semiconductor, superconductor, dielectric, ferroelectric, ceramics) are arranged according to the Fibonacci sequence. A key feature of these man-made materials is the presence of *two kinds of order* in the same sample *at different length scales*: At the atomic level we have the usual crystalline order determined by the periodic arrangement of atoms in each layer, whereas at longer scales we have the quasiperiodic order determined by the sequential deposition of the different layers. This long-range aperiodic order is artificially imposed during the growth process and can be precisely controlled. Since different physical phenomena have their own relevant physical scales, by properly matching the characteristic length scales we can efficiently exploit the aperiodic order we have introduced in the system, hence opening new avenues for technological innovation. In this way, new perspectives in the design of resonating optical devices that exploit the interference possibilities associated to the quasiperiodic order of the substrate, such as optical microcavities or broad multidirectional reflection devices, were discussed in Sections 7.2 and 7.3, respectively.

An important lesson we can learn from the examples presented in these preceding Sections is that several characteristic features of energy spectra of most aperiodic structures are closely related with a number of properties of interest in the design of useful devices. For the sake of illustration we will list the following ones:

- The presence of **forbidden symmetries** in photonic quasicrystals allows for the existence of higher order rotational axes leading to more isotropic complete band-gaps in omnidirectional mirrors,

- the presence of **inflation symmetries** in both photonic QCs and aperiodic multilayers gives rise to the presence of a denser reciprocal space, favoring higher harmonics generation processes in non-linear optics,

- the presence of **critical states** leads to the occurrence of highly localized optical modes in defect-free photonic and phononic QCs of interest for the fabrication of high quality factor resonators as well as coupled-resonator waveguides,

- the presence of a **highly fragmented** energy spectrum becomes particularly useful for a number of applications as laser compression (determined by small group velocities), high spectral resolution (arising from the presence of very narrow transmission peaks for a given spectral range), fine filtering tuning (since reflectance of aperiodic coatings increases with length), selective enhancement of thermal emission for some selected frequencies, a significant efficiency enhancement of photovoltaic cells, or an improved frequency selectivity in acoustic waveguides.

In addition to these properties, directly related to the fractal nature of energy spectra in aperiodic systems (see Chapter 5), one can also benefit from purely structural features of these aperiodic arrangements. For example, both Thue-Morse and Fibonacci multilayers have less interfaces than periodic ones for a given system size, which is convenient in order to reduce losses and dispersion effects. As another illustrative example, we can mention the large number of tetrahedral interstices in the structure of icosahedral QCs, a property of interest for hydrogen-storage purposes.

Nevertheless, along with these useful aspects, quasiperiodic structures also present an important shortcoming of fundamental nature, namely, that both fractal features and long-range quasiperiodicity require a critical size to properly manifest themselves. Accordingly, in practical applications one must reach a balance between those effects related to the presence of energy losses and dispersion effects (requiring relatively small systems) and beneficial aspects stemming from self-similarity and quasiperiodicity related effects (which require a large enough system).

Another interesting conclusion we can draw from the diverse aperiodic systems presented in Chapter 7 concerns the very nature of the geometric patterns adopted in the design of a given aperiodic structure. As we can readily see, most earlier designs were simply based on those aperiodic sequences (in layered systems) or spatial arrangements (in two or three dimensional systems) which had been previously considered in the mathematical literature, namely Fibonacci, Thue-Morse, Rudin-Shapiro, period-doubling sequences, on the one hand, and Penrose tiles, on the other hand. With the exception of the Fibonacci sequence (which can be profusely found in a number of actual systems in Nature, like botanical arrangements, or morphogenetic patterns),[1] the remaining aperiodic structures were originally introduced following more theoretical motivations, basically. Certainly, this is a rather natural procedure from a fundamental point of view, and it has proved very fruitful indeed. However, as the field of aperiodic systems progressively matures, and the

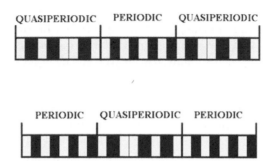

FIGURE 8.1
Scheme of a hybrid-order optical resonating microcavity based on three sub-units. A full transmission periodic dielectric multilayer is encased between two quasiperiodically ordered ones, acting as optical mirrors, and vice versa. Thick, darker layers are made of titanium oxide, while the thin ones are made of silica.

knowledge of fundamental aspects paves the way towards practical applications, it is reasonable to expect that the quest for specific aperiodic orderings, able to yield an improved performance, will gradually grab the limelight.

This line of thought naturally introduces a new way of thinking in the field of aperiodic arrangements of matter, opening the search for those aperiodic orderings which may be more beneficial for some specific technological applications. Accordingly, in this Chapter we aim to take a step ahead in the aperiodic order realm by addressing this appealing issue in some detail.

8.2 One-dimensional designs

8.2.1 Hybrid order multilayers

Up to now we have considered periodic and aperiodic arrangements of matter as two separate categories of order in Nature, each one exhibiting a number of characteristic features, and the spotlight has been focused on aperiodically ordered structures as they substantially widen the possibilities of novel designs. A broader approach considers the construction of *mixed* devices composed of both periodically and quasiperiodically arranged multilayers in order to properly blend in their specific features, thereby obtaining new capabilities. Such devices can be viewed as *hybrid order* systems made of two different kinds of subunits, each one exhibiting a different kind of topological order.[1] Some examples of this kind of arrangement are illustrated in Figs.8.1 and 8.4. The key point here is that one just changes the ordering in the stacking sequence of layers composing each subunit, so that the chemical nature of the different layers remains the same, a remarkable point for industrial implementation. The introduction of these subunits endows the system with an *additional design parameter*, bridging the gap between the atomic level characteristic of the microstructural domain of each layer and the mesoscale level associated to the long-range order of the entire device as a whole. Therefore, since each subunit can sustain a specific kind of order (i.e., periodic or aperiodic) we are able to introduce a *modular design* in the multilayered structure by properly selecting the ordering sequence in each subunit. Noteworthy, this sort of modular design naturally mimics the basic structural principles observed in several macromolecules of biological interest.

These mixed structures were originally proposed in the field of linear optics in order to obtain *complementary* optical responses, ranging from that corresponding to a selective filter to that proper of a reflective coating, by properly choosing the incidence angle geometry.[2, 3] In Fig. 8.1 a sketch illustrating possible designs for optical devices based on the mixed architecture just described is shown. They correspond to resonant-cavities where a high transmission periodic (quasiperiodic) multilayer is encased between high reflectivity Bragg reflectors based on quasiperiodic (periodic) dielectric multilayers. The basic principle can readily be extended to the construction of broad omnidirectional reflection bands on the basis of complementary band gap design. The main idea is that gaps appearing in the highly fragmented quasiperiodic units compensate for the relatively broad band regions appearing in the periodic ones. In this way, one can optimize engineered combinations of quasiperiodic and periodic short multilayered stacks in order to obtain not too thick photonic heterostructures (v.g., $[ABA]^4[BAABA]^2[BA]^7$), exhibiting large omnidirectional reflection.[4, 5] In the same vein, the possible use of a hybrid order structure of the form periodic/Cantor/periodic dielectric multilayer was also proposed for its applications as a polychromatic filter.[6] Though originally proposed for optical applications, the properties of such hybrid order structures have been also considered in the field of acoustics,[?] thermal waves propagation,[7] and phononic crystals as well.[8]

From a mathematical viewpoint the nature of spectra in hybrid order structures is an interesting open problem. A representative example of the obtained numerical results is shown in Fig.8.2 for the frequency spectrum of a Fibonacci/periodic/Fibonacci hybrid structure. For the sake of illustration, in Figs.8.2 a and b the frequency spectra of periodic and Fibonacci chains are respectively presented. In the Fibonacci spectrum one can readily see the characteristic fragmentation scheme, as well as the presence of three allowed bands in the region corresponding to the gap in the binary chain (see Fig.5.11).

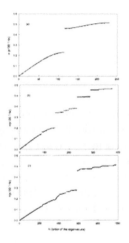

FIGURE 8.2
Frequency eigenvalues versus ordering number for (a) a binary periodic chain
with $N = 244$ atoms; (b) a Fibonacci lattice with $N = 377$ atoms; (c) a
hybrid chain formed by a periodic chain including 116 atoms sandwiched
between two Fibonacci lattices including 377 atoms each ($N = 986$ atoms).
The model parameters are $m_A = 4.805 \times 10^{-23}$ g, $m_B = 1.791 \times 10^{-22}$ g,
$K_{AA} = 4.416 \times 10^4$ dyn/cm, $K_{AB} = 4.724 \times 10^4$ dyn/cm, and they describe
an Al/Ag metallic multilayer. (Courtesy of Victor R.Velasco).

(a)   (b)

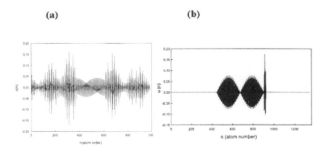

FIGURE 8.3

Normalized atomic displacements $u(n)$ versus the atom order number n
through (a) the Fibonacci/periodic/Fibonacci hybrid structure whose frequency spectrum is shown in Fig.8.2, corresponding to the eigenvalue $\omega =$
15.68 GHz; (b) a hybrid structure formed by the periodic repetition 64 times
of an AB block sandwiched between two seventh-order Thue-Morse multilayers, corresponding to the eigenvalue $\omega = 40.006$ GHz with the same model
parameters as those indicated in Fig.8.2. (Courtesy of Victor R.Velasco).

By inspecting Fig.8.2 one realizes that the spectrum of the hybrid structure
shares some characteristic features with both the periodic and the quasiperiodic frequency spectra. Thus, the hybrid spectrum preserves a high degree of
fragmentations over all the considered frequency range, but the energy bands
located at the intermediate frequency region in the isolated Fibonacci lattice
are pushed towards lower frequency values, nearly closing the gap between 20
and 30 GHz typical of the quasiperiodic lattice.

Intimately related to these modifications in the overall spectral structure,
one may expect that the competition between highly fragmented spectra supporting critical eigenstates (a characteristic feature of quasiperiodic systems)
and continuous spectra possessing Bloch eigenfunctions (typical of periodic
systems) should give rise to some peculiar eigenstates in these hybrid structures. Two illustrative examples are shown in Fig.8.3 for Fibonacci/periodic
and Thue-Morse/periodic hybrid structures. In Fig.8.3a one can appreciate
an extended vibration pattern, corresponding to a frequency which simultaneously belongs to the energy spectra of both Fibonacci and periodic chains. The
vibration pattern looks pretty regular in the periodic portion of the structure
(a reminiscent feature of a typical Bloch-like function), whereas it exhibits
fluctuations at several scales in the quasiperiodic portions of the structure

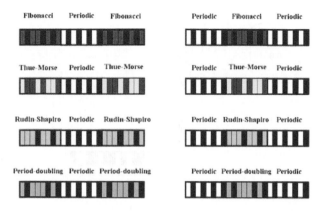

FIGURE 8.4
Diagram showing different possible designs of hybrid systems obtained by combining periodic and aperiodic multilayers.

(showing typical self-similar fingerprints).[7] Conversely, in Fig.8.3b we show the vibration pattern corresponding to a frequency value which belongs to the energy spectrum of the periodic unit, but not to the Thue-Morse one. In this case, the atomic displacements are essentially confined to the periodic block (thus acting as a kind of phononic cavity) and exhibit a notable regularity, as expected for a Block-like function.[9] In this way, the presence of two different kinds of order in the underlying substrate of the hybrid order structure naturally affects the energy spectrum and the spatial distribution of its eigenstates. Though we have focused on the aperiodic/periodic/aperiodic systems it is clear that one should expect similar physical behaviors for other possible combinations of periodic and aperiodic units. In fact, there exist many ways of constructing such combinations, opening a broad avenue for the design of new kinds of multilayered systems. For instance, one may think of reversing the order type in the previously considered structure, thus obtaining periodic/Fibonacci/periodic or periodic/Thue-Morse/periodic multilayers. Similarly, one may mix together different kinds of aperiodic sequences as, for instance, Fibonacci/Thue-Morse/Fibonacci or Fibonacci/Thue-Morse/Rudin-Shapiro, and so on. A graphical account of some possible designs based on hybrid order multilayered structures is shown in Fig.8.4.

8.2.2 Conjugation and mirror symmetries

Following this line of thought one may also consider systems entirely based on quasiperiodic arrangements of layers where the role played by the A and

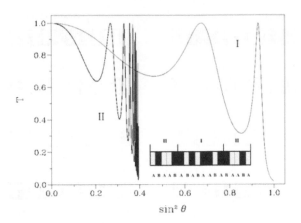

FIGURE 8.5

Dependence of the transmission coefficient with the incidence angle for the conjugated order device sketched in the inset, with $n_A = 1.46$ (SiO$_2$), $n_B = 2.35$ (TiO$_2$) (type I subunit) and $n_A = 2.35$ (TiO$_2$), $n_B = 1.46$ (SiO$_2$) (type II subunit). Inset: Scheme of a conjugated order optical resonating microcavity based on three quasiperiodic subunits, where a Fibonacci dielectric multilayer (with $N = 8$), showing high transmission, is encased between two Fibonacci multilayers (with $N = 5$), acting as optical mirrors. Dark (white) layers correspond to low (high) refractive index materials, respectively. (Adapted from Ref.[3]. Reprinted figure with permission from Maciá E 2001 *Phys. Rev B* **63** 205421 © 2001 by the American Physical Society.)

B layers is respectively interchanged. In that case the characteristic long-range quasiperiodic order is preserved under transformations involving the conjugation operation $A^+ = B$ and $B^+ = A$. This invariance, which follows from the very definition of the Fibonacci sequence, has interesting physical implications which can be exploited in the design of optical devices, as that shown in the inset of Fig.8.5 which consists of two kinds of quasiperiodic subunits. In the first one (labeled I), the A (B) layers are composed of low (high) refractive index materials. In the second one (labelled II), the values of the refractive indices assigned to the layers A and B are reversed, so that the total internal reflection angle condition is achieved when $x_0 \simeq 0.4$ (i.e., $\theta \simeq 40^\circ$, see Section 7.2.2.2). Consequently, the second unit behaves as a perfect mirror for incidence angles verifying $x > x_0$. The key point here is the possibility of combining both kinds of units in order to construct efficient optical microcavities by a judicious choice of the incidence angle geometry. In the main frame of Fig.8.5 the dependence of the transmission coefficient with the incidence angle θ for the optical device just described is shown. The subunits are composed of SiO_2 (TiO_2), whose indices of refraction (at 700 nm) are $n_A = 1.45$ and $n_B = 2.30$, respectively. From this figure one sees that a resonant cavity to work at specific incidence angles could be constructed by making a structure where a Fibonacci multilayer exhibiting full transmission at the incidence angle $x \simeq 0.67 \simeq 55°$ is sandwiched between two Fibonacci multilayers behaving as perfect mirrors at this angle. It is also worth noting that by properly selecting the refraction indices of the layers' materials, it would be possible to achieve broad multidirectional reflection devices based on hybrid order structures.

The basic principle inspiring the combination of conjugated quasiperiodic structures can be readily extended to consider other kinds of aperiodic sequences as well. In fact, the possibility of obtaining a selective confinement of certain normal modes in composite systems based on the combination of Fibonacci/Fibonacci*, Thue-Morse/Thue-Morse*, and Rudin-Shapiro/Rudin-Shapiro* stacks has been recently reported (the symbol * here indicates the conjugation operation $A^+ = B$ and $B^+ = A$).[10] The multilayers are described by nearest-neighbor force constants (see Section 5.4) and the corresponding masses particularized to aluminum and silver layers (respectively containing two or three Al or Ag atoms each). Since the Thue-Morse sequence is invariant under the conjugation transformation by construction (see Section 4.2.1) one realizes that the energy spectrum of the Thue-Morse/Thue-Morse* multilayer does not exhibit any new feature, simply reducing to that corresponding to a higher order generation one. On the contrary, Fibonacci and Rudin-Shapiro based structures exhibit differences in the frequency spectrum (e.g., reduction and partial closing of some primary and secondary gaps) as compared to the original (i.e., not combined) multilayer. The most remarkable feature is the presence of spatial confinement of the atom displacements in just one of the sequences forming the composite structure for certain frequency ranges. This selective confinement of the atomic vibrations is achieved due to the interplay of different aperiodic orders (original and conjugated one) at different scales, and may be useful, at least in principle, for filtering and guiding systems.[10] It could also be exploited to enhance the local interaction of electromagnetic (Kerr effect) or stress (piezoelectric effect) fields in non-linear media.

Thue-Morse sequence has mirror symmetry by construction, which most other substitutional sequences considered so far, as the Fibonacci, period-doubling, or Rudin-Shapiro ones, lack. Nevertheless, one can intentionally introduce an internal mirror symmetry in these sequences by

1. reversing the order of the letters in the original sequence (i.e., $ABAAB \rightarrow BAABA$), and

2. concatenating the original sequence to the reversed one to obtain the string of letters $ABAAB|BAABA$.[11]

It is clear that the resulting sequence has mirror symmetry with respect to the back slash plane by design. In so doing, the structure is endowed with a nested series of dimer-like positional correlations among the letters, starting from the mirror symmetry plane (i.e., $A\{B[A\{A[B/B]A\}A]B\}A$). This structural correlation, in turn, will induce a series of resonant transmission effects (in close analogy to the resonance effects earlier considered in the study of the random dimer model, see Section 1.7). This interesting feature was used to manipulate the resonant transmission properties at specific wavelengths by designing and growing a symmetric Fibonacci/Fibonacci^{-1} multilayer (the symbol $^{-1}$ here indicates the letters reversal operation) based on TiO$_2$ and SiO$_2$ dielectric layers. The measured transmission spectra indicated the potential applications of this kind of aperiodic structures in multiwavelength narrow band filters and wavelength division multiplexing devices.[12]

The vibrational properties of aperiodic systems with artificially imposed mirror symmetry were numerically studied in the case Fibonacci/Fibonacci^{-1}, Thue-Morse/Thue-Morse^{-1}, and Rudin-Shapiro/Rudin-Shapiro^{-1} heterostructures based on aluminum and silver layers.[13] A number of localized modes (absent in the original aperiodic systems) were found in the wide primary gaps and near the band edges of the Fibonacci based mirror structures. These modes only appear when even order Fibonacci generations are considered, as they appear due to the formation of a BB dimer at the mirror plane in that case. Since the presence of BB dimers is forbidden in Fibonacci sequences, its presence can be properly regarded as a structural defect, breaking the ideal quasiperiodicity of the Fibonacci sequence. Accordingly, this defect dimer gives rise to the presence of some isolated modes in gap regions of the frequency spectrum. In the Rudin-Shapiro based structures, analogous localized modes near the band edges were also found, whereas in the Thue-Morse based structures no such features are found, as expected. From a practical viewpoint, the existence of localized modes in the gaps could be used for filter and guiding purposes.[13]

In summary, making use of certain formal operations (letter conjugation, letters reversal), followed by concatenation of the obtained sequences (regarded as structural building blocks), one can exploit additional symmetries in the resulting aperiodic structures. The combination of these systems among them, as well as with periodic ones, would result in a next generation of structurally complex systems based on aperiodic orderings. The possible emergence of unexpected physical properties of possible technological interest will probably deserve further attention on these systems from researchers working in the field in the years to come.

8.2.3   Distorted quasiperiodicity

In Section 2.5 we learnt that quasiperiodic structures can be obtained by means of a suitable projection from periodic arrangements defined in a higher dimensional hyperspace. One can then imagine a further generalization by considering the projection of a quasiperiodic lattice according to a certain

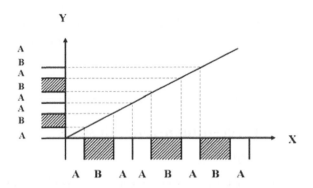

FIGURE 8.6
Sketch showing the principle of introducing a distortion into a one-dimensional quasiperiodic multilayer system by means of a projection scheme.

rule in order to obtain a *distorted quasiperiodic* structure. In so doing, one introduces some additional design parameters in the projected structure, which are directly related to the adopted projection rule. For the sake of illustration let us consider that the original quasiperiodic sequence is arranged along the abscissas of a Cartesian coordinate system, and the projected sequence is obtained along the ordinates (Fig.8.6). The generated structure is then obtained from the original by applying a projection rule of the form $y(x, k_i)$, where k_i are arbitrary parameters characteristic of the adopted rule. This approach was originally introduced to obtain a broad omnidirectional reflection band in one-dimensional photonic QCs based on the Fibonacci sequence. To this end, the power law $y = x^{1+k}$ was used, where k measures the distortion degree introduced to the original Fibonacci multilayer.[14] In this way, the thickness of each layer in the deformed Fibonacci system becomes variable and depends on the sequential order j of the considered layer and the deformation degree according to the expression

$$d'_j = d[j^{k+1} - (j-1)^{k+1}], \quad j \geq 1 \tag{8.1}$$

where d is the thickness of the considered layer in the original Fibonacci sequence. In the particular case of a dielectric multilayer the change in the layer thickness is accompanied by a change in its related optical thickness. This originates a modification of the optical properties in the multilayer which can be precisely controlled by a judicious choice of the projection parameter k. It is clear that this procedure can be immediately generalized to any arbitrary sequence in order to obtain distorted Thue-Morse, Rudin-Shapiro, or period-

doubling sequences for instance. On the other hand, for a given original
aperiodic sequence, one can properly vary the projection rule to tailor the
resulting distorted structure at will. The potential of this approach to design
useful orderings in multilayered devices is certainly extraordinary.

8.3 Two- and three-dimensional designs

8.3.1 Aperiodic squares

As we saw in Chapters 1 and 2, two-dimensional aperiodic tilings are collections of polygons capable of covering a plane with neither gaps nor overlaps in
such a way that the resulting overall pattern lacks any translational symmetry.
Recently, the possible effects of aperiodic order in the electromagnetic radiation properties of antenna arrays arranged according to quasiperiodic tiles
have been theoretically discussed, by considering how various tiling geometries affect directivity, sidelobe level, and bandwidth of the radiated fields.[15]
Quite interestingly, the experimental realization of a small antenna put on top
of a planar structure exhibiting octagonal symmetry has confirmed a very high
directivity at some frequencies as well as the suppression of the transverse electric surface waves at all frequencies on the surface of such a structure.[16] This
suppression can be understood as due to the fact that, in this aperiodic structure, the resonance frequencies of the squares and rhombuses building tiles are
incommensurate. Accordingly, there does not exist any frequency for which
the surface can support surface wave propagation and the overall transmission
of transverse waves is almost suppressed. A completely analogous result has
been recently reported for Penrose phononic crystals (see Section 7.3) made
of polymeric rods ($v_l = 1800$ ms^{-1}, $v_s = 800$ ms^{-1}, $\rho = 1.14$ kg m^{-3}) with
a pentagonal cross section submerged in water at filling fractions of 17%. In
this case, the reduction of the symmetry of the scattering object (from circular in the cylindrical rods to a polygonal one) can affect the symmetry of the
system modes. In particular, for scatters with lower symmetry the resonance
frequencies should change with the direction of the incident wave, probably
leading to wider gaps.[17]

Nonetheless, these applications still rely on well-known aperiodic tiles, such
as Archimedean tiles (see Section 1.7) or Penrose-related tiles (see Section
2.2.2), and do not introduce truly innovative approaches in the field. A first
step forward in the quest for new aperiodic designs comes from the introduction of the so-called square Fibonacci tiling, which was originally proposed
as an illustrative example of quasicrystal without forbidden symmetries.[18]
This square is constructed as follows. Let us consider two identical Fibonacci
grids, each one consisting of an infinite set of lines whose inter-line spacing
follows the Fibonacci sequence of short (unity) and long (τ) distances. By

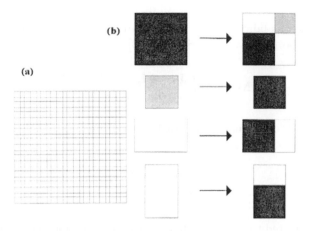

FIGURE 8.7

(a) The square Fibonacci tiling; (b) substitution rules for obtaining the square Fibonacci tiling. (Courtesy of Ron Lifshitz).

superimposing both grids at a right angle one gets a two-dimensional quasi-periodic tiling with tetragonal point group symmetry (Fig.8.7a). This tiling consists of three different tiles: a small square of dimensions 1×1, a large square of dimensions $\tau \times \tau$, and a rectangle of dimensions $1 \times \tau$. This tiling can also be generated by the cut-and-project method from a 4-dimensional space (see Section 2.5) or using the substitution rules illustrated in Fig.8.7b. This construction can be straightforwardly generalized to three dimensions to obtain the cube Fibonacci tiling. It can also be generalized to any quasiperiodic sequence as well.

Although these structures are unlikely to occur spontaneously in Nature, they can be used in the design of aperiodic based devices for optical or acoustical applications, as we saw in Chapter 7. From this perspective the knowledge of their most relevant mathematical properties is worthwhile. The main advantage of these structures (as compared to more realistic structural models for QCs, for instance) is that they are separable, namely electron or phonon problems can be decomposed into separated one-dimensional problems, so that possible dimensionality effects can be more easily addressed. In this way, the nature of both the energy spectra and eigenfunctions of the square Fibonacci lattice has been studied. From the obtained results it has been conjectured that a transition from singular continuous to absolutely continuous spectra may occur in these Fibonacci based two-dimensional lattices in going from one to two dimensions, leading to the emergence of extended wave functions.[19]

The physical properties of photonic aperiodic crystals based on the Thue-

Morse sequence have been recently explored as well. A two-dimensional Thue-Morse grating was fabricated by single beam computer-generated holography according to the recursive rule given by[20]

$$\mathbf{M}_{n+1} = \begin{pmatrix} \mathbf{J}_n - \mathbf{M}_n & \mathbf{M}_n \\ \mathbf{M}_n & \mathbf{J}_n - \mathbf{M}_n \end{pmatrix}, \qquad (8.2)$$

where $\mathbf{J}_n$ is a $2^n \times 2^n$ matrix in which each element is equal to one, and the initial matrices to start iteration are

$$\mathbf{M}_1 = \begin{pmatrix} 1 & 0 \\ 0 & 1 \end{pmatrix}, \quad \mathbf{J}_1 = \begin{pmatrix} 1 & 1 \\ 1 & 1 \end{pmatrix}. \qquad (8.3)$$

Note that $\mathbf{J}_1$ reduces to the Thue-Morse substitution sequence (see Table 4.3), and $\mathbf{M}_1$ is the identity matrix. Making use of Eq.(8.3) into Eq.(8.2) one gets

$$\mathbf{M}_2 = \begin{pmatrix} 0 & 1 & 1 & 0 \\ 1 & 0 & 0 & 1 \\ 1 & 0 & 0 & 1 \\ 0 & 1 & 1 & 0 \end{pmatrix}, \quad \mathbf{M}_3 = \begin{pmatrix} 1 & 0 & 0 & 1 & 0 & 1 & 1 & 0 \\ 0 & 1 & 1 & 0 & 1 & 0 & 0 & 1 \\ 0 & 1 & 1 & 0 & 1 & 0 & 0 & 1 \\ 1 & 0 & 0 & 1 & 0 & 1 & 1 & 0 \\ 0 & 1 & 1 & 0 & 1 & 0 & 0 & 1 \\ 1 & 0 & 0 & 1 & 0 & 1 & 1 & 0 \\ 1 & 0 & 0 & 1 & 0 & 1 & 1 & 0 \\ 0 & 1 & 1 & 0 & 1 & 0 & 0 & 1 \end{pmatrix}, \; \dots \qquad (8.4)$$

It is easily verified that any row and any column of the symmetric $\mathbf{M}_n$ matrix is a Thue-Morse sequence of order n. The photonic properties of this square Thue-Morse lattice, with dielectric cylinders ($\epsilon = 12.25$) located at the "0" sites, was theoretically analyzed confirming that the self-similarity and long-range correlations of this structure give rise to the emergence of a rich bandgap structure as well as to the presence of localized light wave states.[21] On the other hand, studies on the optical behavior of two-dimensional Fibonacci based lattices fabricated by electron-beam lithography on transparent quartz substrates reported on the presence of nearly localized plasmon modes whose exact location can be accurately predicted from purely structural considerations. In this way, the quasiperiodicity of gold nanoparticles distribution on the substrate has a significant impact for the design and fabrication of novel nano-plasmonic devices.[22] In this case, the two-dimensional quasiperiodic lattice was obtained from a seed letter A or B by applying two complementary Fibonacci sequence substitution rules $g_A : A \to AB, B \to A$ and $g_B : A \to B, B \to BA$ (depending on whether the first element encountered in the letter matrix expansion is A or B) along the horizontal and the vertical directions, alternatively. For the sake of illustration the first generations of the two-dimensional Fibonacci structures are shown below in matrix form

$$A \to AB$$

$$\begin{pmatrix} A & B \\ B & A \end{pmatrix} \to \begin{pmatrix} A & B & A \\ B & A & B \end{pmatrix}$$

$$\begin{pmatrix} A & B & A \\ B & A & B \\ A & B & A \end{pmatrix} \to \begin{pmatrix} A & B & A & A & B \\ B & A & B & B & A \\ A & B & A & A & B \end{pmatrix} \cdots .$$

Although the resulting structure does not coincide with that of the square Fibonacci lattice introduced above, both the direct and reciprocal lattices obtained agree very well in both cases. Following this approach two-dimensional generalizations of both Thue-Morse and Rudin-Shapiro sequences have been implemented to design metallic nanoparticles arrays of interest for plasmon nanodetectors.[23]

8.3.2   Spherical stacks and optical lattices

In Section 7.2 we presented several instances of systems based on the aperiodic stacking of many layers. A key feature of those systems is that every layer is geometrically characterized by its thickness along the growth direction (it has also a cross section, of course, but its value is physically irrelevant). This is not the case if one considers non-planar geometries (i.e., cylindrical or spherical ones) for the layers, as it is shown in Fig.8.8. In fact, the preservation of the energy flux in a solid angle along the radial direction requires the consideration of the external (r_{i+1}) and internal (r_i) boundaries of each layer, so that one must consider the order number of the layer, i, to properly express its thickness $d_i = r_{i+1} - r_i$. This is a direct consequence of the existence of a preferential point in the system (its center) which breaks the homogeneity present in the multilayers based on planar slabs. This means that, in general, the properties of wave oscillations depend on the place of a layer with respect to the center of the spherical cavity.

The mathematical treatment of wave propagation through spherical multilayers is similar to that described in Section 7.2 by expressing the Helmholtz Eq.(7.2) in terms of a scalar function called the Debye potential $\Pi(r, \theta, \varphi)$ as

$$\frac{d^2\Pi}{dr^2} + \left[\frac{\omega^2}{c^2} n^2(r) - \frac{l(l+1)}{r^2} \right] \Pi = 0, \tag{8.5}$$

where n is the refractive index, ω is the angular frequency, and l is the angular momentum.[25] Eq.(8.5) can be solved in terms of the spherical Hankel functions and the propagation of the wave can be described in terms of a series of transfer matrices containing these functions. Nevertheless, in the spherical multilayer case these transfer matrices are no longer unimodular, since their

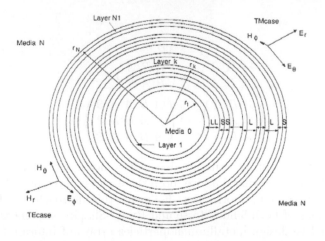

FIGURE 8.8

Geometry of a spherical multilayered system. A stack of multilayers arranged according to the Fibonacci sequence is deposited on the surface of a microsphere. (From ref.[24]. With permission from Elsevier.)

determinant is given by the ratio $(r_{i+1}/r_i)^2 > 1$.[26] Another consequence of the spherical geometry is that the transfer matrix depends not only on the thickness of the layer, but also on the distance of the considered block to the center. As a result, the simple concatenation rule $S_{n+1} = S_n S_{n-1}$, giving the Fibonacci stacking sequence in homogeneous systems (see Section 4.2.1), must be replaced by the following one for the spherical Fibonacci stack,[24]

$$S_{n+1}(n+1) = S_n(n)S_{n-1}(F_{n+2}). \tag{8.6}$$

This result provides a clear illustration of the significant role of geometric three dimensional effects in the very nature of aperiodic order in multilayered systems.

As a final example of new designs based on aperiodic order in three dimensions we will briefly comment on the so-called optical lattices. In previous chapters we have found many examples where the presence of quasiperiodic long-range order in a lattice gives rise to the formation of sharp diffraction peaks due to interference processes coupling waves and matter. In optical lattices several monochromatic travelling waves are combined to create an interference pattern which exhibits long-range order. Then, such a light field is used to trap small particles or atoms inside it. In the resulting structure the atoms adopt a regular arrangement in space, but they are weakly coupled among them, at variance with the usual situation in conventional solid matter. In fact, atomic gases condensates in optical lattices do not present either defects or phonons, offering a powerful tool for investigating the quantum behavior of condensed matter systems under unique control possibilities. The symmetry of the optical potential created by the interference of lasers is completely determined by the geometric arrangement of their beams. Therefore, one can in principle design both periodic and quasiperiodic optical potentials at will. Following this approach, the formation of a three dimensional quasiperiodic optical lattice of Cs atoms by the interference of five or six laser beams was reported some time ago.[27] Quite interestingly, a physical realization of the geometrical cut-and-project method (see Section 2.5.1) was proposed for creating a Fibonacci optical lattice. To this end, four laser beams build a two-dimensional square lattice by their interference. A fifth laser beam is then used to drive Rb atoms along the projection direction at an appropriate angle (see Fig.2.12). By properly arranging the on-site energies of these atoms one would obtain a physical realization of a Fibonacci lattice of atoms.[28] It is then clear that optical quasiperiodic lattices offer dramatic possibilities for designing a wide range of geometrical arrays able to confine Bose-Einstein condensates of both fundamental and practical interest.[29] In this way the notion of aperiodic order spreads over different domains of Condensed Matter, as previous concepts originally restricted to periodic arrangements are properly generalized.

8.4 Beyond quasiperiodic order: Aperiodicity by design

At the beginning it was periodic thinking. Afterwards, quasiperiodic order was discovered; first in mathematical kingdom, subsequently -and partially by chance- in physical sciences as well. From the study of quasiperiodic systems it was progressively realized that the domains of ordered matter could be expanded more and more. First steps in aperiodic condensed matter were mainly guided by previous mathematical knowledge about quasiperiodic functions along with some results obtained from the study of the structural and physical properties of quasicrystals and aperiodic superlattices. In this way, a variety of self-similar systems, some of them exhibiting long-range order based on inflation symmetries and/or substitution rules, other ones exhibiting both inflation and translation symmetries, like fractals, were systematically investigated.

Certainly, there exist many other ways to arrange a series of different layers in an ordered aperiodic way other than the recourse to substitution sequences. For instance, one may consider a multilayer where each layer is labeled by a natural number in such a way that A layers correspond to prime numbers. Since the number of primes progressively decays as one goes to larger integers (the mean density of primes is approximately given by $1/\ln p$) the relative frequency of A layers vanishes as the system grows longer. Therefore, A layers can be regarded as some sort of diluted impurity layers. Nevertheless, an interesting property of primes is that they do not come infrequently in pairs (called twin primes) like the couples (11,13) or (17,19), for instance. In this way, twin primes introduce a short-range correlation among the A layers, which is completely analogous to that characteristic of the random dimer model (see Section 1.7). As a consequence, one should expect that the presence of these correlations gives rise to the emergence of delocalized states in this sort of multilayers.[30]

As the field is coming to reach a certain maturity, and some relationships between spatial structure and physical properties are progressively disclosed, a logical extension naturally appears: it consists in considering systems whose building blocks are intentionally arranged according to certain aperiodic designs in order to tailor their physical properties to achieve some specified requirement. In this case, the guiding principle is not the previous knowledge about a sequence with some well-known mathematical (or even merely aesthetical) properties well deserving to be explored, but a purely technological one. One has in mind a requirement which must be fulfilled, and tries to find out the best arrangement of matter able to fit it. We will refer to the systems obtained in this way as *aperiodic systems by design*. In so doing, appealing aperiodic orderings are created by hand, generally following suitable optimization algorithms. Such a procedure essentially differs from the recursion methods usually considered in the study of aperiodic systems based on the application of substitution rules or self-similar inflation symmetries, thereby substantially enriching the variety of aperiodic orders of interest in Condensed Matter Physics.

A first example of aperiodic systems by design was spurred by the interest in designing quasiperiodic structures able to simultaneously phase match any two non-linear interactions. To this end, a suitable algorithm was used to determine an aperiodic modulation of the non-linear coefficient in ferroelectric devices.[31] This procedure is an example of what is generally known as a reverse engineering problem. Thus, the optimized structures are found by numerically solving an optimization problem where a number of optimization parameters (for instance, the layer thicknesses or the assignment of positive and negative domains in a ferroelectric multilayer) are treated as independent variables subjected to certain physical limitations (e.g., the existence of a thickness small limit for the layers). The merit function to be optimized is defined in terms of the physical magnitude of interest in the considered device. For instance, if one is interested in optimizing the reflectance $R(\lambda)$ of a mirror or a coating over a given spectral range one may use a function of the form

$$f = \frac{1}{N} \sum_{j=1}^{N} [R_0 - R(\lambda_j)]^2 \,, \tag{8.7}$$

where N is the total number of layers and λ the radiation wavelength. Following this approach a number of Mo/Si aperiodic multilayers were designed and grown in order to attain a maximum reflectivity over a specified range of wavelengths or angles of incidence for XUV radiation for their use as mirrors,[32] or analyzers,[33] (as we discussed in Section 7.2.1, Fibonacci multilayers exhibited very promising reflecting properties). Similar approaches have been used to design efficient aperiodic W/Si multilayer mirrors for x-ray plasma diagnostic,[34] narrowband optical filters able to simultaneously tune multiple wavelengths,[35] organic light-emitting diodes with an improved brightness output,[36] or enhanced frequency converters for continuous-wave tunable lasers.[37]

As we learnt in Section 2.6, the discovery of quasicrystalline alloys led to the introduction of a more general definition of crystals, where the focus moves from the physical space to the reciprocal space. We have also seen that the most promising applications of aperiodic multilayers have been demonstrated in the field of optics, where self-similarity provides plentiful reciprocal states for quasi-phase matching in non-linear frequency conversion, or the existence of forbidden, higher order symmetries leads to more isotropic Brillouin surfaces for the design of omnidirectional band gaps. Accordingly, it seems a convenient procedure to move from the physical space to the reciprocal one when trying to search for novel designs of practical interest.[38] This very promising approach has been recently explored, and a general theory to identify those aperiodic structures having most useful band properties has been introduced by means of a Fourier-space based inverse optimization algorithm. In this way, one can manipulate the Fourier components of an aperiodic system to design and engineer any desirable band gap and field localization characteristic.[39]

In addition to its practical interest, this approach properly accounts for a very fundamental feature of quasiperiodic systems. In fact, periodic systems have a unique scale length which is determined by the characteristic period of their unit cell. On the contrary, both fractals and quasiperiodic structures are more precisely defined in an asymptotic way. These asymptotic processes involve longer and longer scales in physical space, which in turn give rise to a rich, nested structure in reciprocal space, hence lacking a well defined scale length. A nice illustration of this rich pattern in reciprocal space is shown in Fig.2.8. As we discussed in previous chapters, the characteristic distribution of nested peaks in reciprocal space, arranged according to a never ending (at least in principle) self-similar pattern, is at the root of a number of both intriguing and useful physical properties of aperiodic structures, like the extremely small thermal conductivity of quasicrystalline alloys or the high symmetry of the Brillouin zone of photonic QCs (see Figure 7.21b). In this way, one may think of a new field for future technological research focusing on the smaller possible features a system possesses in reciprocal space, rather than in physical space. Such a topic research (which we could tentatively refer to as *nano-reciprocal-space science*) closely follows the sequence of theoretical events which were progressively disclosed as the novel paradigm of order without periodicity attained its present status, and I guess it could profitably guide the next generation of Condensed Matter physicists and Materials Science researchers alike, far beyond the still narrow frontiers of knowledge imposed by our current limited understanding on the role of aperiodic order in Science and Technology.

8.5   Some open questions

Our journey is now about completion. I sincerely hope the reading of the preceding pages has provided the reader with some intellectual thrust, spurring his/her interest in giving a closer look on the possible implication of aperiodic order in his/her own research field. To this end, I would like to conclude by listing some appealing topics which, in my opinion, well deserve a deeper consideration in the years to come.

- **What is a crystal?**

At the time being we are still lacking a final definition for that notion. Currently, we only have a provisional definition, distinguishing periodic crystals from aperiodic ones. In addition, in the aperiodic crystals class, one can distinguish among incommensurate composites, modulated phases, and quasiperiodic crystals. So far, we have found just four different symmetry classes among quasicrystals to date, namely, icosahedral, decagonal, dodecagonal, and octogonal ones. Is that all?

Certainly, one can think of many more possible arrangements of matter giving rise to purely discrete diffraction patterns and exhibiting beautiful and subtle spatial symmetries, as compared with those formerly allowed by the, now old-fashioned, crystallographical restriction theorem.[40, 41] Some hints in that direction are provided by the current quest for novel dielectric structures and metamaterials based composites of interest for optical applications and optoelectronics. I suspect that advances in technological innovation will give us a plethora of ingenious complex designs of matter structures during the next decades, most of them based on algorithmic rules far beyond the relatively simple substitution sequences which played the main characters in the aperiodic stage at early times. If so, how will we denominate them?

It is likely that the genealogical tree of aperiodic order will open new branches, and elder ones will fan out in a fractal-like manner in order to accommodate new structures in a coherent fashion. For instance, some time ago the global structure of B_6O oxide was deciphered as a hierarchical packing of icosahedral boron clusters (see Section 1.4).[42] More recently, the x-ray patterns of cubic aperiodic phases observed in melt-quenched Mg-Al and FeNbBSi alloys have been described in terms of a periodic packing of truncated tetrahedra and Frank-Kasper polyhedra hierarchically arranged.[43] Inspired by the example provided by the fruitful recourse to hyperspace crystallography in order to encompass most basic notions of classical crystallography within a unified theoretical framework based on group theory and projection matrices, it is quite possible that the very notion of *hypercrystals* must be then introduced to keep trace of the existence of an increasing hierarchy of complex orderings of matter containing the previous simpler ones (namely, periodic and quasiperiodic crystals!) in a unified way. Thus, the higher dimensional description has recently proved useful in order to describe a *phase transformation* in an incommensurate molecular crystal that essentially affects the interaction between the two composing structures rather than the individual spatial structures themselves.[44] It is then reasonable to expect that, in contrast with periodic materials, many more types of phase transitions may be considered in crystallographic systems belonging to the hyperspace realm.

- ## Organic Quasicrystals?

Closely related to the question regarding the possible existence of quasiperiodic crystals exhibiting new axial symmetries (v.g., 7-fold or 9-fold,[45]) is the question about the very nature of chemical bond in currently known QCs. As we know, the first representatives of quasiperiodic crystals class are alloys composed of atoms belonging to the metallic families in the Periodic Table. Nevertheless, the notion of aperiodic crystal was originally proposed in the context of the debate about the nature of genetic material. Accordingly, there is no fundamental reason preventing the spontaneous emergence of long-range aperiodic order in non-metal based systems. In fact, the occurrence of a quasicrystalline decagonal phase of silicon has been reported,[46] and the possible existence of icosahedral quantum dots containing several hundreds of covalently bounded Si atoms has been theoretically predicted from first principle calculations.[47] In this regard, the possible existence of quasicrystals based on an aperiodic stacking of C_{60} fullerenes has also been explored, with no success so far.[48, 49] On the other hand, the discovery of liquid crystals with quasiperiodic symmetries, in which the basic construction units of the lattice are supramolecular aggregates instead of individual atoms, has been reported and proposed for photonic applications.[50, 51, 52, 53] Therefore, plenty of room is left for the search of new kinds of quasiperiodic arrangements of matter in modern Materials Science.

- ## What about clusters?

A central problem in Condensed Matter Physics is to determine whether quasiperiodicity leads to new physical properties which are significantly different from those of crystalline or amorphous materials. In this context, what physical properties should be expected from a solid mainly consisting of clusters? The possibility of obtaining a semiconducting material by assembling clusters composed of metallic atoms is of both theoretical and practical importance. The question regarding whether or not QCs are typical representatives of cluster-based solids and whether or not their formation, stability, and unusual properties can be explained by employing a cluster-based approach remains open in the field.[54] For instance, one may speculate that quasiperiodicity may give rise to a novel kind of "recurrently" localized covalent bond, characterized by an almost periodic modulation over large-scales (mimicking the delocalized nature of usual metallic bonding) modulating a series of more or less localized bumps at the typical scale of cluster structures (mimicking the oriented geometries typical of covalent bonding networks).

- **Fibonacci genomics?**

As we have mentioned perviously, two kinds of order coexist in DNA, each one related to two separate subsystems in the DNA helix, namely the nucleotide subsystem and the backbone system. Thus, one has periodic structural order at the atomic scale in the sugar-phosphate backbone, which yields discrete Bragg spots in x-ray diffraction patterns, exhibiting a characteristic cross-shaped pattern. On the other hand, one has aperiodic, informative chemical order at the molecular scale, as determined by the base pairs sequence. The chemical order of the bps sequence can be properly characterised by ab-initio quantum chemistry calculations, which nicely highlight the emergence of molecular orbitals beyond the atomic scale. Therefore, from a structural point of view DNA could be classified as a periodic crystal (with helical symmetry!), whereas the nucleobase system electronic structure, defining most basic properties of DNA molecule, is effectively aperiodic. Accordingly, one may think of DNA as a sort of hybrid order system exhibiting both aperiodic (stemming from the nucleobase subsystem) and periodic (corresponding to the sugar-phosphate backbone helix subsystem) order features in its electronic structure. As a consequence, one may then think of probing aperiodic order in DNA (that is reading the bps sequence!) by means of purely physical (as opposed to current chemical based) techniques.[55] In particular, could spectroscopic techniques or transport measurements be able to unveil the aperiodic sequence of bps?

Current bioengineering techniques allow for the growth of oligonucleotide sequences tailored at will. Thus, a nucleic acid arranged according to the Fibonacci sequence, for instance a (G:C)(A:T)(G:C)(G:C)(A:T)(G:C)(A:T)... oligonucleotide, may be easily synthesized. Such a molecule, where complementary Watson-Crick base pairs play the role of *A*s and *B*s in the Fibonacci's substitutional rule, could be properly referred to as a *DNA quasicrystal* approximant. Quite interestingly, the presence of short segments of the Fibonacci sequence has been reported in several biological DNA chains.[56] Along with this result some curious numerical relationships can be obtained by factorizing the total number of base pairs in a given genome sequence. In Table 8.1 we list the total number of base pairs, N, found in the genome of several organisms and their corresponding factorization in prime numbers.

From the data listed in this table we realize that the genome length can be usually expressed as a product involving both *small* and *large* prime factors. Although the biological relevance (if any) of this result is not clear,

TABLE 8.1

Number of base pairs in several genomes and their prime factors decomposition.

ORGANISM	N	PRIME DECOMPOSITION	REF.
Human chromosome 21	$33,127,944$	$2^3 \times 3 \times 72,649$	[57]
Chimpanzee chromosome 22	$32,799,845$	$5 \times 13 \times 53 \times 9,521$	[57]
E. Coli	$4,639,221$	$3^3 \times 171,823$	[58]
Mycobacterium tuberculosis	$4,411,529$	$1,471 \times 2,999$	[59]
Haemophilus influenzae	$1,830,137$	$19 \times 96,323$	[60]
Methanococus jannaschii	$1,739,933$	$13 \times 17 \times 7,873$	[61]
Helicobacter pylori	$1,667,867$	$1,667,867$	[62]
Treponema pallidum	$1,138,006$	$2 \times 569,003$	[63]
Mycoplasma genitalum	$580,070$	$2 \times 5 \times 19 \times 43 \times 71$	[64]

these findings certainly highlight the convenience of gaining a deeper insight into the mathematical structure of the genome design. To this end, suitable mathematical approaches have been recently introduced in order to clarify the evolution of repetitive DNA strings,[65, 66] within the framework of genomic evolution by duplication.[67]

Last but not least, a number of studies have recently reported that DNA based thin films may be used as promising materials for fabricating photonic and optoelectronic devices.[68] Accordingly, a better understanding of the physical bases governing optical transitions in DNA macromolecules[69] appears as a very appealing research topic in the aperiodic systems field.

References

[1] Maciá E 2006 *Rep. Prog. Phys.* **69** 397

[2] Maciá E 1998 *Appl. Phys. Lett.* **73** 3330

[3] Maciá E 2001 *Phys. Rev B* **63** 205421

[4] Dong J W, Han P, and Wang H Z 2003 *Chin. Phys. Lett.* **20** 1963

[5] Kumar Singh S, Thapa K B, and Ojha S P 2007 *Optoelectronics Adv. Mater.-Rapid Commun.* **1** 49

[6] Kanzari M and Rezig B 2001 *J. Opt. A: Pure Appl. Opt.* **3** S201

[7] Montalbán A, Velasco V R, Tutor J, and Fernández-Velicia F J 2004 *Phys. Rev B* **70** 132301

[8] Chen J, Qin B, and Chan H L W 2008 *Solid State Commun.* **146** 491

[9] Montalbán A, Velasco V R, Tutor J, and Fernández-Velicia F J 2007 *Surf. Sci.* **601** 2538

[10] Montalbán A, Velasco V R, Tutor J, and Fernández-Velicia F J 2007 *Physica B* **387** 36

[11] Huang X Q, Jiang S S, Peng R W, and Hu A 2001 *Phys. Rev B* **63** 245104

[12] Peng R W, Huang X Q, Qiu F, Wang M, Hu A, Jiang S S, and Mazzer M 2002 *Appl. Phys. Lett.* **80** 3063; Peng R W, Liu Y M, Huang X Q, Qiu F, Wang M, Hu A, Jiang S S, Feng D, Ouyamg L Z, and Zou J 2004 *Phys. Rev B* **69** 165109

[13] Montalbán A, Velasco V R, Tutor J, and Fernández-Velicia F J 2005 *Surf. Sci.* **594** 174

[14] Ben Abdelaziz K, Zaghdoudi J, Kanzari M, and Rezig B 2005 *J. Opt. A: Pure Appl. Opt.* **7** 544

[15] Pierro V, Galdi V, Castaldi G, Pinto I M, and Felsen L B 2005 *IEEE Trans. Antennas Propag.* **53** 635; Galdi V, Castaldi G, Pierro V, Pinto I M, and Felsen L B 2005 *IEEE Trans. Antennas Propag.* **53** 2044

[16] Li H, Hang Z, Qin Y, Wei Z, Zhou L, Zhang Y, Chen H, and Chan C T 2005 *Appl. Phys. Lett.* **86** 121108

[17] Sutter-Widmer D, Neves P, Itten P, Sainidou R, and Steurer W 2008 *Appl. Phys. Lett.* **92** 073308

[18] Lifshitz R 2002 *J. Alloys and Compounds* **342** 186

[19] Ilan R, Liberty E, Even-Dar Mandel S, and Lifshitz R 2004 *Ferroelectrics* **305** 15; Even-Dar Mandel S and Lifshitz R 2006 *Phil. Mag.* **86** 759

[20] Zito G, Piccirillo B, Santamato E, Marino A, Tkachenko V, and Abbate G 2008 *Optics Express* **16** 5164

[21] Moretti L and Mocella V 2007 *Optics Express* **15** 15314

[22] Dallapiccola R, Gopinath A, Stellacci F, and Dal Negro L 2008 *Optics Express* **16** 5544

[23] Dal Negro L, Feng N N, and Gopinath A 2008 *J. Opt. A: Pure Appl. Opt.* **10** 064013

[24] Burlak G N and Díaz-de-Anda A 2008 *Opt. Commun.* **281** 181

[25] Chew W 1996 *Waves and Fields in Inhomogeneous Media* (IEEE Press, New York)

[26] Burlak K G, Koshevaya S, and Sánchez-Mondragón J 2000 *Opt. Commun.* **180** 49

[27] Guidoni L, Triché C, Verkek P, and Grynberg G 1997 *Phys. Rev. Lett.* **79** 3363

[28] Eksioglu Y, Vignolo P, and Tosi M P 2004 *Opt. Commun.* **243** 175

[29] Sánchez-Palencia L and Santos L 2005 *Phys. Rev. A* **72** 053607

[30] Ryu C S, Kim I M, Oh G Y, and Lee M H 1994 *Phys. Rev B* **49** 14991

[31] Fradkin-Kashi K and Arie A 1999 *IEEE J. Quantum Electron.* **35** 1649; Fradkin-Kashi K, Arie A, Urenski P, and Rosenman G 2002 *Phys. Rev. Lett.* **88** 023903

[32] Levashov V E, Mednikov K N, Pirozhkov A S, and Ragozin E N 2006 *Radiation Phys. Chem.* **75** 1819

[33] Wang H, Zhu J, Wang Z, Zhang Z, Zhang S, Wu W, Chen L, Michette A G, Powell A K, Pfauntsch S J, Schäfers F, and Gaupp A 2006 *Thin Solid Films* **515** 2523

[34] Champeaux J Ph, Troussel Ph, Villier B, Vidadl V, Khachroum T, Vidal B, and Krumrey M 2007 *Nuclear Instr. Methods Phys. Res. A* **581** 687

[35] Gu X, Chen X, Chen Y, Zeng X, Xia Y, and Chan Y 2004 *Opt. Commun.* **237** 53

[36] Agrawal M, Sun Y, Forrest S R, and Peumans P 2007 *Appl. Phys. Lett.* **90** 241112

[37] Capmany J, Pereda J A, Bermúdez V, Callejo D, and Diéguez E 2001 *Appl. Phys. Lett.* **79** 1751

[38] García-Moliner F 2005 *Microelectron. J* **36** 870

[39] Chakraborty S, Hasko D G, and Mears R J 2004 *Microelec. Eng.* **73/74** 392; Chakraborty S, Parker M C, and Mears R J 2005 *Photonics and Nanostructures Fundam. Appl.* **3** 139

[40] Steurer W and Deloudi S 2008 *Acta Cryst.* **A64** 1

[41] Senechal M 2007 *Z. Kristallogr.* **222** 311; Lifshitz R 2007 *Z. Kristallogr.* **222** 313; Zimmermann H 2007 *Z. Kristallogr.* **222** 318

[42] Haubert H, Devouard B, Garvie L A J, O'Keefe M O, Buseck P R, Petuskey W T, and McMillan P F 1998 *Nature (London)* **391** 376

[43] Kraposhin V S, Talis A L, and Thanh Lam H 2008 *J. Phys.: Condens. Matter* **20** 114115

[44] Toudic B, García P, Odin C, Rabiller Ph, Ecolivet C, Collet E, Bourges Ph, Mclntyre G J, Hollingsworth M D, and Breczewski T 2008 *Science* **319** 69

[45] Gogotsi Y, Libera J A, Kalashnikov N, and Yoshimura M 2000 *Science* **290** 317

[46] Kamalakaran R, Singh A K, and Srivastava O N 2000 *Phys. Rev. B* **61** 12686

[47] Zhao Y, Kim Y H, Du M H, and Zhang S B 2004 *Phys. Rev. Lett.* **93** 015502

[48] Fleming R M, Kortan A R, Siegrist T, Thiel F A, Marsh P, Haddon R C, Tycko R, Dabbagh G, Kaplan M L, and Mujsce A M 1991 *Phys. Rev. B* **44** 888

[49] Michaud F, Barrio M, Toscani S, López D O, Tamarit J Ll, Agafonov V, Szwarc H, and Céolin R 1998 *Phys. Rev. B* **57** 10351

[50] Zeng X, Ungar G, Liu Y, Pereec V, Dulcey A E, and Hobbs J K 2004 *Nature (London)* **428** 157

[51] Zeng X 2005 *Current Opinion in Colloid & Interface Sci.* **9** 384

[52] Gorkhali S P, Qi J, and Crawford G P 2005 *Appl. Phys. Lett.* **86** 011110

[53] Lifshitz R and Diamant H 2007 *Phil. Mag.* **87** 3021

[54] Steurer W 2006 *Phil. Mag.* **86** 1105; Henley C L, de Boissieu M, and Steurer W 2006 *Phil. Mag.* **86** 1131

[55] Zwolak M and Di Ventra M 2008 *Rev. Mod. Phys.* **80** 140

[56] Mezquinta J et al. 1985 *J. Mol. Evol.* **21** 209

[57] International Chimpanzee Chromosome 22 Consortium 2004 *Nature (London)* **429** 382

[58] Blattner F R et al. 1997 *Science* **277** 1453

[59] Cole S T et al. 1998 *Nature (London)* **393** 537

[60] Fleischmann R D et al. 1995 *Science* **269** 496

[61] Bult C J et al. 1996 *Science* **273** 1058

[62] Tomb J F et al. 1997 *Nature (London)* **388** 539

[63] Fraser C M et al. 1998 *Science* **281** 375

[64] Fraser C M et al. 1995 *Science* **270** 397

[65] Dress A, Giegerich R, Grünewald S, and Wagner H 2003 *Annals of Combinatorics* **7** 259

[66] Beleza Yamaguchi M E and Shimabukuro A I 2008 *Bull. Math. Biol.* **70** 643

[67] Ohno S 1984 *J. Mol. Evol.* **20** 313

[68] Steckl A J 2007 *Nature Photonics* **1** 3

[69] Díaz E, Malyshev A V, and Domínguez-Adame F 2007 *Phys. Rev. B* **76** 205117

9

Mathematical tools

9.1 Almost periodic and quasiperiodic functions

From a mathematical viewpoint quasiperiodic functions are a special case of *almost periodic functions*. The theory of these functions was developed by Harald Bohr (1887-1951, brother of the well-known physicist Niels Bohr).[1] A function $f(\mathbf{x})$ is called almost periodic if for any arbitrary small number $\varepsilon > 0$, there are almost-periods $\mathbf{P}$ such that the shifted function differs less than ε from the unshifted one, namely, $|f(\mathbf{x}) - f(\mathbf{x} + \mathbf{P})| < \varepsilon$, for all $\mathbf{x} \in \mathbb{R}^n$.[2, 3] In general, the smaller the value of ε, the larger becomes the required translation P, although they are relatively dense in $\mathbb{R}^n$. By this we mean that there are, for each ε, values R_1 and R_2 such that every sphere of radius R_2 contains at least one $\mathbf{P}$ satisfying the above condition, and in every sphere of radius R_1 around any translation $\mathbf{P}$ satisfying the condition, there is no other translation but $\mathbf{P}$.

Almost periodic functions can be uniformly approximated by Fourier series containing a countable infinity of pairwise incommensurate frequencies. When the set of frequencies required can be generated from a finite-dimensional basis, the resulting function is referred to as a *quasiperiodic* one. Let us consider a quasiperiodic function given by its discrete Fourier decomposition

$$f(\mathbf{x}) = \sum_{\mathbf{k}} \tilde{f}(\mathbf{k}) e^{i\mathbf{k} \cdot \mathbf{x}}, \tag{9.1}$$

where the reciprocal vectors are defined by

$$\mathbf{k} = \sum_{j=1}^{N} n_j \mathbf{b}_j. \tag{9.2}$$

If the minimal number of basis vectors $\mathbf{b}_j$ is larger than three (i.e., $N > 3$), then a higher dimensional description (see Section 2.5) is needed to describe the reciprocal lattice, and the related structure is an aperiodic crystal. Otherwise, we obtain a periodic crystal, which indicates that, in turn, periodic functions are just a particular case of quasiperiodic ones.

As an illustrative counter-example we can consider the function

$$f(x) = \lim_{N \to \infty} \sum_{n=1}^{N} \frac{1}{n!} \sin\left(\frac{x}{2^n}\right), \tag{9.3}$$

which is almost periodic but not quasiperiodic, since its Fourier spectrum does not have a finite basis. The simplest one-dimensional example of a quasiperiodic function can be written as

$$f(x) = \cos(x) + \cos(\alpha x), \tag{9.4}$$

where α is an irrational number. It is interesting to note that this quasiperiodic function can be obtained as the one-dimensional projection of a related *periodic function* in two dimensions

$$f(x, y) = \cos x + \cos y, \tag{9.5}$$

through the restriction $y = \alpha x$. This property is the basis of the so-called cut and project method, which is widely used in the study of quasiperiodic crystals (see Section 2.5). In fact, since any quasiperiodic function can be thought of as deriving from a periodic function in a space of higher dimension, most of the basic notions of classical crystallography can be properly extended to the study of quasicrystals in appropriate hyper-spaces.[4, 5]

9.2 Transfer matrix technique

Many problems of physical interest are described by linear ordinary second-order differential equations for which different types of transfer matrices can be introduced.[6, 7, 8, 9] In this book we will focus on the description of elementary excitations in low dimensional systems which can be reduced, as a first approximation, to the study of one-dimensional lattice models defined by the following general equation[10, 11, 12, 13]

$$v_n \phi_n = t_{n,n-1}\phi_{n-1} + t_{n,n+1}\phi_{n+1}, \tag{9.6}$$

along with an appropriate set of boundary conditions. In Eq.(9.6), ϕ_n is the amplitude of the elementary excitation at the nth lattice position, and v_n depends on the excitation energy (frequency), E (or ω), as well as on other characteristic physical magnitudes of the system, like atomic masses m_n, elastic constants $K_{n,n+1}$, or electronic binding energies ε_n, as it is illustrated in Table 9.1. The transfer integrals $t_{n,n\pm1}$ describe the excitation transfer from site n to its neighboring sites $n \pm 1$ (hence $t_{n,n\pm1} = t_{n\pm1,n}$), and will

TABLE 9.1
Values adopted by the different coefficients appearing
in the general Eq.(9.6) depending on the considered
elementary excitation.

Parameters	Electrons	Phonons
ϕ_n	ψ_n	u_n
v_n	$E - \varepsilon_n$	$K_{n,n-1} + K_{n,n+1} - m_n\omega^2$
$t_{n,n\pm1}$	$t_{n,n\pm1}$	$K_{n,n\pm1}$

generally depend on the excitation energy. It is convenient to cast Eq.(9.6) in
the following matrix form

$$\begin{pmatrix} \phi_{n+1} \\ \phi_n \end{pmatrix} \equiv \mathbf{T}_n \begin{pmatrix} \phi_n \\ \phi_{n-1} \end{pmatrix}, \tag{9.7}$$

where

$$\mathbf{T}_n(E) \equiv \begin{pmatrix} \frac{v_n}{t_{n,n+1}} & -\frac{t_{n,n-1}}{t_{n,n+1}} \\ 1 & 0 \end{pmatrix} \tag{9.8}$$

is the so-called transfer matrix at site n. Let N be the number of sites of the
considered lattice. By iterating Eq.(9.7) N times one obtains

$$\begin{pmatrix} \phi_{N+1} \\ \phi_N \end{pmatrix} \equiv \mathcal{M}_N \begin{pmatrix} \phi_1 \\ \phi_0 \end{pmatrix}, \tag{9.9}$$

where ϕ_1 and ϕ_0 define the initial conditions, and the global transfer matrix
is defined by the product

$$\mathcal{M}_N(E) = \prod_{n=N}^{1} \mathbf{T}_n(E). \tag{9.10}$$

From the knowledge of the $\mathcal{M}_N(E)$ matrix elements, several magnitudes of
physical interest, like the density of states (see Section 9.5.1), the transmission
coefficient (see Section 9.5.2), or the localization length (see Section 9.5.4),
can be readily evaluated. In this way, the transfer matrix formalism provides
a simple mathematical tool allowing for a unified treatment of such diverse
problems as electron and phonon dynamics in both periodic and aperiodic lattices (Chapter 5), charge transport through DNA chains (Chapter 6), optical
properties of dielectric multilayers (Chapter 7), the propagation of acoustic
waves in semiconductor heterostructures and metallic superlattices,[14, 15, 16]
or localization of elastic waves in heterogeneous media.[17]

In order to explicitly evaluate the global transfer matrix one usually must
calculate some power matrices. To this end, the recourse to the Cayley-
Hamilton theorem (named after the mathematicians Arthur Cayley (1821-
1895) and William Rowan Hamilton (1805-1865)) has proved very useful. To

start with, we recall that the characteristic polynomial of a $n \times n$ matrix $\mathbf{M}$ is given by the equation $\det(\mathbf{M} - \lambda \mathbf{I}) = 0$, where $\mathbf{I}$ is the identity matrix and λ are the matrix eigenvalues. For 2×2, 3×3, and 4×4 matrices this general condition respectively reads

$$\lambda^2 - a_1 \lambda + \det \mathbf{M} = 0, \tag{9.11}$$

$$\lambda^3 - a_1 \lambda^2 + \lambda a_2 - \det \mathbf{M} = 0, \tag{9.12}$$

$$\lambda^4 - a_1 \lambda^3 + a_2 \lambda^2 - \lambda a_3 + \det \mathbf{M} = 0, \tag{9.13}$$

where $a_2 = [(\mathrm{tr}\mathbf{M})^2 - \mathrm{tr}(\mathbf{M}^2)]/2$, $a_3 = [(\mathrm{tr}\mathbf{M})^3 - 3\mathrm{tr}(\mathbf{M}^2)\mathrm{tr}\mathbf{M} + 2\mathrm{tr}(\mathbf{M}^3)]/6$, and $a_1 = \mathrm{tr}\mathbf{M}$ stands for the trace of matrix $\mathbf{M}$. According to the Cayley-Hamilton theorem any $n \times n$ square matrix over the real or complex field is a root of its own characteristic polynomial,[18] so that Eq.(9.11) can be written in the matrix form

$$\mathbf{M}^2 - 2z\mathbf{M} + \mathbf{I}\det \mathbf{M} = \mathbf{0}, \tag{9.14}$$

where $z \equiv \frac{1}{2}\mathrm{tr}\mathbf{M}$ is usually referred to as the semi-trace of matrix $\mathbf{M}$. Analogous expressions are derived from Eqs.(9.12) and (9.13) for higher dimensional 3×3 and 4×4 matrices. If $\mathbf{M}$ belongs to the Sl(2,$\mathbb{R}$) group (i.e., $\det \mathbf{M} = 1$, so that $\mathbf{M}$ is referred to as an unimodular matrix), one can readily exploit these matrix equations in order to properly express any higher power of $\mathbf{M}$ as a linear combination of matrices $\mathbf{I}$ and $\mathbf{M}$ itself. For instance, in the case of 2×2 unimodular matrices one can make use of Eq.(9.14) to obtain by induction the expression

$$\mathbf{M}^N = U_{N-1}(z)\mathbf{M} - U_{N-2}(z)\mathbf{I}, \tag{9.15}$$

where

$$U_N \equiv \frac{\sin(N+1)\varphi}{\sin \varphi}, \tag{9.16}$$

with $\varphi \equiv \cos^{-1} z$, are Chebyshev polynomials of the second kind satisfying the recursion relation

$$U_{n+1} - 2zU_n + U_{n-1} = 0, \quad n \geq 1 \tag{9.17}$$

with $U_0(z) = 1$ and $U_1(z) = 2z$. In the more general case, when $\det \mathbf{M} = \Delta \neq 1$, one gets[19]

$$\mathbf{M}^N = \Delta^{(N-1)/2}U_{N-1}(\tilde{z})\mathbf{M} - \Delta^{N/2}U_{N-2}(\tilde{z})\mathbf{I}, \tag{9.18}$$

where $\tilde{z} \equiv \frac{1}{2}\Delta^{-1/2}\mathrm{tr}\mathbf{M}$. Eq.(9.18) properly reduces to Eq.(9.15) when $\Delta = 1$. The above results can be generalized to obtain arbitrary powers of any dimensional square matrix by using polynomials of many arguments, which generalize the Chebyshev polynomials given by Eq.(9.16).[19]

In a series of works devoted to the study of periodic superlattices it was reported that Chebyshev polynomials play, for *finite* systems, a similar role to the one played by Bloch functions in the description of transport properties of infinite periodic systems.[20] In addition, the ability of these polynomials to properly describe the propagation of both quantum and classical waves in *locally periodic* media (namely, systems having only a relatively small number of repeating elements) in a compact way has been recently illustrated,[7] as well as the convenience of their use when describing the presence of extended states in *correlated random* systems.[21, 22] Chebyshev polynomials are also very useful to perform numerical calculations in a broad collection of one-dimensional quasiperiodic systems. In particular, closed analytical expressions for several diagnostic tools (see Section 9.5) can be derived in terms of Chebyshev polynomials (see Chapters 5, 6, and 7).

The dispersion relation of periodic approximants of quasiperiodic lattices can be obtained by imposing the cyclic boundary conditions $\phi_{N+1} = e^{iqNa_0}\phi_1$ and $\phi_0 = e^{-iqNa_0}\phi_N$ to Eq.(9.9), where q is the wave number and a_0 is the lattice constant. The motion equation then reads

$$\begin{pmatrix} e^{iqNa_0}\phi_1 \\ \phi_N \end{pmatrix} \equiv \begin{pmatrix} M_{11} & M_{12} \\ M_{21} & M_{22} \end{pmatrix} \begin{pmatrix} \phi_1 \\ e^{-iqNa_0}\phi_N \end{pmatrix}, \tag{9.19}$$

where M_{ij} are the elements of the global transfer matrix $\mathcal{M}_N(E)$. Equation (9.19) leads to the system

$$e^{iqNa_0}\phi_1 = M_{11}\phi_1 + M_{12}e^{-iqNa_0}\phi_N \tag{9.20}$$

$$\phi_N = M_{21}\phi_1 + M_{22}e^{-iqNa_0}\phi_N, \tag{9.21}$$

and solving for ϕ_N in Eq.(9.21) and plugging the obtained expression in Eq.(9.20) one gets

$$e^{iqNa_0} + e^{-iqNa_0}\det\mathcal{M}_N = \text{tr}\mathcal{M}_N. \tag{9.22}$$

Eq.(9.22) significantly simplifies when $\mathcal{M}_N$ is unimodular, in which case one obtains the dispersion relation

$$\cos(qNa_0) = \frac{1}{2}\text{tr}\mathcal{M}_N(E). \tag{9.23}$$

Alternatively, one can derive Eq.(9.22) by adopting the value $\lambda = e^{iqNa_0}$ for the eigenvalues appearing in the characteristic polynomial given by Eq.(9.11). This approach can be extended to consider ladder quantum models described in terms of 4×4 transfer matrices (see Section 6.4.2). In that case, by assuming periodic boundary conditions, the motion equation reads

$$\begin{pmatrix} \lambda\psi_{1,A} \\ \psi_{0,A} \\ \lambda\psi_{1,B} \\ \psi_{0,B} \end{pmatrix} = \begin{pmatrix} M_{11} & M_{12} & M_{13} & M_{14} \\ M_{21} & M_{22} & M_{23} & M_{24} \\ M_{31} & M_{32} & M_{33} & M_{34} \\ M_{41} & M_{42} & M_{43} & M_{44} \end{pmatrix} \begin{pmatrix} \psi_{1,A} \\ \lambda^{-1}\psi_{0,A} \\ \psi_{1,B} \\ \lambda^{-1}\psi_{0,B} \end{pmatrix}, \tag{9.24}$$

where M_{ij} are the elements of the global 4×4 transfer matrix $\mathcal{M}_N(E)$. Making use of Eq.(9.13) its characteristic polynomial can be expressed as

$$\lambda^4 - a_1\lambda^3 + a_2\lambda^2 - a_3\lambda + a_4 = 0, \tag{9.25}$$

where $a_4 = \det \mathcal{M}_N$. Now, if $\lambda = e^{iqNa_0}$ is a solution of the characteristic polynomial describing a Bloch wave propagation towards the right through the ladder, then the wave propagating with reversed momentum $-q$ is also a solution represented by λ^{-1}.[23, 24, 25] Accordingly, λ^{-1} must also be an eigenvalue of Eq.(9.25) so that,

$$\lambda^{-4} - a_1\lambda^{-3} + a_2\lambda^{-2} - a_3\lambda^{-1} + a_4 = 0. \tag{9.26}$$

Multiplying Eq.(9.26) by λ^4 and comparing the resulting polynomial with (9.25) we get the additional conditions $a_1 = a_3$, and $a_4 = 1$, so that $\mathcal{M}_N$ is unimodular too. Making use of these results and dividing Eq.(9.25) by λ^2 we can reduce the original quartic equation to the following quadratic one:

$$u^2 - a_1 u + a_2 - 2 = 0, \tag{9.27}$$

where $u \equiv \lambda + \lambda^{-1} = 2\cos(qNa_0)$. Finally, by solving Eq.(9.27) we obtain the dispersion relation,[24]

$$\cos(qNa_0) = \frac{1}{4}\left[\operatorname{tr}\mathcal{M}_N \pm \sqrt{2\operatorname{tr}(\mathcal{M}_N^2) - (\operatorname{tr}\mathcal{M}_N)^2 + 8}\right], \tag{9.28}$$

which properly generalizes Eq.(9.23) for 4×4 matrices. The double sign indicates that we have two branches describing the bonding (antibonding) states between the two parallel chains composing the ladder.

We note that Eqs.(9.25) and (9.26) impose the condition $\mathcal{M}_N^\dagger \mathbf{J}\mathcal{M}_N$ on the matrix $\mathcal{M}_N$, where $\mathcal{M}_N^\dagger$ stands for the Hermitian conjugate and

$$\mathbf{J} = \begin{pmatrix} 0 & -1 & 0 & 0 \\ 1 & 0 & 0 & 0 \\ 0 & 0 & 0 & -1 \\ 0 & 0 & 1 & 0 \end{pmatrix}. \tag{9.29}$$

This property characterizes the so-called symplectic property of $\mathcal{M}_N$.[24, 26, 27]

9.3   Trace map formalism

In the previous Section we have seen that the semi-trace plays an important role in the study of dynamical systems characterized by unimodular transfer matrices. The physical reason why these traces are so important is due to its presence in the dispersion relations [Eqs.(9.23) and (9.28)] which, in turn, determine the structure of the energy spectrum. This property can be fruitfully exploited by means of the so-called trace map formalism, which was introduced by Mahito Kohmoto on the basis of the following theorem.

Consider a set of matrices $\mathbf{M}_n$ belonging to the Sl(2,$\mathbb{R}$) group and satisfying the concatenation rule $\mathbf{M}_{n+1} = \mathbf{M}_{n-1}\mathbf{M}_n$, then[28]

$$\mathrm{tr}\mathbf{M}_{n+1} = \mathrm{tr}\mathbf{M}_n \mathrm{tr}\mathbf{M}_{n-1} - \mathrm{tr}\mathbf{M}_{n-2}, \quad n \geq 2. \tag{9.30}$$

By defining $z_n \equiv \mathrm{tr}\mathbf{M}_n/2$, Eq.(9.30) is rewritten as the dynamical map

$$z_{n+1} = 2z_n z_{n-1} - z_{n-2}, \quad n \geq 2, \tag{9.31}$$

usually referred to as the trace map. This map has the constant of motion[28]

$$I = z_{-1}^2 + z_0^2 + z_1^2 - 2z_{-1}z_0 z_1 - 1, \tag{9.32}$$

determined by the initial conditions $z_{-1} = \mathrm{tr}M_0/2$, $z_0 = \mathrm{tr}M_1/2$, and $z_1 = \mathrm{tr}M_2/2$. The trace map formalism has been extended to other aperiodic binary chains, such as those based in the Thue-Morse sequence, in which case it reads[29, 30]

$$z_{n+1} = 4z_{n-1}^2(z_n - 1) + 1, \quad n \geq 2. \tag{9.33}$$

The trace map formalism was also generalized to the case of ternary and quaternary quasiperiodic lattices,[31] as well as to the case of continuous potentials,[32] and generalized Fibonacci sequences.[33] More recently, the trace map formalism has been further extended in order to describe Fibonacci superlattices as well (see Chapter 4). The key feature of these aperiodic structures is the coexistence of two kinds of order in the same sample at different length scales as it is illustrated in Fig.9.1. As a first approximation, the physical description is substantially simplified by assuming all the force constants to be equal, i.e., $K_A = K_B = K_{AB} \equiv K$. Within the transfer matrix formalism the dynamic response of a Fibonacci superlattice composed of N layers can then be expressed in terms of the global transfer matrix

$$\mathcal{M}_N = ...\mathbf{L}_B\mathbf{L}_A\mathbf{L}_A\mathbf{L}_B\mathbf{L}_A, \tag{9.34}$$

where the *layer matrices*

$$\mathbf{L}_A \equiv \mathbf{Q}_A^{n_A} = \begin{pmatrix} 2-\Omega & -1 \\ 1 & 0 \end{pmatrix}^{n_A}, \quad \mathbf{L}_B \equiv \mathbf{Q}_B^{n_B} = \begin{pmatrix} 2-\alpha\Omega & -1 \\ 1 & 0 \end{pmatrix}^{n_B} \tag{9.35}$$

describe the phonon propagation through layers A and B as a product of *atomic matrices* $\mathbf{Q}_A$ and $\mathbf{Q}_B$, respectively. The atomic matrices are characterized by the normalized frequency $\Omega \equiv m_A\omega^2/K$, where ω is the phonon frequency, the mass ratio $\alpha \equiv m_B/m_A$, and the number of atoms in each layer $n_{A(B)}$, respectively. In a Fibonacci superlattice the layers are arranged

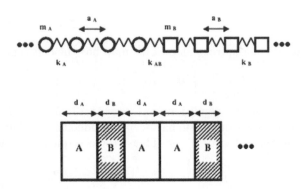

FIGURE 9.1

Sketch illustrating the hierarchical arrangement of a Fibonacci heterostructure. At the atomic scale (top frame) the system can be modelled as a lattice chain composed of two kinds of atoms m_A and m_B coupled via force constants K_A (layer A), K_B (layer B), and K_{AB} (interfaces). At a larger scale (bottom frame) the system is described in terms of a sequence of layers of different composition and width $d_A = n_A a_A$ and $d_B = n_B a_B$, respectively, where n_ν is the number of atoms composing the layer and a_ν is its lattice constant. ([34] Reprinted figure with permission from Maciá E 2006 *Phys. Rev. B* **73** 184303 © 2006 by the American Physical Society.)

according to the Fibonacci sequence $ABAAB...$, which determines their order of appearance in the matrix product given by Eq.(9.34). Making use of the Cayley-Hamilton theorem the power matrices given by Eq.(9.35) can be readily expressed as[34]

$$\mathbf{L}_r = \begin{pmatrix} U_{n_r}(x_r) & -U_{n_r-1}(x_r) \\ U_{n_r-1}(x_r) & -U_{n_r-2}(x_r) \end{pmatrix}, \quad r = \{A, B\} \tag{9.36}$$

where $U_{n_r}(x_r)$ are Chebyshev polynomials of the second kind, $x_r \equiv tr\mathbf{Q}_r/2$, and we have explicitly used Eq.(9.17). Since $\mathbf{L}_r$ is a product of Sl(2,$\mathbb{R}$) group elements, the layer matrices are unimodular themselves, and one can exploit this fact in order to extend Kohmoto's theorem, originally introduced to describe a Fibonacci lattice of atoms, to the Fibonacci superlattice case. In fact, since $\det(\mathbf{L}_{v_r}) = 1$, the set of *superlattice transfer matrices* $\mathcal{M}_N$ given by Eq.(9.34) satisfies the conditions of that theorem as well. Accordingly, the dynamical map given by Eq.(9.31) can be properly applied to the Fibonacci superlattice,[35] provided the initial conditions

$$z_{-1} = \frac{1}{2}tr\mathbf{L}_B = T_{n_B}(x_B), \tag{9.37}$$

$$z_0 = \frac{1}{2}tr\mathbf{L}_A = T_{n_A}(x_A), \tag{9.38}$$

$$z_1 = \frac{1}{2}tr(\mathbf{L}_B\mathbf{L}_A) = T_{n_A}(x_A)T_{n_B}(x_B) + (x_Ax_B - 1)U_{n_A-1}(x_A)U_{n_B-1}(x_B), \tag{9.39}$$

where $T_{n_r}(x_r) = \cos[n_r \cos^{-1}(x_r)]$ are Chebyshev polynomials of the first kind, and we have used the relationship $U_n - U_{n-2} = 2T_n$. Making use of Eqs.(9.37)-(9.39) the invariant of the dynamical map is now given by

$$I = (x_A - x_B)^2 \, U_{n_A-1}^2(x_A)U_{n_B-1}^2(x_B). \tag{9.40}$$

In this way, the trace map formalism can be extended to discuss the phonon propagation through a Fibonacci superlattice characterized by the presence of two relevant physical scales. In fact, the dynamical map given by Eqs.(9.31) and (9.37-9.39) can be physically interpreted as follows. By equating $z_{-1} = T_{n_B}(x_B) \equiv \cos(qd_B)$ and $z_0 = T_{n_A}(x_A) \equiv \cos(qd_A)$, where q is the wave vector, we readily obtain the dispersion relation corresponding to the A or B layer

$$\omega^2 = \frac{4K}{m_{A(B)}} \sin^2\left(\frac{qa_{A(B)}}{2}\right), \tag{9.41}$$

respectively. Analogously, the equation $z_1 \equiv \cos[q(d_A + d_B)]$ leads to the dispersion relation corresponding to the binary periodic superlattice with unit cell AB.[36, 37] Accordingly, the initial conditions implementing the generalized trace map are directly related to the phonon dispersion relations corresponding to the constituent layers (z_{-1} and z_0) and the lowest order periodic approximant to the Fibonacci superlattice (z_1). Consequently, the expression

$$z_n \equiv \cos(qD), \quad (n \geq 2),$$ (9.42)

with $D = F_n d_A + F_{n-1} d_B$, can be properly regarded as the dispersion relation corresponding to successive Fibonacci superlattice approximants obtained from a continued iteration of the trace map.[38]

The phonon spectrum of the Fibonacci superlattice can then be obtained as the asymptotic limit of a series of approximants whose dispersion relations are determined by the successive application of the trace map recursion relation given by Eq.(9.31). In so doing, one realizes that the dispersion relation of a Fibonacci superlattice approximant can be generally split into two complementary contributions. The first one describes a periodic binary lattice, where the layers alternate in the form $ABABABA....$ The other one includes the effects related to the emergence of quasiperiodic order in the system.[34] A similar splitting was discussed in order to describe the general structure of Fibonacci quasicrystals, where it was shown that their quasilattice can be seen as an average periodic structure plus quasiperiodic fluctuations.[39]

In order to properly describe the phonon dynamics in finite Fibonacci heterostructures where two kinds of order (periodic at the atomic scale and quasiperiodic beyond the layer scale) are present in the same sample at different scale lengths, it is convenient to express the trace map in terms of nested Chebyshev polynomials of the form $T_{F_\nu}[T_{n_r}(x_r)]$ and $U_{F_\nu-1}[T_{n_r}(x_r)]$, where the variable x_r describes the atomic scale physics and the function $T_{n_r}(x_r)$ describes the dynamics at the layer scale. This representation can be regarded as describing a *scale transformation,* formally expressed as

$$T_{n_r}(x_r) \rightarrow X_r.$$ (9.43)

Since the trace map itself can be interpreted as giving the dispersion relation of a given Fibonacci superlattice realization in terms of the dispersion relations corresponding to lower order approximants, this nested structure provides a suitable unified description of the phonon dynamics in Fibonacci superlattices, able to encompass their characteristic hierarchical structure in a natural way. By applying this transformation to Eqs.(9.37)-(9.39) the trace map initial conditions now read[34]

$$z_{-1} = X_B,$$
$$z_0 = X_A,$$
$$z_1 = X_A X_B + \sqrt{I + Y_A^2 Y_B^2},$$ (9.44)

where $Y_r = \sqrt{1 - X_r^2}$, and we have made use of the constant of motion I given by Eq.(9.40). This result provides a direct link between the topological self-similarity of these quasiperiodic heterostructures and the dynamics of the elementary excitations propagating through them. In this regard, the emergence of specific features related to the quasiperiodic order imposed to the heterostructure can be properly described in terms of the scale transformation given by Eq.(9.43). The inclusion of interface effects renders a much more involved mathematical description, although it does not significantly affect the underlying physics of the approach.[34]

9.4 Renormalization matrix methods

The formalism we are going to introduce is based on the transfer matrix technique discussed in Section 9.2. Real-space renormalization group approaches, based on decimation schemes, have proved themselves very successful in order to study the energy spectrum of aperiodic systems based on the application of substitution rules.[40, 41, 42, 43, 44, 45] The convenience of such procedures stems from the fact that, by properly decimating the original chain into successively longer blocks, one is able to describe the elementary excitation state corresponding to sites more and more farther apart. In order to fully exploit the inflation/deflation symmetry characteristic of most aperiodic systems considered in this book it is convenient to renormalize *the set of transfer matrices* instead of the *lattice itself.* Since these matrices contain all the relevant information concerning the dynamics of the elementary excitations, this approach becomes specially well suited to deal with those characteristic features stemming from the long-range order of the underlying aperiodic system for, as we will see below, it preserves the original quasiperiodic order of the lattice at any stage of the renormalization process.

For the sake of illustration, let us first consider the electronic problem in the general Fibonacci lattice given by Eq.(5.3), in which both diagonal and off-diagonal terms are present in the Hamiltonian. The corresponding motion equation can be cast in terms of the following matrices,[44]

$$X \equiv \begin{pmatrix} \frac{E-\varepsilon_B}{t_{AB}} & -1 \\ 1 & 0 \end{pmatrix}, \quad Y \equiv \begin{pmatrix} \gamma^{-1} \frac{E-\varepsilon_A}{t_{AB}} & -\gamma^{-1} \\ 1 & 0 \end{pmatrix},$$

$$Z \equiv \begin{pmatrix} \frac{E-\varepsilon_A}{t_{AB}} & -\gamma \\ 1 & 0 \end{pmatrix}, \quad W \equiv \begin{pmatrix} \frac{E-\varepsilon_A}{t_{AB}} & -1 \\ 1 & 0 \end{pmatrix}, \quad (9.45)$$

relating three consecutive sites (ABA, AAB, BAA, and BAB) along the lattice, where E is the electron energy, ε_A and ε_B are the on-site energies, $t_{AB} = t_{DA}$ and t_{AA} are the corresponding transfer integrals and $\gamma \equiv t_{AA}/t_{AB} > 0$. In the study of the phonon problem given by Eq.(5.2) the corresponding transfer matrices are given by

$$X \equiv \begin{pmatrix} 2 - \alpha\lambda & -1 \\ 1 & 0 \end{pmatrix}, \quad Y \equiv \begin{pmatrix} 1 + \gamma^{-1}(1-\lambda) & -\gamma^{-1} \\ 1 & 0 \end{pmatrix},$$

$$Z \equiv \begin{pmatrix} 1+\gamma-\lambda & -\gamma \\ 1 & 0 \end{pmatrix}, \qquad W \equiv \begin{pmatrix} 2-\lambda & -1 \\ 1 & 0 \end{pmatrix}, \tag{9.46}$$

where $\alpha \equiv m_B/m_A$, $\gamma \equiv K_{AA}/K_{AB}$, and $\lambda \equiv m_A \omega^2/K_{AB}$.

The global transfer matrix given by Eq.(9.10) then *translates* the atomic sequence $ABAAB\ldots$ describing the topological order of the Fibonacci lattice to the transfer matrix sequence $\ldots\mathbf{XZYXZYXWXZYXW}$ describing the behavior of electrons moving through it. In spite of its greater apparent complexity, we realize that by renormalizing this transfer matrix sequence according to the blocking scheme $\mathbf{R}_A \equiv \mathbf{ZYX}$ and $\mathbf{R}_B \equiv \mathbf{WX}$, we get the considerably simplified sequence $\ldots\mathbf{R}_B\mathbf{R}_A\mathbf{R}_A\mathbf{R}_B\mathbf{R}_A$.[44] The subscripts in the $\mathbf{R}$s matrices are introduced to emphasize the fact that the renormalized transfer matrix sequence is also arranged according to the Fibonacci one and, consequently, the *topological order* present in the original lattice is *preserved* by the renormalization process. Let $N = F_n$ be the number of lattice sites, where F_n is a Fibonacci number (see Section 2.4). It can then be shown by induction that the renormalized sequence contains $n_A \equiv F_{n-3}$ matrices R_A and $n_B \equiv F_{n-4}$ matrices R_B.

During the last two decades a significant amount of work has been devoted to the study of systems based on two simple kinds of transfer matrices, namely, the so-called on-site and transfer models. In the on-site model one assumes all the transfer integrals to be equal, so that $\mathbf{T}_n$ becomes unimodular at every site of the chain. In the transfer model, all the on-site energies are assumed to be identical (and usually set to zero), so that we have $\det(\mathbf{T}_n) = -t_{n,n-1}/t_{n,n+1} \neq 1$, in this case. From a physical point of view, one expects that the value of the transfer integral, coupling two neighbor atoms in the lattice, will be determined by the chemical nature of these atoms which, in turn, define certain distribution of on-site energies along the chain. Therefore, in most physical situations of interest, one must consider the so-called mixed models, where both the on-site energies and the transfer integrals explicitly appear in Eq.(9.8). In that case, one must usually deal with non-unimodular transfer matrices at every lattice site.

Nevertheless, as we have seen in the previous Section, unimodular matrices belonging to the Sl(2,$\mathbb{C}$) group have a number of appealing mathematical properties, rendering the study of on-site models much more easy than the study of mixed ones. This fact has spurred the interest in searching for a suitable transformation, able to reduce the general motion equation (9.6) to the on-site form

$$u_n \varphi_n - t\varphi_{n-1} - t\varphi_{n+1} = 0, \tag{9.47}$$

where φ_n, u_n, and t are determined by the transformation rule. In this way, the dynamics of elementary excitations in the original mixed system could be described in terms of an effective on-site model, while still preserving its generality. Aiming at this goal, the local transformation $\phi_n \rightarrow \varphi_n/\lambda_n$ was introduced into Eq.(9.6).[46] This transformation can be physically interpreted as a rescaling of the elementary excitation amplitude at each lattice site, where the parameter $\lambda_n \in \mathbb{C}$ plays the role of a local scale factor, which is determined from the relationship $\lambda_{n,n\pm1}\lambda_n^* \equiv t_{n,n\pm1}^{-1}$. Making use of this transformation into Eq.(9.6) we obtain $v_n|\lambda_n|^2\varphi_n - \varphi_{n-1} - \varphi_{n+1} = 0$, which has the form of Eq.(9.47) with $t = 1$ and $u_n \equiv v_n|\lambda_n|^2$. Accordingly, both the the original on-site energies, ε_n, and the excitation energies E (included in the u_n coefficients) are affected by the transformation. Following a different approach a unitary transformation satisfying the condition $\mathbf{H} = \mathbf{U}\mathbf{H}_0\mathbf{U}$, where $\mathbf{H}$ and $\mathbf{H}_0$ are tridiagonal Hamiltonian matrices related to Eqs.(9.6) and (9.47), respectively, and $\mathbf{U}$ is a unitary diagonal matrix, was subsequently introduced.[47] In this case, the energies of the eigenstates are not affected by the transformation and the elements of the transformation matrix are recursively obtained from the knowledge of the original transfer integrals.

Inspired by these previous results, a local similarity transformation which acts at two different scale lengths has been recently introduced. At the atomic scale the transformation adopts the form

$$\mathbf{M}_n\mathbf{T}_n\mathbf{M}_n^{-1} = e^\lambda\tilde{\mathbf{T}}_n, \tag{9.48}$$

transforming a non-unimodular transfer matrix $\mathbf{T}_n$ into a unimodular one, $\tilde{\mathbf{T}}_n$, where the phase factor e^λ plays the role of an effective transfer term. Making use of this transformation, the global transfer matrix given by Eq.(9.10) can be expressed as a product of unimodular matrices, so that it becomes unimodular itself. When acting on larger scale lengths, corresponding to short segments of the original lattice, the main effect of this transformation is to map the original mixed lattice into an effective on-site lattice which is related to the original one by means of a renormalization process. Accordingly, the similarity transformation can be properly regarded as a renormalization operator acting on one-dimensional lattice models in this case.[48]

One starts making use of the decomposition property[49, 50]

$$\mathbf{T}_n \equiv \begin{pmatrix} \frac{v_n}{t_{n,n+1}} & -\frac{t_{n,n-1}}{t_{n,n+1}} \\ 1 & 0 \end{pmatrix} = \begin{pmatrix} t_{n,n+1}^{-1} & 0 \\ 0 & 1 \end{pmatrix}\begin{pmatrix} v_n & -1 \\ 1 & 0 \end{pmatrix}\begin{pmatrix} 1 & 0 \\ 0 & t_{n,n-1} \end{pmatrix}, \tag{9.49}$$

so that, by properly rearranging the product given in Eq.(9.10), the global transfer matrix can be expressed as

$$\mathcal{M}_N(E) = \mathbf{\Lambda}_{N+1}\left[\prod_{n=N}^{1}\mathbf{Q}_n\right]\mathbf{\Lambda}_0^{-1}, \tag{9.50}$$

where

$$\mathbf{Q}_n \equiv \begin{pmatrix} v_n t_{n,n-1}^{-1} & -t_{n,n-1} \\ t_{n,n-1}^{-1} & 0 \end{pmatrix} \tag{9.51}$$

are unimodular matrices and the boundary conditions are given by the non-unimodular matrices

$$\mathbf{\Lambda}_{N+1} \equiv \begin{pmatrix} t_{N,N+1}^{-1} & 0 \\ 0 & 1 \end{pmatrix}, \quad \mathbf{\Lambda}_0 \equiv \begin{pmatrix} t_{0,1}^{-1} & 0 \\ 0 & 1 \end{pmatrix}. \tag{9.52}$$

Then, one introduces a similarity transformation which acts locally in order to express every $\mathbf{Q}_n$ matrix in the form

$$\mathbf{Q}_n = \mathbf{M}_n^{-1} \mathbf{P}_n \mathbf{M}_n, \tag{9.53}$$

where

$$\mathbf{P}_n = \begin{pmatrix} v_n t_{n,n-1}^{-1} & -e^{-\lambda_n} \\ e^{\lambda_n} & 0 \end{pmatrix} \tag{9.54}$$

and

$$\mathbf{M}_n = \begin{pmatrix} e^{-\frac{\lambda_n}{2}} t_{n,n-1}^{-1/2} & 0 \\ 0 & e^{\frac{\lambda_n}{2}} t_{n,n-1}^{1/2} \end{pmatrix}, \tag{9.55}$$

where λ_n is a local phase factor which will be subsequently determined. By plugging Eq.(9.53) into Eq.(9.50) we get

$$\mathcal{M}_N(E) = \mathbf{\Lambda}_{N+1} \mathbf{M}_N^{-1} \left[\prod_{n=N}^{1} \mathbf{B}_n \right] \mathbf{M}_0 \mathbf{\Lambda}_0^{-1}, \tag{9.56}$$

where

$$\mathbf{B}_n = e^{\frac{\lambda_{n-1}+\lambda_n}{2}} \sqrt{\frac{t_{n-1,n-2}}{t_{n,n-1}}} \begin{pmatrix} \frac{v_n}{t_{n,n-1}} e^{-\lambda_n} & -\frac{t_{n,n-1}}{t_{n-1,n-2}} e^{-\lambda_n-\lambda_{n-1}} \\ 1 & 0 \end{pmatrix}. \tag{9.57}$$

At this point one exploits the degrees of freedom associated to the local phase factor λ_n in order to further simplify Eq.(9.56). To this end, we impose the condition

$$e^{\lambda_n+\lambda_{n-1}} \equiv t_{n,n-1}/t_{n-1,n-2}, \tag{9.58}$$

so that Eq.(9.57) reduces to

$$\mathbf{B}_n = \begin{pmatrix} u_n & -1 \\ 1 & 0 \end{pmatrix}, \tag{9.59}$$

which can be regarded as a unimodular matrix adopting the standard on-site model form with

$$u_n \equiv v_n e^{-\lambda_n} t_{n,n-1}^{-1}. \tag{9.60}$$

Now, making use of Eq.(9.59) into Eq.(9.56) the global transfer matrix can be expressed as

$$\mathcal{M}_N(E) = \mathbf{L}_{N+1} \left[\prod_{n=N}^{1} \mathbf{B}_n \right] \mathbf{L}_0, \tag{9.61}$$

where the boundary matrices are given by

$$\mathbf{L}_{N+1} \equiv \boldsymbol{\Lambda}_{N+1}\mathbf{M}_N^{-1} = e^{-\lambda_N/2}t_{N,N-1}^{-1/2} \begin{pmatrix} e^{-\lambda_{N+1}} & 0 \\ 0 & 1 \end{pmatrix}, \qquad (9.62)$$

$$\mathbf{L}_0 \equiv \mathbf{M}_0\boldsymbol{\Lambda}_0^{-1} = e^{\lambda_0/2}t_{0,-1}^{1/2} \begin{pmatrix} e^{\lambda_1} & 0 \\ 0 & 1 \end{pmatrix}, \qquad (9.63)$$

and we have made explicit use of Eq.(9.58). By adopting periodic boundary conditions we have $t_{0,-1} = t_{N,N-1}$, $\lambda_{N+1} = \lambda_1$, and $\lambda_0 = \lambda_N$, so that Eq.(9.61) adopts the simple form

$$\mathcal{M}_N(E) = \mathbf{L}_1^{-1} \left[\prod_{n=N}^{1} \mathbf{B}_n \right] \mathbf{L}_1, \qquad (9.64)$$

where

$$\mathbf{L}_1 \equiv \begin{pmatrix} e^{\lambda_1} & 0 \\ 0 & 1 \end{pmatrix}. \qquad (9.65)$$

From a physical viewpoint, Eq.(9.64) shows that the dynamics of elementary excitations in any arbitrary system, originally described by means of transfer matrices of the form given by Eq.(9.6), can be properly expressed in terms of an equivalent on-site model given by the transfer matrices set $\{\mathbf{B}_n\}$.

The local phase factors λ_n appearing in Eq.(9.60) exhibit a very remarkable feature, namely, their value at a given lattice site is determined by the values of all the transfer integrals which precede it along the chain. In fact, by taking logarithms in Eq.(9.58) and substituting successive terms into each other, one obtains

$$\lambda_n = (-1)^n \left[\ln\left(t_{N,N-1} \prod_{k=1}^{n-1} t_{k,k-1}^{2(-1)^k} \right) + \lambda_N \right] + \ln t_{n,n-1}, \quad n \geq 2, \quad (9.66)$$

along with the boundary relation

$$\lambda_1 = \ln t_{1,N} - \ln t_{N,N-1} - \lambda_N. \qquad (9.67)$$

According to Eq.(9.66), the value of the phase at a given site is a cumulative magnitude expressing the correlations among different transfer terms in an explicit form. Therefore, λ_n values will generally depend on the possible presence of long-range correlations in the system. This general treatment is valid for any arbitrary topological order of the lattice, as determined by the sequence of appearance of the different transfer matrices in Eq.(9.10). In order to illustrate the main features of this approach we will first consider the periodic, ternary mixed model corresponding to the unit cell ABC, which is illustrated in Fig.9.2. Taking $t_{AB} \equiv t$ as a reference value we can express $t_{BC} = bt$ and $t_{CA} = at$, without loss of generality. Accordingly, the boundary conditions read $t_{N,N-1} = t_{BC} = bt$, and $t_{N,1} \equiv t_{1,0} = t_{AC} = at$ in this

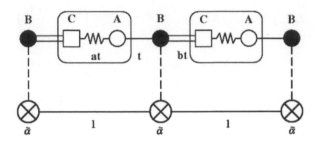

FIGURE 9.2

A mixed ternary lattice is mapped into an equivalent renormalized monatomic chain by decimating the AC dimers. ([48] Reprinted figure with permission from Maciá E and Rodríguez-Oliveros R 2006 *Phys. Rev. B* **74** 144202 © 2006 by the American Physical Society.)

case. The translation symmetry ensures we have just three kinds of sites v_A, v_B, and v_C, periodically arranged along the chain, satisfying $\lambda_A = \lambda_{1\pm3k}$, $\lambda_B = \lambda_{2\pm3k}$, and $\lambda_C = \lambda_{3\pm3k}$, with $k = 0, 1, \dots$. Making use of the condition $\lambda_4 = \lambda_1 \equiv \lambda_A$ into Eqs.(9.66) and (9.67) we obtain $\lambda_A = -\ln b$, $\lambda_B = \ln(b/a)$, and $\lambda_C = \ln a$, so that the transformed on-site energies given by Eq.(9.60) read

$$u_A = v_A \frac{b}{at}, \quad u_B = v_B \frac{a}{bt}, \quad u_C = v_C \frac{1}{abt}. \tag{9.68}$$

Making use of the Cayley-Hamilton theorem we obtain the power matrix

$$(\mathbf{B}_C\mathbf{B}_B\mathbf{B}_A)^m = \begin{pmatrix} U_m + u_B U_{m-1} & (1 - u_B u_C)U_{m-1} \\ (u_A u_B - 1)U_{m-1} & -u_B U_{m-1} - U_{m-2} \end{pmatrix}, \tag{9.69}$$

where $m \equiv N/3$, and $U_m(z)$ are Chebyshev polynomials of the second kind with

$$z \equiv \frac{v_A v_B v_C - t^2(b^2 v_A + a^2 v_B + v_C)}{2abt^3}. \tag{9.70}$$

Plugging Eq.(9.69) into Eq.(9.64) we obtain

$$\mathcal{M}_N(E) = \begin{pmatrix} U_m + u_B U_{m-1} & e^{-\lambda_1}(1 - u_B u_C)U_{m-1} \\ e^{\lambda_1}(u_A u_B - 1)U_{m-1} & -u_B U_{m-1} - U_{m-2} \end{pmatrix}, \tag{9.71}$$

so that we get $\mathrm{tr}[\mathcal{M}_N(E)]/2 = (U_m - U_{m-2})/2 = T_m(z)$, where $T_m(z)$ is a Chebyshev polynomial of the first kind. Then, making use of Eq.(9.23), we finally obtain the dispersion relation

$$2t_{AB}t_{BC}t_{CA}\cos(3qa_0) = u_A u_B u_C - u_A t_{BC}^2 - u_B t_{CA}^2 - u_C t_{AB}^2, \qquad (9.72)$$

which is invariant under cyclic permutations of the atoms in the unit cell. This procedure can be straightforwardly extended to obtain the dispersion relation of periodic lattices with arbitrary unit cells.[48]

We note that the mixed ternary lattice is characterized by three local transfer matrices, namely,

$$\mathbf{T}_A = \begin{pmatrix} \frac{v_A}{t} & -a \\ 1 & 0 \end{pmatrix}, \quad \mathbf{T}_B = \begin{pmatrix} \frac{v_B}{bt} & -b^{-1} \\ 1 & 0 \end{pmatrix}, \quad \mathbf{T}_C = \begin{pmatrix} \frac{v_C}{at} & -\frac{b}{a} \\ 1 & 0 \end{pmatrix}, \qquad (9.73)$$

none of which is unimodular. Quite remarkably, however, the product

$$\mathbf{Q} \equiv \mathbf{T}_C \mathbf{T}_B \mathbf{T}_A = \frac{1}{bt^2} \begin{pmatrix} \frac{1}{at}\left(v_A v_B v_C - b^2 t^2 v_A - t^2 v_C\right) b^2 t^2 - v_B v_C \\ v_A v_B - t^2 & -atv_B \end{pmatrix} \qquad (9.74)$$

belongs to the Sl(2,$\mathbb{C}$) group, for any choice of the system parameters and for any value of the elementary excitation energy. Accordingly, we can transform $\mathbf{Q}$ to the form $\mathbf{MQM}^{-1} = e^\lambda \tilde{\mathbf{T}}$,[48] where

$$\tilde{\mathbf{T}} = \begin{pmatrix} \frac{\text{tr}\mathbf{Q}}{e^\lambda} & -\frac{e^{-\lambda}}{e^\lambda} \\ 1 & 0 \end{pmatrix}. \qquad (9.75)$$

At this point we note that by defining $\text{tr}\mathbf{Q} \equiv E - \tilde{\alpha}$, the matrix $\tilde{\mathbf{T}}$ can be properly regarded as a transfer matrix describing the elementary excitation propagation through a monatomic lattice, composed of atoms of on-site energy $\tilde{\alpha}$ coupled to its neighbors through transfer integrals $\tilde{t}_{k,k\pm1} = e^{\pm\lambda}$. Then, it is tempting to think of Eq.(9.48) as describing an effective renormalization of the original mixed ternary lattice leading to the monatomic one. To confirm this physical scenario we shall consider the lattice pentamers $ACBAC$ in the ternary chain and decimate the AC sites, as it is illustrated in Fig.9.2. In so doing, we obtain $\tilde{t}_{n,n\pm1} = 1$, and

$$E - \tilde{\alpha} = \frac{v_A v_B v_C - t^2(b^2 v_A + a^2 v_B + v_C)}{abt^3}, \qquad (9.76)$$

which coincides with $\text{tr}\mathbf{Q}$, as given by Eq.(9.74). Therefore, we conclude that the similarity transformation given by Eq.(9.48) describes a *local renormalization* transformation, acting on certain segments of the original ternary lattice in order to transform it into an effective monatomic lattice. Since the renormalized transfer integrals trivially reduce to the unity, all the relevant physical information is now contained in the renormalized on-site energies $\tilde{\alpha}$.

This approach can be straightforwardly extended to aperiodic lattices as well. For instance, making use of Eqs.(9.45) we obtain

$$\mathbf{R}_A = \gamma^{-1} \begin{pmatrix} qy - x\gamma^2 & \gamma^2 - y^2 \\ q & -y \end{pmatrix}, \qquad \mathbf{R}_B = \begin{pmatrix} q & -y \\ x & -1 \end{pmatrix}, \qquad (9.77)$$

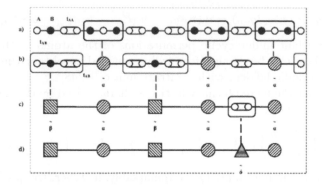

FIGURE 9.3

Renormalization scheme mapping a mixed Fibonacci lattice model into an effective on-site model, which proceeds according to the following steps: (a) decimation of the B sites in the BAB trimers to obtain the renormalized on-site energy sites $\tilde{\alpha}$; (b) decimation of the AA dimers belonging to the $AABAA$ pentamers to obtain the renormalized on-site energy sites $\tilde{\beta}$; and (c) decimation of the remaining AA dimers to obtain the on-site energy sites $\tilde{\delta}$. The resulting aperiodic ternary lattice is shown in (d). ([48] Reprinted figure with permission from Maciá E and Rodríguez-Oliveros R 2006 *Phys. Rev. B* **74** 144202 © 2006 by the American Physical Society.)

where $x \equiv E + \varepsilon_A$, $y \equiv E - \varepsilon_A$, and $q \equiv xy - 1$. For the sake of simplicity we set the origin of energies in such a way that $\varepsilon_B = -\varepsilon_A$, and fix the energy scale so that $t_{AB} \equiv 1$ without loss of generality. It is readily checked that both $\mathbf{R}_A$ and $\mathbf{R}_B$ matrices are unimodular for *any* choice of the system parameters and for *any* value of the electron energy. Accordingly, we can transform them to the form $\mathbf{M}_A \mathbf{R}_A \mathbf{M}_A^{-1} = e^{\lambda_A} \tilde{\mathbf{T}}_A$ and $\mathbf{M}_B \mathbf{R}_B \mathbf{M}_B^{-1} = e^{\lambda_B} \tilde{\mathbf{T}}_B$, respectively, where

$$\tilde{\mathbf{T}}_A = \begin{pmatrix} \frac{\mathrm{tr}\mathbf{R}_A}{e^{\lambda_A}} & -\frac{e^{-\lambda_A}}{e^{\lambda_A}} \\ 1 & 0 \end{pmatrix}, \qquad \tilde{\mathbf{T}}_B = \begin{pmatrix} \frac{\mathrm{tr}\,\mathbf{R}_B}{e^{\lambda_B}} & -\frac{e^{-\lambda_B}}{e^{\lambda_B}} \\ 1 & 0 \end{pmatrix}. \qquad (9.78)$$

Now, by analogy with the treatment introduced in the study of periodic lattices, we note that by defining $\mathrm{tr}\mathbf{R}_A \equiv E - \tilde{\alpha}$ and $\mathrm{tr}\,\mathbf{R}_B \equiv E - \tilde{\beta}$ the matrices $\tilde{\mathbf{T}}_A$ and $\tilde{\mathbf{T}}_B$ can be properly regarded as transfer matrices describing the electron propagation through a lattice composed of atoms of on-site energy $\tilde{\alpha}$ (alternatively $\tilde{\beta}$) coupled to their neighbors through transfer integrals $\tilde{t}_{k,k\pm1} = e^{\pm\lambda_A}$ (alternatively $\tilde{t}_{k,k\pm1} = e^{\pm\lambda_B}$). To confirm this physical picture we shall consider the lattice trimers BAB in the Fibonacci chain and decimate the B sites, as it is illustrated in Fig.9.3a. In so doing, we obtain $E - \tilde{\alpha} =$

$xy - 2$, which coincides with tr $\mathbf{R}_B = q - 1$, as given by Eq.(9.77). In a similar way, we consider the lattice pentamers $AABAA$ and decimate the AA dimers, as indicated in Fig.9.3b, to obtain $E - \tilde{\beta} = \gamma^{-1}(x(y^2 - \gamma^2) - 2y)$, which coincides with tr$\mathbf{R}_A = \gamma^{-1}(y(q - 1) - x\gamma^2)$ as given by Eq.(9.77). Therefore, we conclude that the similarity transformation given by Eq.(9.48) describes a *renormalization transformation* in the quasiperiodic lattice case as well. By simultaneously applying the renormalization transformations shown in Figs.9.3a and 9.3b the original Fibonacci chain is mapped into the lattice shown in Fig.9.3c, where some AA dimers, connected via γ transfer terms, still remain. In order to complete the renormalization of the original chain, we decimate these dimers, obtaining the fully renormalized lattice shown in Fig.9.3d, where $E - \tilde{\delta} = y - \gamma$. As we can see, the resulting lattice is now composed of three different atoms, and all the transfer integrals coupling them have the same value, $\tilde{t} = t_{AB} \equiv 1$. Therefore, the original binary Fibonacci chain has been transformed in an equivalent ternary lattice which can be properly described in terms of an on-site model. Broadly speaking we can say that the reduction of the mixed model to an on-site one is obtained at the cost of increasing the system's chemical complexity (i.e., we now have three different atomic flavors, rather than the original two). In addition, the topological order present in the renormalized lattice is no longer described by the Fibonacci sequence, although it still is an aperiodic one.

9.5 Diagnostic tools

The absence of a well suited mathematical framework to obtain general analytical results on the behavior of quasiperiodic systems has led to the progressive introduction of what we have referred to as diagnostic tools.[51] These include the density of states (DOS), the transmission coefficient, the Landauer conductance, or the Lyapunov coefficient. Although the information that any of these tools can supply by itself is not conclusive as rigorous proof, when grouped together they can provide quite compelling evidence about the nature of the spectrum and its related states.

9.5.1 Density of states

The density of states is an important quantity for the understanding of many phenomena in a large number of physical systems. The interpretation of experimental data is directly related to the DOS in many cases. For numerical purposes it is convenient to introduce the DOS in terms of a closely related magnitude, namely, the integrated density of states (IDOS), which is defined as the number of states whose energy is smaller than E according to the expression

$$I(E) = \int_{-\infty}^{E} D(E')dE', \tag{9.79}$$

where $D(E)$ is the density of states (DOS). This magnitude is always well defined from a strict mathematical point of view, the energy spectrum being singular continuous or not (see Section 5.2).[52] Eq.(9.79) can be numerically determined by the node counting method proposed by Dean.[53] Plots of the IDOS as a function of the energy are very useful in describing the global as well as local structure of the energy (or frequency) spectrum of self-similar systems (see Figs.5.9-5.12). The pattern of the IDOS of such systems corresponds to the so called devil staircase, whose steps indicate the position of gaps, thus clearly showing the fragmentation scheme of the spectrum. Moreover, the difference of the relative heights between successive steps indicate the number of states between gaps. This allows one to extract valuable information regarding the level populations.[54] From the knowledge of the IDOS, the DOS can be obtained by numerical derivation.

Alternatively, closed analytical expressions for the DOS can be derived from the knowledge of the global transfer matrix given by Eq.(9.10) according to the differential expression,[24]

$$D(E)dE = \frac{1}{N}d\left[\cos^{-1}\left(\frac{1}{2}\mathrm{tr}\mathcal{M}_N\right)\right]. \tag{9.80}$$

For the sake of illustration, let us consider the global transfer matrix corresponding to a double-stranded DNA chain given by Eq.(6.38) in Chapter 6, whose semi-trace is given by

$$\frac{1}{2}\mathrm{tr}\mathcal{M}_N = \frac{U_{N/2} - U_{N/2-2}}{2} = T_{N/2}(z) = \cos\left(\frac{N}{2}\cos^{-1}z\right), \tag{9.81}$$

where

$$z(E) = \frac{[E - \alpha(E)][E - \beta(E)]}{2t_0^2} - 1, \tag{9.82}$$

and $\alpha(E)$ and $\beta(E)$ are given by Eq.(6.40). By plugging Eq.(9.81) into Eq.(9.80) one gets

$$D(E) = -\frac{1}{2\sqrt{1 - z^2}}\frac{dz}{dE}, \tag{9.83}$$

and making use of the expression given in Eq.(9.82) one readily obtains Eq.(6.42).

9.5.2   Transmission coefficient

Let us consider a lattice composed of N atoms which is sandwiched between two periodic chains (playing the role of metallic contacts), each one with on-site energy ε and transfer integral t. The dispersion relation of the contacts is then given by $E = \varepsilon + 2t \cos \kappa$. From the knowledge of the global transfer matrix $\mathcal{M}_N(E)$ elements, M_{ij}, one can obtain the transmission coefficient, $T_N(E)$, from the expression,[55]

$$
t e^{i\kappa N} \begin{pmatrix} e^{i\kappa} \\ 1 \end{pmatrix} = \begin{pmatrix} M_{11} & M_{12} \\ M_{21} & M_{22} \end{pmatrix} \begin{pmatrix} e^{i\kappa} + r e^{-i\kappa} \\ 1 + r \end{pmatrix},
$$

where r and t are the reflection and transmission amplitudes, respectively. Thus, one has

$$
t = -\frac{2i \sin k \; e^{-i\kappa N}}{M_{11}e^{-i\kappa} + M_{12} - M_{21} - M_{22}e^{i\kappa}} \det \mathcal{M}_N ,
$$

and after some algebra the transmission coefficient $T_N(E) = |t|^2$ reads[10]

$$
T_N(E) = \frac{4(\det \mathcal{M}_N)^2 \sin^2 k}{[M_{12} - M_{21} + (M_{11} - M_{22}) \cos k]^2 + (M_{11} + M_{22})^2 \sin^2 k}. \tag{9.84}
$$

The transmission coefficient is a useful quantity to describe the transport efficiency in quantum systems. Nonetheless, $T_N(E)$ is usually difficult to be directly measured experimentally. Relevant information on the transmission properties can be gained by measuring current-voltage characteristics. However applying a voltage bias in between conducting leads contacting the chain has also some influence on the scattering properties inside the system, and direct information on intrinsic effects on transmission should thus be considered with care. In experiments on molecular wires the electric current I through the molecule is measured as a function of the voltage, V, and it can be calculated from the expression,[56]

$$
I(V) = \frac{2e}{h} \int T_N(E, \delta) \; [f_L(E, T, V) - f_R(E, T, V)] \; dE, \tag{9.85}
$$

where $f_{L(R)}$ is the Fermi-Dirac distribution at the left (right) contacts, respectively, and we assume the charges propagate from left to right. In this expression the transmission coefficient depends on the system's length, N, the charge carrier energy, E, and a lead coupling factor, δ. If δ were a function of the applied voltage then $T = T_N(E, V)$ as well. However, one usually assumes the transmission coefficient is evaluated a zero bias. This low bias approach is a very good description for short molecules connected to metallic leads. Nevertheless, in the case of large molecules the possible modification of the macromolecule electronic structure in the vicinity of the leads should be considered for moderated biases.

9.5.3 Landauer conductance

Closely related to the transmission coefficient, the Landauer resistance formula appears as a very simple expression of the residual resistance of a one-dimensional system in terms only of its scattering properties at the Fermi energy.[57] Accordingly, it is a useful magnitude in the context of molecular electronics, which allows for a direct comparison with experimental current-voltage curves, since one reasonably expects the energy interval under the $T_N(E)$ curve to be proportional to the electrical current flowing through the molecule.[58] The Landauer formula establishes the relationship between the resistance of a finite chain, $\rho_N(E)$, and the transmission coefficient, $T_N(E)$, as follows

$$\rho_N(E) = \frac{h}{2e^2} \frac{1 - T_N(E)}{T_N(E)}. \tag{9.86}$$

Such formula should not depend on external leads and it has a range of applicability limited to the regimes where $T_N(E) \ll 1$. In the case of resonant (or ballistic) transmission, $\rho_N(E)$ obviously tends to zero, but in that regime, the resistance is fixed by the contacts plus molecule system, and will be limited to the so-called Landauer conductance $G(E) = G_0 T_N(E)$, where $G_0 \equiv 2e^2/h \simeq 12906^{-1} \ \Omega^{-1}$ is the conductance quantum. The use of the Landauer resistance is specially relevant in the study of electron transport in one-dimensional finite systems, since Eq.(9.86) can be easily combined with the transfer matrix formalism, considering that electrons enter in the system by one of the edges and exit through the other side according to the expression given by Eq.(9.84). The generalization of such concept to more than one dimension is not a trivial issue as it has been discussed by Landauer and co-workers.[59]

The Landauer resistance has been widely used to characterize the localization properties of one-dimensional disordered systems.[60] Thus, for exponentially localized states the Landauer resistance increases exponentially with the system size. On the contrary, for critical states the resistance fluctuates and shows self-similar patterns. According to this result it was suggested that, in the case of Fibonacci systems, the Landauer resistance should show an algebraic dependence of the form $\rho_N \sim N^\delta$ with $\delta > 0$.[61] This conjecture, originally based on numerical results, was later confirmed by means of rigorous mathematical proofs.[62] Therefore, the power-law scaling of the resistance might be regarded as a fingerprint of critical states. On the other side, for a given system size, the plot of the Landauer conductance as a function of energy displays the global structure of the energy spectrum (see Fig.5.24). Therefore, for ideal aperiodic systems (without structural defects) one can reasonably expect that the Landauer conductance will show self-similar fluctuations as a consequence of the highly fragmented structure of their energy spectrum.[10, 63]

To derive Eq.(9.86) it was assumed that the temperature is low enough, so that inelastic scattering (responsible for thermalization, energy dissipation, and phase randomization) takes place inside the metallic contacts only. However, in actual systems the effects of finite temperature as well as the defects present in the sample can hide finer details of the energy spectrum, thus making it difficult to experimentally determine fractal features. This fact is of great relevance if those features are to be applied in technological devices (see

Figs.5.25 and 7.1). In this case it is convenient to make use of the formula[64]

$$\kappa(T,\mu) = \frac{\int \left(-\frac{\partial f}{\partial E}\right) T_N(E) dE}{\int \left(-\frac{\partial f}{\partial E}\right) [1 - T_N(E)] dE} \qquad (9.87)$$

for the dimensionless conductance of the system at finite temperature, where the integration is extended over the entire energy spectrum, f is the Fermi-Dirac distribution and μ denotes the chemical potential of the sample.

9.5.4 Lyapunov coefficient

The Lyapunov coefficient, $\Gamma(E)$, provides the growth ratio of the wave function of the eigenstate with energy E along the system. Accordingly, its inverse, $\Gamma^{-1}(E)$, determines the localization length of the eigenstate. Therefore, this magnitude is quite useful in order to establish a relationship between the energy spectrum and the transport properties of the system.

There exist several definitions of this parameter in the literature. Thouless pointed out the relation between the Lyapunov coefficient and the local structure of the spectrum, determined by the DOS (see Section 9.5.1),[65]

$$\Gamma(E) = \int \ln \left| \frac{4(E - E')}{W} \right| D(E') dE' - \ln |Z|, \qquad (9.88)$$

where W is the conduction-band width and Z is the geometric mean of the off-diagonal elements of the Hamiltonian. Gaps in the energy spectrum are characterized by maxima of Γ, whereas allowed bands correspond to minima of Γ. For systems arranged according to the Fibonacci sequence the energy spectrum is singular continuous and, for Fibonacci tight-binding transfer-like Hamiltonians, the condition $\Gamma(E) = 0$ for every state in the spectrum has been rigorously proven.[66]

From a numerical point of view,[55, 66] it is more convenient to use the expression

$$\Gamma(E) = \lim_{N \to \infty} \frac{1}{N} \ln \|\mathcal{M}_N(E)\| \qquad (9.89)$$

where $\mathcal{M}_N(E)$ is the global transfer matrix of the system given by Eq.(9.10) and $\|...\|$ stands for a suitable matrix norm. One can consider the matrix norm

$$\|\mathcal{M}_N(E)\| = \sup \frac{\|\mathcal{M}_N \psi\|}{\|\psi\|}, \qquad (9.90)$$

associated with the euclidean vector norm $\|\psi\| = \sqrt{|\psi_1|^2 + |\psi_0|^2}$, where $\psi \equiv (\psi_1, \psi_0)^t$ is the initial vector. At first sight, the Lyapunov exponent seems to depend on the norm of this initial vector, but it has been shown that, under very general circumstances, Eq.(9.89) can be expressed as[31, 55]

$$\Gamma(E) = \lim_{N \to \infty} \frac{1}{2N} \ln(\text{tr}\mathcal{M}_N \mathcal{M}_N^\dagger) = \lim_{N \to \infty} \frac{1}{2N} \ln \left(\sum M_{ij}^2 \right), \qquad (9.91)$$

where † denotes the Hermitian conjugate. This expression is very useful in numerical studies where the elements of the global transfer matrix, $M_{ij}(E, N)$, can be evaluated in a recursive way. It allows for the derivation of closed analytical expressions as well. For instance, plugging the matrix elements given in Eq.(6.38) into Eq.(9.91), and making use of the relationship $U_n^2 + U_{n-1}^2 - 2zU_nU_{n-1} = 1$, along with Eq.(9.17), one readily obtains Eq.(6.43).

Alternatively, one may use the expression[55, 67]

$$\Gamma(E) = \lim_{N \to \infty} \frac{1}{2N} \ln \lambda_N, \tag{9.92}$$

where

$$\lambda_N = \frac{\|\mathcal{M}_N\|_*^2}{2} + \sqrt{\frac{\|\mathcal{M}_N\|_*^4}{4} - (\det \mathcal{M}_N)^2} \tag{9.93}$$

is the largest eigenvalue of the matrix $\mathcal{M}_N \mathcal{M}_N^\dagger$, and $\|\mathcal{M}_N\|_* \equiv \sqrt{\sum M_{ij}^2}$ is the so-called trace matrix norm.

Several alternative definitions of the Lyapunov coefficient can be found in the literature aimed to relate it with several magnitudes of interest for the study of the transport properties. For example, the following expressions relating the Lyapunov coefficient to the transmission coefficient $T_N(E)$,[67, 68]

$$\Gamma(E) = -\lim_{N \to \infty} \frac{1}{2N} \ln T_N(E), \tag{9.94}$$

or to the wavefunction values at the ends of the chain,

$$\Gamma(E) = \lim_{N \to \infty} \frac{1}{N} \ln \left| \frac{\psi_N}{\psi_0} \right|, \tag{9.95}$$

have been extensively considered. When considering extended states belonging to the allowed bands of periodically ordered crystalline systems one gets $\Gamma(E) = 0$. In fact, in that case (i) the global transfer matrix elements oscillate between two bounded values [i.e., we have $|\mathcal{M}_N| \leq 2$ in Eq.(9.89)], (ii) we have a unity transmission coefficient, hence vanishing Eq.(9.94), and (iii) the Bloch wave functions satisfy the boundary condition $\psi_N \equiv \psi_0$, hence vanishing Eq.(9.95) as well. In the case of exponentially localized eigenstates Eq.(9.95) can be related to the localization length as follows. Let ψ be an exponentially localized state. Then, the wavefunction amplitude at site n can be expressed as $\psi_n \simeq e^{-|n-n_0|/\xi}$, where n_0 indicates the site where the state exhibits its maximum amplitude along the chain and ξ measures its localization length. Accordingly, $\psi_{n+1} \simeq e^{-|n-n_0|/\xi - 1/\xi}$, so that

$$\xi^{-1} = -\ln \left| \frac{\psi_{n+1}}{\psi_n} \right|. \tag{9.96}$$

By comparing Eq.(9.96) with Eq.(9.95) we see the Lyapunov exponent is related to the inverse localization length in the straightforward way $\Gamma(E) = \xi^{-1}$

in the limit $n \to 0$. Therefore, the Lyapunov exponent provides a direct measure of the inverse localization length for exponentially localized states. Works dealing with the so-called random dimer model (see Section 1.7), where a number of impurity dimers are randomly interspersed in a host chain, have considered the Lyapunov coefficient as defined by Eq.(9.94), and used it to discuss the localization length of resonant states. The criterion for the presence of delocalized states was the condition of $\Gamma \gg N^{-1}$ for these states.[51] Can this interpretation be directly extended to critical states characteristic of quasiperiodic systems?

It is known that in the particular case of Fibonacci on-site models (see Chapter 5) the norm of the global transfer matrix is polynomially bounded for every state in the spectrum, i.e., $\|\mathcal{M}_N(E)\| \leq N^\beta$, when N goes to infinity.[62] This result prevents a naive extension of the argument presented when dealing with exponentially localized states in order to interpret the Lyapunov exponent as the inverse localization length. In fact, in this case we have critical wavefunctions whose amplitudes are roughly modulated by scaling exponents β describing a power law behavior of an envelope as $|\psi_n| \simeq |n - n_k|^{-\beta}$. This relation applies to the sites n_k where the wave function has a local maximum. These sites, in turn, are distributed in a self-similar way along the chain. Then, for each one of these bumps in the wave function we get

$$\ln \left| \frac{\psi_{n+1}}{\psi_n} \right| = -\beta \ln \left| 1 + \frac{1}{n - n_0} \right|, \tag{9.97}$$

and the limit $n \to 0$ yields vanishing Lyapunov exponents for any eigenvalue in the spectrum.[66] The values for the exponents β corresponding to the center ($E = 0$) and edges of the main subbands in the energy spectrum of the Fibonacci chain (transfer model) can be analytically obtained.[28] Numerical studies have shown that the function $\Gamma(E)$, determined by means of Eqs.(9.91) and (9.94), exhibits a self-similar pattern, reflecting the highly fragmented, fractal structure of the energy spectrum.[10] The multifractal nature of the electronic spectrum structure probably requires the use of a set of exponents β_n, rather than a single one.[69]

References

[1] Bohr H 1952 *Collected Mathematical Works. II Almost Periodic Functions* (Copenhagen: Dansk Matematisk Forening)

[2] Janssen T, Chapuis G, and de Boissieu M 2007 *Aperiodic Crystals: From Modulated Phases to Quasicrystals* (Oxford University Press, Oxford)

[3] Baake M 2002 in *Quasicrystals: An Introduction to Structure, Physical Properties, and Applications*, Eds. Suck J B, Schreiber M, and Häussler P (Springer, Berlin) p 17

[4] Elser V 1985 *Phys. Rev. B* **32** 4892

[5] Elcoro L and Pérez-Mato J M 1996 *Phys. Rev. B* **54** 12115

[6] Pérez-Álvarez R, Trallero-Herrero C, and García-Moliner F 2001 *Eur. J. Phys.* **22** 1

[7] Griffiths D J and Steinke C A 2001 *Am. J. Phys.* **69** 137

[8] Bendickson J M, Dowling J P, and Scalora M 1996 *Phys. Rev. E* **53** 4107

[9] Wu H, Sprung D W L, and Martorell J 1993 *J. Phys. D: Appl. Phys.* **26** 798

[10] Maciá E and Domínguez-Adame F 2000 *Electrons, Phonons and Excitons in Low Dimensional Aperiodic Systems* (Colección Línea 300, Ed. Complutense, Madrid)

[11] Maciá E 2006 *Rep. Prog. Phys.* **69** 397

[12] Albuquerque E L and Cottam M G 2003 *Phys. Rep.* **376** 225; Albuquerque E L and Cottam M G 2004 in *Polaritons in Periodic and Quasiperiodic Structures* (Elsevier, Amsterdam)

[13] Velasco V R 2001 in *Some Contemporary Problems of Condensed Matter Physics*, Eds. Vlaev S J and Gaggero-Sager L M (New York, Nova-Scientiae); Pérez-Álvarez R, Trallero-Herrero C, and García-Moliner F 2001 *ibid.* p.1-37

[14] Tamura S and Wolfe J P 1987 *Phys. Rev. B* **36** 3491

[15] Fernández-Álvarez L and Velasco V R 1998 *Phys. Rev. B* **57** 14141; Zárate J E, Fernández-Álvarez L, and Velasco V R 1999 *Superlatt. Microst.* **25** 529; Velasco V R and Zárate J E 2001 *Prog. Surf. Sci.* **67** 383

[16] Aynaou H, Velasco V R, Nougaoui A, El Boudouti E H, Djafari-Rouhani B, and Bria D 2003 *Surf. Sci.* **538** 101

[17] Shahbazi F, Bahraminasab A, Mehdi Vaez Allaei S, Sahimi M, and Reza Rahimi Tabar M 2005 *Phys. Rev. Lett.* **94** 165505

[18] Gantmacher F R 1974 *The Theory of Matrices* **2** (Chelsea, New York)

[19] Peisakhovich Y G 1996 *J. Phys. A: Math. Gen.* **29** 5103

[20] Pereyra P 1998 *Phys. Rev. Lett.* **80** 2677; 1998 *J. Phys. A: Math. Gen.* **31** 4521; 2000 *Phys. Rev. Lett.* **84** 1772

[21] Ojeda P, Huerta-Quintanilla R, and Rodríguez-Achach M 2002 *Phys. Rev. B* **65** 233101

[22] Cao L S, Peng R W, Zhang R L, Zhang F, Wang M, Huang X Q, Hu A, and Jiang S S 2005 *Phys. Rev. B* **72** 214301

[23] Mochán W L and del Castillo-Mussot M 1988 *Phys. Rev. B* **37** 6763

[24] Iguchi K 1997 *Int. J. Mod. Phys. B* **11** 2405; 2001 *J. Phys. Soc. Jpn.* **70** 593

[25] Monsivais G, Otero J A, and Calás H 2005 *Phys. Rev. B* **71** 064101

[26] Sedrakyan T and Ossipov A 2004 *Phys. Rev. B* **70** 214206

[27] Díaz E, Sedrakayan A, Sedrakayan D, and Domínguez-Adame F 2007 *Phys. Rev. B* **75** 014201

[28] Kohmoto M, Kadanoff L P, and Tang C 1983 *Phys. Rev. Lett.* **50** 1870; Kohmoto M, Sutherland B, and Tang C 1987 *Phys. Rev. B* **35** 1020

[29] Axel F and Peyriére J 1989 *J. Stat. Phys.* **57** 1013

[30] Ghosh A and Karmakar S N 1999 *Physica A* **274** 555

[31] Iguchi K 1994 *Phys. Rev. B* **49** 12633; 1994 *J. Math. Phys.* **35** 1008

[32] Baake N, Joseph D, and Kramer P 1992 *Phys. Lett. A* **168** 199

[33] Baake M, Grimn U, and Joseph D 1993 *Int. J. Mod. Phys. B* **7** 1527; Roberts J A C and Baake M 1994 *J. Stat. Phys.* **74** 829

[34] Maciá E 2006 *Phys. Rev. B* **73** 184303

[35] The similarity between the trace maps corresponding to discrete and continuous Fibonacci models was discussed by MacDonald A H and Aers G C 1987 *Phys. Rev. B* **36** 9142

[36] Simkin M V and Mahan G D 2000 *Phys. Rev. Lett.* **84** 927

[37] Velasco V R, García-Moliner F, Miglio L, and Colombo L 1988 *Phys. Rev. B* **38** 3172

[38] Holzer M 1988 *Phys. Rev. B* **38** 1709

[39] Naumis G G, Aragón J L, and Torres M 2002 *J. Alloys Compounds* **342** 210; Naumis G G, Wang Ch, Thorpe M F, and Barrio R A 1999 *Phys. Rev. B* **59** 14302

[40] Niu Q and Nori F 1990 *Phys. Rev. B* **42**, 10 329; 1986 *Phys. Rev. Lett.* **57**, 2057

[41] Piéchon F, Benakli M, and Jagannathan A 1995 *Phys. Rev. Lett.* **74**, 5248

[42] Yndurain F, Barrio R A, Elliot R J, and Thorpe M F 1983 *Phys. Rev. B* **28** 3576

[43] You J Q, Yang Q B, and Yan J R 1990 *Phys. Rev. B* **41** 7491

[44] Maciá E and Domínguez-Adame F 1996 *Phys. Rev. Lett.* **76** 2957

[45] Maciá E 1999 *Phys. Rev. B* **60** 10032; 2000 *Phys. Rev. B* **61** 6645; 1998 *Appl. Phys. Lett.* **73** 3330

[46] Flores J C 1989 *J. Phys. C: Condens. Matter* **1** 8471

[47] Lindquist B and Riklund R 1998 *Phys. Stat. Sol. (b)* **209** 353

[48] Maciá E and Rodríguez-Oliveros R 2006 *Phys. Rev. B* **74** 144202

[49] Lamb J S W and Wijnands F 1998 *J. Stat. Phys.* **90** 261

[50] Kroon L and Riklund R 2003 *J. Phys. A: Math. Gen.* **36** 4519

[51] Sánchez A, Maciá E, and Domínguez-Adame F 1994 *Phys. Rev. B* **49** 147

[52] Luck J M 1989 *Phys. Rev. B* **39** 5834

[53] Dean P 1972 *Rev. Mod. Phys.* **44** 127

[54] Maciá E, Domínguez-Adame F, and Sánchez A 1994 *Phys. Rev. E* **50** 679

[55] Sütö A 1995 in *Beyond Quasicrystals* Eds. Axel F and Gratias D (Les Editions de Physique, Les Ullis) p.481

[56] Mugica V, Kemp M, Roitberg and Ratner M A 1996 *J. Chem. Phys.* **104** 7296

[57] Landauer R 1970 *Phil. Mag.* **21** 683

[58] S. Datta 1995 in *Electronic Transport in Mesoscopic Systems* (Cambridge University Press, Cambridge)

[59] Büttiker M, Imry Y, Landauer R, and Pinhas S 1985 *Phys. Rev. B* **31** 6207; Imry Y and Landauer R 1999 *Rev. Mod. Phys.* **71** S306

[60] Thouless D 1974 *Phys. Rep.* **13** 93

[61] Sutherland B and Kohmoto M 1987 *Phys. Rev. B* **36** 5877

[62] Iochum B and Testard D 1991 *J. Stat. Phys.* **65** 715

[63] Das Sarma S and Xie X C 1988 *Phys. Rev. B* **37** 1097

[64] Engquist H L and Anderson P W 1981 *Phys. Rev. B* **24** 1159

[65] Thouless D 1972 *J. Phys. C: Solid State Phys.* **5** 77

[66] Sütö A 1989 *J. Stat. Phys.* **56** 25

[67] Pichard J L 1986 *J. Phys. C: Solid State Phys.* **19** 1519

[68] Kirkman P D and Pendry J B 1984 *J. Phys. C: Solid State Phys.* **17** 4327

[69] Naumis G G 1999 *Phys. Rev. B* **59** 11315

Index